Springer Theses

Recognizing Outstanding Ph.D. Research

Aims and Scope

The series "Springer Theses" brings together a selection of the very best Ph.D. theses from around the world and across the physical sciences. Nominated and endorsed by two recognized specialists, each published volume has been selected for its scientific excellence and the high impact of its contents for the pertinent field of research. For greater accessibility to non-specialists, the published versions include an extended introduction, as well as a foreword by the student's supervisor explaining the special relevance of the work for the field. As a whole, the series will provide a valuable resource both for newcomers to the research fields described, and for other scientists seeking detailed background information on special questions. Finally, it provides an accredited documentation of the valuable contributions made by today's younger generation of scientists.

Theses are accepted into the series by invited nomination only and must fulfill all of the following criteria

- They must be written in good English.
- The topic should fall within the confines of Chemistry, Physics, Earth Sciences, Engineering and related interdisciplinary fields such as Materials, Nanoscience, Chemical Engineering, Complex Systems and Biophysics.
- The work reported in the thesis must represent a significant scientific advance.
- If the thesis includes previously published material, permission to reproduce this must be gained from the respective copyright holder.
- They must have been examined and passed during the 12 months prior to nomination.
- Each thesis should include a foreword by the supervisor outlining the significance of its content.
- The theses should have a clearly defined structure including an introduction accessible to scientists not expert in that particular field.

Indexed by zbMATH.

More information about this series at http://www.springer.com/series/8790

Marie-Christine Zdora

X-ray Phase-Contrast Imaging Using Near-Field Speckles

Doctoral Thesis accepted by
University College London, London,
United Kingdom

Springer

Author
Dr. Marie-Christine Zdora
School of Physics and Astronomy
University of Southampton
Southampton, UK

Supervisors
Dr. Irene Zanette
School of Physics and Astronomy
University of Southampton
Southampton, UK

Prof. Pierre Thibault
Department of Physics
University of Trieste
Trieste, Italy

Prof. Alessandro Olivo
Department of Medical Physics
and Biomedical Engineering
University College London
London, UK

ISSN 2190-5053 ISSN 2190-5061 (electronic)
Springer Theses
ISBN 978-3-030-66331-5 ISBN 978-3-030-66329-2 (eBook)
https://doi.org/10.1007/978-3-030-66329-2

This Springer imprint is published by the registered company Springer Nature Switzerland AG
The registered company address is: Gewerbestrasse 11, 6330 Cham, Switzerland

Supervisors' Foreword

The principle of phase-contrast imaging, exploiting the wave properties of light, was introduced for the visible-light regime in the 1930s by Frits Zernike. For the first time, phase shifts induced by the imaged object, which carry important information on the sample's properties, could be visualised in a microscopic image. Phase-contrast imaging revolutionised the field of visible-light microscopy and was recognised with the Nobel Prize in Physics in 1953. More than a decade later, the concept was translated to the X-ray regime by Ulrich Bonse and Michael Hart. They developed an X-ray crystal interferometer, today known as Bonse-Hart interferometer, to translate X-ray phase effects into intensity modulations recorded by a detector. As for visible light, the X-ray phase-contrast signal leads to significantly improved image contrast, in particular for samples with small density differences, which can hardly be visualised by the conventional absorption-based modality. At first not widely applied due to the limitations in the available instrumentation, X-ray phase-contrast imaging was further pursued in the 1990s with the advent of high-brilliance X-ray synchrotron sources and more advanced X-ray optics. A number of groups started working on X-ray phase-contrast imaging during this time and introduced various other X-ray phase-contrast imaging techniques, such as analyser-based imaging, propagation-based imaging, the edge-illumination approach and Talbot(-Lau) grating interferometry. X-ray phase-contrast imaging methods have since seen increasing interest in the last decade for a wide range of applications, one of the most promising being (bio-) medical imaging.

In recent years, efforts have increasingly been directed towards the development and simplification of X-ray phase-contrast methods, also focussing on their translation from synchrotron facilities to lower brilliance conventional laboratory X-ray sources. X-ray speckle-based imaging, introduced in 2012, is a simple yet very sensitive and quantitative method to measure the phase shift induced by the sample and can be adapted to conventional X-ray sources. The beauty of the technique lies in the use of a simple optical element, such as a piece of sandpaper, to modulate the X-ray wavefront and create interference effects, from which the phase information is extracted. The great interest of the X-ray imaging community in speckle-based imaging has resulted in its rapid development, to which Marie-Christine Zdora's Ph.D. project has contributed significantly. Her work is a multidisciplinary project spanning from theoretical advances and algorithmic developments to applications of the method to relevant research areas.

Marie-Christine Zdora put X-ray speckle-based imaging in context with another powerful X-ray phase-contrast method named grating interferometry and unified these two methods with an innovative algorithmic approach for phase extraction from interferometric data, the Unified Modulated Pattern Analysis, applicable also to other wavelengths. She subsequently focussed on demonstrating the potential of her method to various scientific applications, a range of which she presents in this thesis.

By exploiting the high accuracy and high precision of the phase shift measured with X-ray speckle-based imaging, Marie-Christine Zdora performed in-situ metrology of X-ray optics, such as refractive X-ray lenses, commonly used at synchrotron facilities. The results are relevant for the X-ray synchrotron community to accurately characterise the optical components of synchrotron beamlines.

When combined with tomography, X-ray speckle-based imaging allows for the visualisation of the inner structure of the sample and directly measures its mass density distribution. Marie-Christine Zdora used X-ray speckle-based tomography to answer relevant questions in materials science and geology, also extending the technique to higher X-ray energies to image denser and thicker samples.

Marie-Christine Zdora pioneered the development of X-ray speckle-based tomography for three-dimensional virtual histology of biomedical soft tissue (biopsies from human tissues and full organs of small animals) in near-native state. The data obtained in this way complement and advance the images from conventional histopathology, while preserving the real three-dimensional connectivity information and mapping in detail even the tiniest and most localised inhomogeneities in the sample.

Another crucial development was the translation of the Unified Modulated Pattern Analysis to conventional laboratory-based X-ray systems, making it accessible to a wider range of users.

When Marie-Christine Zdora was only 2 years into her Ph.D. project, she had already been recognised by the X-ray imaging community as a pioneer of speckle-based imaging. The impact of her Ph.D. work has been awarded by an important recognition in our field, the Werner Meyer-Ilse Award for excellence in X-ray microscopy in 2018. The expertise and deep understanding that Marie-Christine Zdora has demonstrated is also reflected in the first review article in X-ray speckle-based imaging that she has published as a single author during her Ph.D. project.

This thesis will serve as a handbook of X-ray speckle-based imaging, providing a comprehensive introduction to the technique and a guide on successfully implementing it at synchrotron beamlines and laboratory-based systems. The high quality and the diligent, detailed analysis of the experimental results presented in the following pages speak for themselves. This work will be highly valuable not only to the X-ray imaging community, but also disciplines that benefit from high-contrast imaging as well as accurate quantitative phase sensing and density measurements, such as metrology and optics characterisation, the biomedical and clinical fields, materials science, geology, palaeontology and archaeology, among others. We anticipate the uptake of X-ray speckle-based imaging in these fields in the near future.

Southampton, UK
September 2020

Dr. Irene Zanette
Prof. Pierre Thibault

Abstract

In the last decades, X-ray phase-contrast imaging has proven to be a powerful method for unveiling the inner structure of samples and is capable of visualising even minute density differences. Recently, speckle-based imaging (SBI), the youngest X-ray phase-sensitive technique, has received great interest due to its high sensitivity, quantitative character and relaxed requirements on the setup components and beam properties.

This thesis is focussed on the development, experimental optimisation and applications of SBI, with the aim of simplifying its implementation, increasing its flexibility and expanding its potential.

For this, a robust, flexible data acquisition and reconstruction approach, the unified modulated pattern analysis (UMPA), was developed, which lifts previous constraints of SBI. UMPA allows for tuning of the sensitivity and spatial resolution by adjusting the scan and reconstruction parameters. It is applicable not only to random speckle but also periodic interference patterns, bridging the gap and improving the performance of both speckle- and single-grating-based techniques.

Following the first demonstration of UMPA, its potential for a range of applications is illustrated in this thesis. It is shown that UMPA can be employed for X-ray optics characterisation to quantify aberrations in the focussing behaviour of X-ray refractive lenses with high precision and accuracy. UMPA phase tomography is applied to the field of biomedical imaging for high-sensitivity three-dimensional (3D) virtual histology of unstained, hydrated soft tissue, giving unprecedented structural and quantitative density information.

Further developments of SBI explored in this thesis include the testing of novel customisable speckle diffusers, the extension of SBI to higher X-ray energies for geology and materials science applications and the demonstration of UMPA at a laboratory X-ray source. These progresses promise new possibilities of SBI for high-sensitivity, robust and high-throughput imaging in previously inaccessible fields and make SBI accessible to a wider range of users in research and industry.

Acknowledgements

This Ph.D. project would not have been possible without the contributions and support of my supervisors, collaborators and colleagues. It has been a pleasure working with so many wonderful and bright people from whom I have gained a lot of knowledge and inspiration.

First and foremost, I would like to sincerely thank my main supervisors Irene Zanette and Pierre Thibault who have supported and guided me at all times throughout this project. I have learnt so much from you and I am very lucky to have had such knowledgeable and at the same time kind and caring supervisors. It was just wonderful to work with you and I hope we will continue to collaborate in the future! Thank you for always being available for support, answering my numerous questions and giving me the freedom to work on topics that interest me.

I would also like to thank my supervisor at Diamond Light Source, Christoph Rau, for giving me the opportunity to work at his beamline and providing me with plenty of resources throughout the last years, including financial support for several conference trips. I would, furthermore, like to acknowledge Diamond Light Source for funding my joint Ph.D. studentship with UCL.

Thanks to my secondary supervisor at UCL, Alessandro Olivo, who provided valuable feedback during the upgrade viva that was very helpful for the writing of this thesis. I am, furthermore, grateful to Desmond McMorrow for having been my second examiner for the transfer exam.

A special thanks goes to my former Ph.D. colleagues in Pierre's group at UCL: Simone Sala, Stephanos Chalkidis and Charan Kuppili. Thanks for sharing many beamtime and conference experiences—it was always great fun with you guys! I would also like to thank the former and current members of Pierre's group at Southampton, Hans Deyhle, Sharif Ahmed, Toby Walker and Ronan Smith, for their support during beamtimes at the synchrotron and in the lab.

I would like to acknowledge the whole group at Diamond I13 beamline, especially Kaz Wanelik and Malte Storm who helped with various problems and always had a kind word for me.

My gratitude goes to all my collaborators who I have had the pleasure to work with over the course of my Ph.D. project. Special thanks to the group at ID19 at the

ESRF, especially Alexander Rack and Margie Olbinado, for the many successful and interesting beamtimes and for supporting my work. I have learnt a lot at ID19! Thanks to the group of Bert Müller at the University of Basel for inviting me to many of their beamtimes, which taught me a lot about grating interferometry: Bert Müller, Georg Schulz, Anna Khimchenko, Griffin Rodgers, Christos Bikis and Peter Thalmann. Special thanks also to Willy Kuo, University of Zurich, who provided many of the biomedical samples and was always available to help and explain biological things to me. Thanks to the people at the Faculty of Medicine, University of Southampton, Peter Lackie, Orestis Katsamenis and Matthew Lawson, who helped substantially with the histology of the kidney sample. Thanks also to Vincent Fernandez, Natural History Museum London, for his help with the kidney project, in particular, his amazing VGStudio skills and his appreciation of Bavarian beer. Thanks to Joan Vila-Comamala for introducing me to I13 and for the collaborations and beamtime support on speckle imaging. Thanks to Frieder Koch and Arndt Last from KIT for the collaboration on lens characterisation and Jenny Romell from KTH Stockholm for beamtime support. Thanks to my collaborators at Diamond, Tunhe Zhou and Nghia Vo, for the great beamtimes at I13 and I12 on various topics (and lots of popcorn). Furthermore, thanks to Beverley Coldwell and Fei Yang for providing samples. I would like to thank also collaborators on several projects that were not directly related to my Ph.D. project, but nonetheless extremely interesting: Patrik Vagovič from DESY and XFEL, Irvin Teh and Jürgen Schneider from the School of Medicine, University of Leeds (previously University of Oxford), Roger Benson, Armin Schmitt and Donald Davesne from the Earth Sciences Department, University of Oxford, Carles Bosch Piñol and Andreas Schaefer from the Francis Crick Institute, London.

Moreover, I would like to express my gratitude to some inspiring role models: Franz Pfeiffer who first introduced me to X-ray imaging 8 years ago as a Bachelor student. Without your encouragement and support I would not be working in this field today. David Paganin who dedicated many hours discussing speckle imaging with me, which was very inspirational and encouraging.

I would like to thank all the kind people who helped with proofreading of this thesis: Irene, Pierre, Nick, Xiaowen, Nghia, Malte and Vincent. Thanks a lot, your comments substantially helped to improve the thesis.

Finally, I would like to thank my family and friends in Germany, the UK and various other places in the world, for their ongoing support, love and encouragement during these past years that have not always been easy. I wouldn't have made it to this point without you! A special thanks goes to my boyfriend for the great support during the last stages of preparing this thesis. Thanks for dealing with all my stressfulness, tears and late nights working. I wouldn't have gotten through these tough past weeks without your help!

Contents

Abbreviations

1D	One dimenisional
2D	Two dimensional
3D	Three dimensional
AO	Aorta
AV	Aortic valve
C	Carbon
Ca	Calcium
CCD	Charge-coupled device
CDI	Coherent diffractive imaging
CMOS	Complementary metal-oxide-semiconductor
COR	Cortex
CRL	Compound refractive lens
CT	Computed tomography
CTF	Contrast transfer function
DCM	Double-crystal monochromator
DCMM	Double-crystal multilayer monochromator
ED	Estimated embryonic day
ERFC	Gaussian error function
ESRF	European Synchrotron Radiation Facility
ESRF-EBS	ESRF extremely brilliant source
FBP	Filtered back-projection
FCC	Fluid catalytic cracking
FPS	Fourier power spectrum
FWHM	Full width at half maximum
Ga	Gallium
GBI	X-ray grating-based imaging
GPU	Graphics processing unit
H&E	Haematoxylin and eosin

H_2O	Water
HREM	High-resolution episcopic microscopy
ID	Insertion device
IM	Inner medulla
In	Indium
ISOM	Inner stripe of the outer medulla
IVS	Interventricular septum
LIGA	X-ray lithography and electroplating
linac	Linear accelerator
LSF	Line spread function
LV	Left ventricle
MACE	Metal-assisted chemical etching
microCT	Micro-computed tomography
MLM	Multilayer monochromator
MRI	Magnetic resonance imaging
MTF	Modulation transfer function
MTRI	Masson's trichrome
MuCLS	Munich compact light source
MV	Mitral valve
NOM	Nanometer optical metrology
OSOM	Outer stripe of the outer medulla
PA	Pulmonary artery
PAS	Periodic acid-Schiff
PBS	Phosphate buffered saline
PET	Polyethylene terephthalate
PFA	Paraformaldehyde
PM	Phase modulator
PMMA	Polymethyl methacrylate
PS	Polystyrene
PSF	Point spread function
PV	Pulmonary valve
PVC	Polyvinyl chloride
RA	Right atrium
ROI	Region of interest
RV	Right ventricle
SBI	X-ray speckle-based imaging
sCMOS	Scientific complementary metal-oxide-semiconductor
Si	Silicon
Sn	Tin
TIE	Transport of intensity
Tukey	Tapered cosine
UMPA	Unified modulated pattern analysis

XFEL	X-ray free electron laser
XGI	X-ray grating interferometry
XPCI	X-ray phase-contrast imaging
XSS	X-ray speckle scanning
XST	X-ray speckle tracking
XSVT	X-ray speckle-vector tracking

Chapter 1
Introduction

We shall see what we shall see. We have the start now; the developments will follow in time. Dam (1896)

This is the answer W. C. Röntgen reportedly gave when asked about the future of X-rays, which he had just discovered in 1895. And he was right. Since these early days, X-rays have been subject to rapid developments in terms of imaging techniques as well as applications, both areas of intense ongoing research to this date. This Ph.D. thesis is part of this actively progressing field. It presents work contributing to the development and optimisation of emerging X-ray imaging techniques, namely X-ray speckle-based and grating-based phase-contrast imaging, as well as demonstrations of the potential of these methods for existing and new areas of applications.

1.1 Background, Motivation and Present Work

The X-ray images taken by Röntgen were based on exploiting the absorption of X-rays by the object to visualise its inner structure, which is nowadays called absorption-based X-ray imaging. This conventional way is still the main workhorse of X-ray imaging and used in a large number of applications such as medical diagnostic imaging, security screening, non-destructive testing, foreign body detection in medicine and food production, and many more.

Absorption-based X-ray imaging relies on the fact that for a fixed wavelength the absorption of X-rays in a specimen depends on its composition, density and thickness. This, in fact, had already been observed by Röntgen in his very first experiments (Röntgen 1898). As a consequence, high-density materials such as metals or bone lead to strong X-ray absorption while low-density materials such as plastics or biomedical soft tissue only attenuate the X-ray beam very weakly, in particular at higher energies. It is hence relatively easy to distinguish materials with large differences in density based on X-ray absorption, but small density variations in a specimen are difficult

M.-C. Zdora, *X-ray Phase-Contrast Imaging Using Near-Field Speckles*,
Springer Theses, https://doi.org/10.1007/978-3-030-66329-2_1

to visualise. A prominent example is the visualisation of biomedical specimens. Bones can easily be distinguished from surrounding soft tissue, as observed early in Röntgen's images of his assistant's and wife's hands, whereas differences in the soft tissue itself cannot be visualised with sufficient contrast.

This limitation is addressed by X-ray phase-contrast imaging, introduced 70 years after Röntgen's early work (Bonse and Hart 1965). Instead of measuring the absorption in the sample, the X-ray phase-contrast imaging approach exploits the phase shift of the X-rays as they travel through the specimen. It has been demonstrated that utilising the phase information can significantly increase image contrast between features of similar densities, in particular for biomedical soft tissue (Fitzgerald 2000; Momose et al. 1996). Furthermore, X-ray phase-contrast imaging has the potential to lead to better dose efficiency, a crucial factor for medical imaging applications (Lewis 2004). This is due to the fact that the X-ray phase-shift cross-section drops less rapidly with the X-ray energy than the absorption cross-section. This allows for the use of higher X-ray energies, which leads to a reduction in the dose absorbed by the sample. Since it was first introduced, X-ray phase-contrast imaging has found a large range of applications originally mainly for medical and biomedical imaging, but later also in other areas such as materials science, geology and archaeology, as well as metrology (for the characterisation of X-ray optics) and wavefront sensing (for the analysis of the X-ray beam itself).

Different methods have been proposed to extract the X-ray phase-shift information about the sample. They all rely on translating the phase shift into intensity variations in the observation plane, which can be recorded by a detection system, as will be explained in more detail in the next chapter of this thesis. Some of the methods not only deliver the phase-contrast image but also complementary X-ray transmission and small-angle scattering information from the same data set. The latter is commonly referred to as dark-field signal in this context (Nesterets 2008; Yashiro et al. 2010).

Among these multimodal techniques, X-ray grating interferometry (David et al. 2002; Momose et al. 2003; Weitkamp et al. 2005) has gained popularity in the imaging community during the last decade due to its quantitative character, high phase sensitivity and compatibility with low-brilliance polychromatic X-ray sources that allowed for its translation to the laboratory (Pfeiffer et al. 2006) and raises hopes for its future implementation in the clinics. The principle of X-ray grating interferometry is to use an X-ray beam-splitting grating to create a periodic interference pattern downstream in the detection plane, which is then used as a wavefront marker. The information on the specimen is encoded in modulations of this reference pattern arising when the sample is inserted into the beam path. These are subsequently decoded computationally to extract the transmission, refraction and small-angle scattering signals of the sample from the change in intensity, the lateral displacement and the change in visibility of the reference pattern, respectively.

A similar idea is the basis for the most recently proposed phase-sensitive (and multimodal) imaging method, namely X-ray speckle-based imaging (Bérujon et al. 2012a, b; Morgan et al. 2012). The periodic grating pattern is replaced by a random pattern, known as X-ray near-field speckle pattern (Cerbino et al. 2008). The latter is produced by placing a diffuser, i.e. a material containing small randomly distributed particles, into the X-ray beam, which leads to X-ray scattering and interference effects. X-ray speckle-based imaging has raised great attention in the last years as it can reach a very high phase sensitivity down to a few nanoradians angular resolution and can be operated with a simple setup that does not require additional specialised equipment and is compatible with polychromatic and divergent beams. Commonly, cheap and widely available sandpaper is used as a diffuser, making this method significantly less costly than many other methods.

These two phase-sensitive imaging methods are the subject of this Ph.D. thesis, which explores X-ray phase-sensitive imaging with a single, periodic or random, phase modulator (grating or diffuser) as a wavefront marker to access the phase-contrast and other complementary multimodal signals of a sample under investigation. The approach of using a single phase modulator allows for an easily implemented, flexible experimental setup that has the potential for wider uptake by the user communities, also at X-ray laboratory sources and in clinical environments. The main focus of the project is the further development and optimisation of the relatively young X-ray speckle-based imaging technique, exploring advanced data acquisition and reconstruction approaches, but also demonstrating its potential for a range of applications. The latter include high-contrast, quantitative phase-contrast and multimodal imaging in the fields of biomedical research, geology and materials science, in particular in three-dimensional (3D) tomographic implementation, as well as X-ray optics characterisation. In addition to the studies and developments on the speckle-based technique, X-ray grating interferometry using a single grating was investigated and optimised for biomedical imaging applications. Although the process for creating the interference pattern and the commonly applied algorithms for signal extraction differ for X-ray grating interferometry and speckle-based imaging, both share the same basic principles of signal generation. In fact, it will be shown in this thesis that the two approaches can be unified in a single data acquisition and reconstruction method, which was developed during this Ph.D. project. This bridges the gap between grating- and speckle-based methods and generalises the concepts discussed in this thesis to any kind of reference pattern, making it transferable to most existing setups designed for X-ray phase-contrast imaging. Moreover, the approach can be extended to laboratory sources, which makes it widely accessible for research applications and future clinical and industrial use.

1.2 Outline of this Thesis

This thesis contains the results of newly developed concepts and experimental validations of X-ray single-grating and X-ray speckle-based phase-sensitive imaging. It is organised as follows.

Chapter 2 gives an overview of the fundamental principles of X-ray imaging that are essential for the work presented later in the thesis. Starting from the basics of the interaction of X-rays with matter, the concepts and implications of temporal and spatial coherence are explained, followed by a summary of different X-ray phase-contrast imaging methods. For the latter, a more detailed overview of the techniques relevant to this thesis is given. In the last sections of the chapter, the principles of computed tomography as well as a summary of different types of X-ray sources, in particular the ones used during this Ph.D. project, is given.

Chapter 3 provides some basic information on the layout, specifications, instrumentation and equipment of the two synchrotron beamlines at which most of the experiments during this Ph.D. project were carried out: beamline I13 at Diamond Light Source (UK) and beamline ID19 at the European Synchrotron Radiation Facility (France). At the end of the chapter, the contributions to these beamlines resulting from this Ph.D. project are summarised.

Chapter 4 presents results on X-ray grating interferometry with a single grating. It starts with the theoretical background on the working principles, signal generation and extraction and developments of X-ray grating interferometry, followed by the demonstration of X-ray single-grating interferometry at Diamond I13 beamline for high-contrast 3D biomedical imaging. In the last section of the chapter, more recent results on the implementation of single-grating interferometry at I13 are shown.

Chapter 5 contains a comprehensive literature review of the concepts and state of the art of X-ray speckle-based imaging. Starting from the principles of X-ray near-field speckle and its use as a wavefront marker, the image formation and reconstruction processes are explained and the existing experimental implementations are outlined. Furthermore, an overview of the developments and applications of X-ray speckle-based imaging to date is given, including some most recently reported advances.

Chapter 6 introduces the unified modulated pattern analysis (UMPA), which unifies the X-ray grating- and speckle-based imaging techniques in a single approach and is one of the main developments achieved during this Ph.D. project. The chapter starts with a section exploring the performance and limitations of existing operational modes for speckle-based imaging. This is followed by the first demonstration of UMPA, which is shown to address some of the main limitations of previous implementations of the speckle- and grating-based techniques. After introducing the principles of UMPA data acquisition and analysis, the potential of the method for multimodal imaging is experimentally demonstrated and quantitatively analysed on a test sample and a more complex specimen. It is, moreover, shown that UMPA can be applied not only to random speckle but also periodic reference patterns. The last section of the chapter contains a more detailed study of the tunable character of UMPA and an analysis of the effects of different scan and reconstruction parameters on the image quality.

In Chap. 7, the UMPA approach implemented with both a random and a periodic reference pattern is applied to X-ray optics characterisation for the analysis of two different X-ray refractive lenses. Aberrations in the focussing behaviour due to previous beam damage and fabrication errors are successfully identified and quantified using UMPA phase-sensitive imaging.

Chapter 8 presents another major contribution of this project to the field of X-ray speckle-based imaging: the first speckle-based phase tomography of a scientifically relevant specimen using UMPA, demonstrating its potential for biomedical imaging.

UMPA phase tomographies of various unstained biomedical soft-tissues samples, such as a mouse testicle, a mouse kidney and human brain tissue, are shown and evaluated qualitatively as well as quantitatively. An in-depth analysis of the results on the murine kidney illustrates the great potential of X-ray speckle-based phase tomography for 3D virtual histology, giving unprecedented insights into the interrelationship and connectivity of features within fully hydrated biomedical specimens without the need for contrast agents.

Chapter 9 reports on some of the recent and ongoing work conducted during this Ph.D. project that is aimed at making X-ray speckle-based imaging adaptable to various different experimental conditions and setups and widening its accessibility to the user community. This includes the optimisation of setup components and the extension of the method to higher X-ray energies as well as polychromatic, low-brilliance laboratory X-ray sources. A new type of customisable speckle diffuser is presented, which has the potential to optimise the imaging setup by adapting the speckle properties to specific experimental conditions. Furthermore, high-energy speckle imaging with the UMPA technique is demonstrated on volcanic rock and mortar samples, extending the technique to new areas of research. In the last section of the chapter, the translation of UMPA to a laboratory X-ray system is reported and its performance is illustrated in a proof-of-principle measurement of a test sample and a bug.

The thesis ends with Chap. 10, which contains a summary and conclusions of the work presented in the previous chapters and a guide of which phase-sensitive imaging method might be most suitable for given experimental conditions. The chapter closes with a discussion on future developments and perspectives.

1.3 Contributions

The main work conducted during this Ph.D. project and presented in this thesis, is based on ideas and concepts conceived by the author and her primary supervisors Dr. Irene Zanette and Prof. Pierre Thibault. Further work was performed in collaboration with a number of European research groups, some on topics pursued in this thesis, some not directly related. The major part of the research within this project is focussed on X-ray grating- and speckle-based imaging. This involved collaborations with several people, in particular for beamtime support and for the supply and preparation of samples. Their contributions are mentioned in the relevant sections. For the parts of this thesis based on previously published papers, collaborators are not explicitly mentioned but can be found in the author list of the related publications.

Specifically, the contributions of the Ph.D. candidate to the main projects presented in this thesis are summarised in the following:

- *Implementation and applications of X-ray grating interferometry at Diamond I13-2 (Chap.* 4*)*: The project was initiated by Dr. Irene Zanette and continued by the Ph.D. candidate. Planning and experiments were led by the Ph.D. candidate and measurements were performed with the assistance of beamline staff and collaborators. The specimens were prepared and provided by collaborators. All data analysis was performed by the Ph.D. candidate.

- *Development of the unified modulated pattern analysis (UMPA) for speckle- and grating-based imaging (Chap.* 6*)*: The initial idea of the UMPA data acquisition and analysis method and the first version of the basic Python code were conceived by Prof. Pierre Thibault and Dr. Irene Zanette. The Ph.D. candidate parallelised and optimised the code and carried out first performance tests. All experiments using UMPA were initiated, planned and led by the candidate with input from her supervisors. Beamline staff and collaborators provided beamtime support and samples. Data analysis was performed by the candidate.
- *Optics characterisation with the unified modulated pattern analysis (Chap.* 7*)*: The initial idea for characterising refractive lenses was proposed by the Ph.D. candidate in discussion with collaborators Dr. Frieder Koch and Dr. Arndt Last. Experiment planning and measurements were led by the candidate with support from beamline staff and collaborators. Collaborators provided the samples and information on them. All analysis was performed by the candidate.
- *3D virtual histology with the unified modulated pattern analysis (Chap.* 8*)*: The idea was conceived by the Ph.D. candidate. The experiments were planned and led by the candidate with support from beamline staff and collaborators. The data analysis was performed by the candidate. Some analysis steps, such as the 3D visualisation of the reconstructed data, videos and the conventional histology procedure, were carried out with support from collaborators, as indicated in the relevant sections. Samples were prepared and provided by collaborators.
- *Development of customisable phase modulators (Chap. 9, Sect.* 9.2*)*: The principle of customising phase modulators for speckle-based imaging was conceived by the Ph.D. candidate together with the collaborator Dr. Joan Vila-Comamala. Dr. Joan Vila-Comamala had the idea of using the technique of metal-assisted chemical etching and produced the phase modulators. The experiments were planned and carried out by the Ph.D. candidate and the collaborator with support from beamline staff. The sample was provided by Dr. Johannes Ihli. Data reconstruction was performed by the Ph.D. candidate.
- *Investigation of geological and materials science samples (Chap. 9, Sect.* 9.3*)*: The experiment was initiated and planned by collaborators Dr. Tunhe Zhou and Dr. Beverley Coldwell. The measurements were carried out by the Ph.D. candidate together with the collaborators and with support from beamline staff. The samples and information on them were provided by Dr. Beverley Coldwell and Dr. Fei Yang. Analysis of the data presented in this thesis was performed by the Ph.D. candidate.
- *First implementation of the unified modulated pattern analysis at a laboratory source (Chap. 9, Sect.* 9.4*)*: The laboratory setup with the liquid-metal-jet X-ray source was conceived by the Ph.D. candidate's supervisor Prof. Pierre Thibault. The experimental arrangement and procedure for lab-based UMPA speckle imag-

ing were developed and planned by the candidate in discussion with Dr. Irene Zanette and Prof. Pierre Thibault. The experiment was led by the candidate and carried out with support from collaborators as listed in Sect. 9.4. All data analysis was performed by the candidate.

In addition to the measurements and results presented in the following chapters, collaborative work on X-ray grating interferometry was performed at Diamond I13 and ESRF ID19 beamlines with the group of Prof. Bert Müller, Biomaterials Science Centre at the University of Basel (Switzerland), mainly for imaging of brain tissue. This resulted in a number of co-authored conference proceedings, see Schulz et al. (2016, 2017), Khimchenko et al. (2016). Further work on the development of X-ray speckle-based imaging was conducted in collaboration with Dr. Tunhe Zhou and Jenny Romell, first at the liquid-metal-jet laboratory X-ray source at KTH Stockholm (Sweden) in the group of Prof. Hans Hertz and later at Diamond I13 and Diamond I12 beamlines (UK), leading to co-authored publications, see Zanette et al. (2014), Zanette et al. (2015), Zhou et al. (2015), Zhou et al. (2016), Romell et al. (2017). In another collaboration, both speckle- and grating-based imaging were used for high-speed differential phase-contrast imaging at ESRF ID19 beamline for the visualisation of fast processes. This research was carried out with Dr. Patrik Vagovič, DESY and XFEL (Germany) and Dr. Margie Olbinado, ESRF (France), as evidenced in Olbinado et al. (2018); Vagovič et al. (2019).

Although not discussed in this thesis, the author was also involved in experiments and data analysis using X-ray single-distance propagation-based (inline) phase-contrast imaging to investigate biomedical and palaeontological specimens. Propagation-based phase-contrast measurements were performed in a collaboration with the group of Prof. Bert Müller, Biomaterials Science Centre at the University of Basel (Switzerland), for the high-resolution visualisation of human brain tissue and further analysis such as cell quantification, as reported in resulting articles, see Hieber et al. (2016); Khimchenko et al. (2016, 2017). The aim of another project in collaboration with Dr. Irvin Teh and Prof. Jürgen Schneider, previously Radcliffe Department of Medicine, University of Oxford (UK), now Leeds Institute of Cardiovascular & Metabolic Medicine, University of Leeds (UK), was the high-contrast, high-resolution visualisation of rat and mouse heart tissue for the validation of diffusion-tensor magnetic resonance imaging (MRI) data. The results have been published as Teh et al. (2017, 2018). The third biomedical application of propagation-based imaging during this Ph.D. project was a collaboration with Dr. Carles Bosch Piñol and Prof. Andreas Schaefer, Francis Crick Institute London (UK), on the visualisation of the murine olfactory tube in the brain, which is an essential part of the olfactory sensory system.

Apart from biomedical imaging, contributions in the scope of this Ph.D. project include propagation-based phase-contrast imaging of palaeontological specimens in collaboration with the group of Prof. Roger Benson, Department of Earth Sciences, University of Oxford (UK). In this project, X-ray phase-contrast imaging at high spatial resolution was used to visualise osteocytes, i.e. bone cells, in a large number of fossilised fish bone samples from different points in time to study the evolution of the cell size in bony fishes.

Moreover, the author of this thesis was involved in work on absorption-based micro computed tomography of biomedical specimens based on staining with contrast agents, which she had initiated during her Bachelor project in the group of Prof. Franz Pfeiffer at Technical University Munich (Germany). This resulted in a co-authored journal publication, see Bidola et al. (2019).

A list of the first and co-authored publications derived from the main work and collaborative side projects conducted during this Ph.D. project can be found in the author's Curriculum Vitae at the end of this book. It also includes a list of invited and contributed talks by the Ph.D. candidate given at international conferences, workshops and meetings.

References

Bérujon S, Wang H, Sawhney K (2012a) X-ray multimodal imaging using a random-phase object. Phys Rev A 86(6):063813

Bérujon S, Ziegler E, Cerbino R, Peverini L (2012b) Two-dimensional x-ray beam phase sensing. Phys Rev Lett 108(15):158102

Bidola P, Martins de Souza e Silva J, Achterhold K, Munkhbaatar E, Jost PJ, Meinhardt A-L, Taphorn K, Zdora M-C, Pfeiffer F, Herzen J (2019) A step towards valid detection and quantification of lung cancer volume in experimental mice with contrast agent-based X-ray microtomography. Sci Rep 9(1):1325

Bonse U, Hart M (1965) An X-ray interferometer. Appl Phys Lett 6:155–156

Cerbino R, Peverini L, Potenza MAC, Robert A, Bösecke P, Giglio M (2008) X-ray-scattering information obtained from near-field speckle. Nat. Phys 4(3):238–243

Dam H (1896) The new marvel in photography. A visit to Professor Röntgen at his laboratory in Würzburg.-His own account of his great discovery.-Interesting experiments with the cathode rays.-Practical uses of the new photography. McClure's Mag. 6(5)

David C, Nöhammer B, Solak HH, Ziegler E (2002) Differential x-ray phase contrast imaging using a shearing interferometer. Appl Phys Lett 81(17):3287–3289

Fitzgerald R (2000) Phase-sensitive X-ray imaging. Phys Today 53(7):23–26

Hieber SE, Bikis C, Khimchenko A, Schulz G, Deyhle H, Thalmann P, Chicherova N, Rack A, Zdora M-C, Zanette I, Schweighauser G, Hench J, Müller B (2016) Computational cell quantification in the human brain tissues based on hard X-ray phase-contrast tomograms. Proc SPIE 9967:99670K

Khimchenko A, Bikis C, Schulz G, Zdora M-C, Zanette I, Vila-Comamala J, Schweighauser G, Hench J, Hieber SE, Deyhle H, Thalmann P, Müller B (2017) Hard X-ray submicrometer tomography of human brain tissue at Diamond Light Source. J Phys Conf Ser 849(1):012030

Khimchenko A, Schulz G, Deyhle H, Thalmann P, Zanette I, Zdora M-C, Bikis C, Hipp A, Hieber SE, Schweighauser G, Hench J, Müller B (2016) X-ray micro-tomography for investigations of brain tissues on cellular level. Proc SPIE 9967:996703

Lewis RA (2004) Medical phase contrast x-ray imaging: current status and future prospects. Phys Med Biol 49(16):3573–3583

Momose A, Kawamoto S, Koyama I, Hamaishi Y, Takai K, Suzuki Y (2003) Demonstration of X-ray talbot interferometry. Jpn J Appl Phys 42:L866–L868

Momose A, Takeda T, Itai Y, Hirano K (1996) Phase-contrast X-ray computed tomography for observing biological soft tissues. Nat Med 2:473–475

Morgan KS, Paganin DM, Siu KKW (2012) X-ray phase imaging with a paper analyzer. Appl Phys Lett 100(12):124102

Nesterets YI (2008) On the origins of decoherence and extinction contrast in phase-contrast imaging. Opt Comm 281(4):533–542

Olbinado M, Grenzer J, Pradel P, Resseguier TD, Vagovic P, Zdora M-C, Guzenko V, David C, Rack A (2018) Advances in indirect detector systems for ultra high-speed hard X-ray imaging with synchrotron light. J Instrum 13(04):C04004

Pfeiffer F, Weitkamp T, Bunk O, David C (2006) Phase retrieval and differential phase-contrast imaging with low-brilliance X-ray sources. Nat Phys 2:258–261

Romell J, Zhou T, Zdora M, Sala S, Koch F, Hertz H, Burvall A (2017) Comparison of laboratory grating-based and speckle-tracking x-ray phase-contrast imaging. J Phys Conf Ser 849(1):012035

Röntgen WC (1898) Über eine neue Art von Strahlen (Erste Mittheilung). Ann Phys (Berl) 300(1):1–11

Schulz G, Götz C, Deyhle H, Müller-Gerbl M, Zanette I, Zdora M-C, Khimchenko A, Thalmann P, Rack A, Müller B (2016) Hierarchical imaging of the human knee. Proc SPIE 9967:99670R

Schulz G, Götz C, Müller-Gerbl M, Zanette I, Zdora M-C, Khimchenko A, Deyhle H, Thalmann P, Müller B (2017) Multimodal imaging of the human knee down to the cellular level. J Phys Conf Ser 849(1):012026

Teh I, McClymont D, Zdora M-C, Whittington H, Gehmlich K, Rau C, Lygate CA, Schneider JE (2018) Validation of diffusion tensor imaging in diseased myocardium. In: Proceedings of the joint annual meeting ISMRM-ESMRMB 2018, international society for magnetic resonance in medicine. Joint Annual Meeting ISMRM-ESMRMB 2018

Teh I, McClymont D, Zdora M-C, Whittington HJ, Davidoiu V, Lee J, Lygate CA, Rau C, Zanette I, Schneider JE (2017) Validation of diffusion tensor MRI measurements of cardiac microstructure with structure tensor synchrotron radiation imaging. J Cardiovasc Magn Reson 19(1):31

Vagovič P, Sato T, Mikeš L, Mills G, Graceffa R, Mattsson F, Villanueva-Perez P, Ershov A, Faragó T, Uličný J, Kirkwood H, Letrun R, Mokso R, Zdora M-C, Olbinado MP, Rack A, Baumbach T, Schulz J, Meents A, Chapman HN et al (2019) Megahertz x-ray microscopy at x-ray free-electron laser and synchrotron sources. Optica 6(9):1106–1109

Weitkamp T, Diaz A, David C, Pfeiffer F, Stampanoni M, Cloetens P, Ziegler E (2005) X-ray phase imaging with a grating interferometer. Opt Express 13(16):6296–6304

Yashiro W, Terui Y, Kawabata K, Momose A (2010) On the origin of visibility contrast in x-ray Talbot interferometry. Opt Express 18(16):16890–16901

Zanette I, Zdora M-C, Zhou T, Burvall A, Larsson DH, Thibault P, Hertz HM, Pfeiffer F (2015) X-ray microtomography using correlation of near-field speckles for material characterization. Proc Natl Acad Sci USA 112(41):12569–12573

Zanette I, Zhou T, Burvall A, Lundström U, Larsson DH, Zdora M-C, Thibault P, Pfeiffer F, Hertz HM (2014) Speckle-based X-ray phase-contrast and dark-field imaging with a laboratory source. Phys Rev Lett 112(25):253903

Zhou T, Zanette I, Zdora M-C, Lundström U, Larsson DH, Hertz HM, Pfeiffer F, Burvall A (2015) Speckle-based x-ray phase-contrast imaging with a laboratory source and the scanning technique. Opt Lett 40(12):2822–2825

Zhou T, Zdora M-C, Zanette I, Romell J, Hertz HM, Burvall A (2016) Noise analysis of speckle-based x-ray phase-contrast imaging. Opt Lett 41(23):5490–5493

Chapter 2
Principles of X-ray Imaging

X-rays are electromagnetic radiation emitted by electrons outside the nucleus of atoms. They typically have energies in the range of 100 eV to 500 keV, corresponding to a wavelength range from 2.5 pm to 10 nm (Als-Nielsen and McMorrow 2011). Compared to visible light, X-rays have a much higher penetration power through dense materials and in particular hard X-rays (with an energy of >10 keV) have the ability to penetrate deep into matter. This was first observed by Röntgen when he discovered X-rays in 1895 (Röntgen 1895, 1898). X-rays were immediately used to investigate the inner structure of materials as well as the human body and they have been exploited for various imaging applications ever since.

While the first applications of X-rays were based on their absorption in the material, e.g. for medical imaging of bones (Codman 1896; Editorial 1896b, a; Spiegel 1995), it was discovered later that, analogous to the visible light case (Zernike 1942, 1955), also their phase shift can be exploited for signal generation (Bonse and Hart 1965). However, the extraction of phase-contrast information is not as straightforward as for absorption imaging as detectors can only measure the beam intensity, and not the phase shift. Therefore, methods were developed to encode this information in intensity variations that could be recorded by the detector. The first X-ray phase-contrast setup was proposed in 1965 by Bonse and Hart (Bonse and Hart 1965) who built a crystal interferometer to visualise the X-ray phase shift. However, X-ray phase contrast only gained increased interest with the development of powerful and coherent X-ray sources at the end of the 20th century. In particular, the discovery of X-ray propagation-based phase-contrast imaging at the European Synchrotron Radiation Facility (ESRF) was a major milestone in the popularity of the X-ray phase-contrast modality (Snigirev et al. 1995, 1996; Raven et al. 1996). An interesting fact here is that both the first discovery of X-rays by Röntgen and the first discovery of propagation-based phase contrast happened by chance when the researchers were investigating on a different topic. Röntgen was performing experiments with a Crookes tube and noticed the green fluorescence light on the phosphor screen covering the tube, which was caused by X-rays generated in the tube. X-ray

M.-C. Zdora, *X-ray Phase-Contrast Imaging Using Near-Field Speckles*,
Springer Theses, https://doi.org/10.1007/978-3-030-66329-2_2

propagation-based phase contrast was first discovered when Snigirev and co-workers at the ESRF observed fringes around an object that was accidentally left in the beam. These edge-enhancement fringes are created by free-space propagation of X-rays which translates the phase modulation at the edges of features into an intensity modulation, as will be explained later in this chapter.

Since then, X-ray phase-contrast imaging has seen rapidly growing interest and is the subject of ongoing research and new developments, including those reported in this thesis. Additionally, it was later recognised that a complementary image signal related to small-angle scattering of the X-rays in the sample, known as dark-field signal (Nesterets 2008; Yashiro et al. 2010; Pfeiffer et al. 2008), can be obtained with some X-ray imaging techniques including the methods discussed in this thesis.

The principles underlying the generation of X-ray absorption-based and phase-contrast imaging are outlined in this chapter to build a basis for the following content, which focusses on the X-ray imaging techniques developed and applied during this Ph.D. project. For a more in-depth treatment of the theoretical background of X-ray imaging, the reader is referred to popular X-ray science handbooks, e.g. Als-Nielsen and McMorrow (2011), Paganin (2006) and Willmott (2011), which are the main sources of the following sections.

Starting from the interactions of X-rays with matter, their implications for X-ray imaging and important X-ray properties, the main X-ray phase-sensitive imaging methods relevant for this Ph.D. project are introduced. It is then explained how complementary three-dimensional (3D) information on the sample can be extracted via tomographic imaging exploiting the absorption, phase-contrast and dark-field signals. The chapter ends with a short overview of the different available X-ray sources and their properties.

2.1 X-ray Interactions with Matter

There are three main ways X-rays with an energy in the kiloelectronvolts range may interact with the atoms of a material that it is travelling through: the photoelectric effect, Rayleigh scattering and Compton scattering. The photoelectric effect describes the absorption of an X-ray photon by an atom of the sample material leading to the expulsion of an outer-shell electron, which triggers the emission of fluorescent X-rays or an Auger electron. The cross-section σ_{pe} of photoelectric absorption is strongly dependent on the atomic number Z of the material and the photon energy E:

$$\sigma_{pe} \propto \frac{Z^4}{E^{3.5}}. \tag{2.1}$$

Apart from photoelectric absorption, X-rays can also interact with the material via coherent, elastic Rayleigh scattering (Thomson scattering for the case of a free electron that is not bound in an atom) and incoherent, inelastic Compton scattering. Both Rayleigh and Thomson scattering scale with the square of the classical electron radius r_0, which is also known as Thomson scattering length. While at higher energies Compton scattering is dominant and almost inversely proportional to the photon energy ($\sigma_{inel} \propto 1/E$), at lower energies elastic scattering has a higher interaction cross-section due to binding effects. The scattering cross-section σ_{el} for elastic Rayleigh scattering approaches $\sigma_{el} \propto Z^2$ for low energies and the inelastic Compton scattering cross-section σ_{inel} approaches $\sigma_{inel} \propto Z$ for high energies.

2.2 Complex Index of Refraction, Attenuation and Phase Shift

We have seen that photoelectric absorption, Rayleigh scattering and Compton scattering are the three main ways X-rays interact with a material. When performing an X-ray imaging experiment, these processes lead to an attenuation of the X-ray beam after passing through the specimen due to the absorption of photons and scattering out of the field of view of the detector.

The propagation of electromagnetic waves can be expressed mathematically by Maxwell's equations describing the full vector wave field. However, under the conditions of monochromatic, unpolarised X-rays, it is sufficient to assume a scalar wave field $E(x, y, z)$. Here, z is the propagation direction of the beam and x, y are the coordinates in the transverse plane. Figure 2.1 visualises the modulations of the X-ray wave $E_2(x, y, z)$ when passing through a homogeneous sample compared to the undisturbed wave field $E_1(x, y, z)$. They are caused by absorption and scattering interactions as discussed previously.

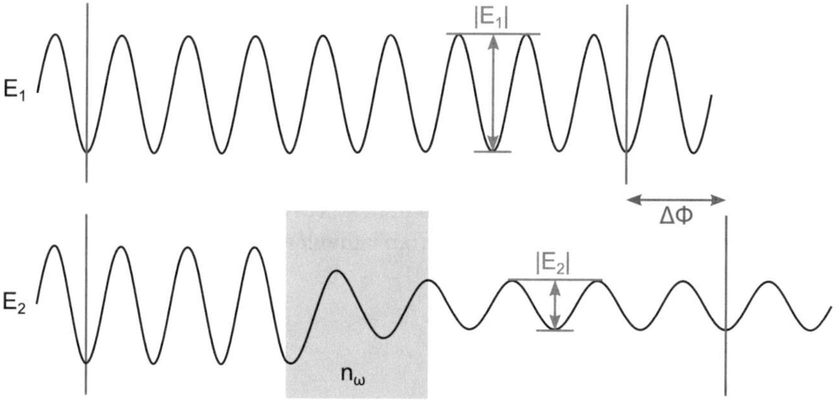

Fig. 2.1 X-ray absorption and phase shift in matter. Propagation of two monochromatic plane waves E_1 and E_2 with the same wavelength λ (frequency ω) and amplitude $|E_1| = |E_2|$. While E_1 travels through vacuum, E_2 passes through a sample of homogeneous material with complex-valued refractive index $n_\omega = 1 - \delta_\omega + i\beta_\omega$, which induces a phase shift $\Delta\Phi$ of the wave and a decrease in its amplitude

The quantity that describes how a material affects an incoming X-ray wave of frequency ω is the refractive index n_ω. We can express the wave field $E(x, y, z)$ directly after passing through a specimen placed at $z = 0$ as follows:

$$\begin{aligned} E(x, y, z) &= E_0(x, y, 0)e^{ik\int n_\omega(x,y,z)dz} \\ &= E_0(x, y, 0)\mathscr{T}(x, y, z), \end{aligned} \tag{2.2}$$

where E_0 is the incident wave field, $k = 2\pi/\lambda$ the wave number and $\mathscr{T}(x, y, z)$ the so-called object transmission function, which describes how the object modulates the wavefront.

The refractive index n_ω can be written as a complex number:

$$n_\omega(x, y, z) = 1 - \delta_\omega(x, y, z) + i\beta_\omega(x, y, z). \tag{2.3}$$

Its imaginary part β_ω quantifies the attenuating properties of the sample leading to a reduction in the amplitude of the wavefront (green arrows in Fig. 2.1), while the decrement δ_ω of the real part of the refractive index is associated with a phase shift $\Delta\Phi$ of the incoming wave (red arrow in Fig. 2.1). Combining Eqs. (2.2) and (2.3), we can write:

$$E(x, y, z) = E_0(x, y, 0)e^{ik\int(1-\delta_\omega(x,y,z))dz}e^{-k\beta_\omega(x,y,z)dz}. \tag{2.4}$$

Alternatively, the complex refractive index[1] n can be expressed by considering scattering of a wave using the atomic scattering length $f(\mathbf{Q}) = f^0(\mathbf{Q}) + f' + if''$, where $\mathbf{Q}$ is the scattering vector, i.e. the wave vector transfer $\mathbf{Q} = \mathbf{k} - \mathbf{k'}$ from the incident (wave vector $\mathbf{k}$) to the scattered (wave vector $\mathbf{k'}$) wave. The factors f' and f'' are the real and imaginary parts, respectively, of the atomic dispersion correction to $f^0(\mathbf{Q})$ (Als-Nielsen and McMorrow 2011):

$$n = 1 - \frac{2\pi\rho_a r_0}{k^2}\left(f^0(\mathbf{Q}) + f' + if''\right). \tag{2.5}$$

Here, ρ_a is the atomic number density, $r_0 = 2.82 \times 10^{-6}$ Å the Thomson scattering length and k the wave vector. We can reduce this to the case of forward scattering as a result of the scattering contributions from individual atoms and hence use $\mathbf{Q} = 0$ and $f^0(\mathbf{0}) = Z$:

$$n = 1 - \frac{2\pi\rho_a r_0}{k^2}\left(Z + f' + if''\right), \tag{2.6}$$

where Z is the atomic number.

The factor f' corrects for the fact that the electrons in the atom are not free. It is only significant close to the absorption edges of the material. At energies far from

[1]The subscript ω is dismissed from this point for the sake of readability.

the absorption edges, the electrons can practically be considered free and f' can be approximated by zero.

From Eqs. (2.6) and (2.3) we see that the refractive index decrement δ can be described as:

$$\delta = \frac{2\pi \rho_a r_0}{k^2}(Z + f') = \frac{\rho_a}{k}\sigma_{ph}, \tag{2.7}$$

where $\sigma_{ph} = \frac{2\pi r_0}{k}(Z + f')$ represents the phase-shift interaction cross-section. Furthermore, the imaginary part β of the refractive index can be written as:

$$\beta = -\frac{2\pi \rho_a r_0}{k^2} f'' = \frac{\rho_a \sigma_a}{2k} \tag{2.8}$$

with the attenuation cross-section $\sigma_a = -\frac{4\pi r_0}{k} f''$.

The above equations are valid for the interaction with atoms of a single element. For the case of a compound material, the sum over all the elements l present in the sample is taken:

$$\delta = \frac{2\pi r_0}{k^2} \sum_l \rho_{a;l}(Z_l + f_l') = \frac{1}{k} \sum_l \rho_{a;l}\sigma_{ph;l} \tag{2.9}$$

and

$$\beta = -\frac{2\pi r_0}{k^2} \sum_l \rho_{a;l} f_l'' = \frac{1}{2k} \sum_l \rho_{a;l}\sigma_{a;l}. \tag{2.10}$$

For low-Z materials, such as biomedical soft tissue, δ can be up to three orders of magnitude larger than β, as shown in Fig. 2.2. It has been demonstrated that phase-contrast imaging is much more sensitive to small density variations in the specimen than conventional absorption X-ray imaging, with the potential for a dramatic improvement in image contrast. Furthermore, it can be observed in Fig. 2.2 that at low X-ray energies the attenuation cross-section drops more rapidly with the energy than the phase-shift cross-section, which as a consequence makes phase-contrast imaging more dose efficient.

For a practical implementation of X-ray imaging, it is important to note that the signal obtained from a measurement is the beam intensity $I(x, y, z) = |E(x, y, z)|^2$. This means that the absorption of X-rays in a sample can be directly measured by the detector, but the phase information is not directly accessible from the recorded intensity. As it will be shown in Sect. 2.4, it is, however, possible to translate the phase shift into intensity variations in the detector, which can subsequently be decoded computationally to retrieve the phase-contrast signal.

The attenuation signal, i.e. the decrease in amplitude of the wave field due to the presence of the specimen, can be determined directly by measuring the intensities of the wavefront $E(x, y, z)$ after passing through the sample and the incident wavefront $E_0(x, y, 0)$:

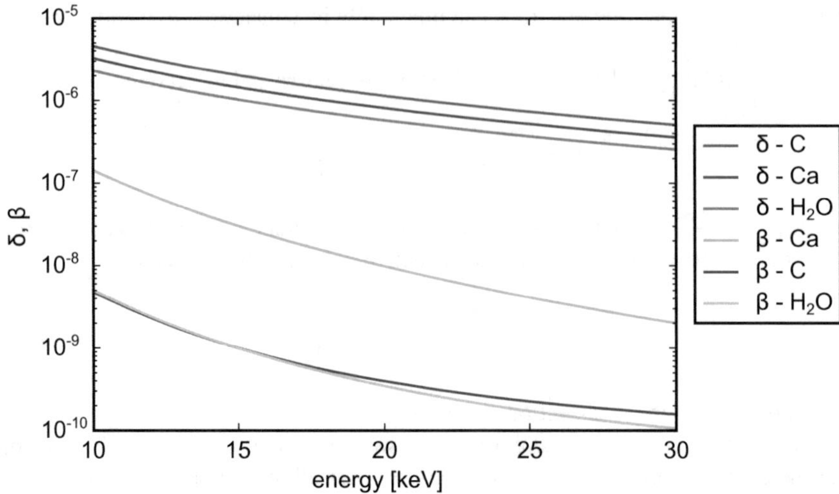

Fig. 2.2 Absorption and phase-shifting properties of different materials over an X-ray energy range of 10–30 keV, far from the absorption edges. Refractive index decrement δ and absorption coefficient β of carbon (C), calcium (Ca) and water (H_2O), which are the main elements present in biomedical specimens. It can be observed that δ is several orders of magnitude larger than β for low-Z materials (C, H_2O) and drops less rapidly with the X-ray energy. Data from Henke et al. (1993)

$$\begin{aligned}A(x, y, z) = 1 - T = 1 - \left|\frac{E(x, y, z)}{E_0(x, y, 0)}\right|^2 = 1 - \left|e^{ik\int n(x,y,z)dz}\right|^2 = 1 - e^{-2k\int \beta(x,y,z)dz} \\ = 1 - e^{-\int \mu(x,y,z)dz},\end{aligned} \tag{2.11}$$

where T is the transmission of X-rays through the sample and μ the linear attenuation coefficient, which is defined as:

$$\mu(x, y, z) = 2\,k\,\beta(x, y, z). \tag{2.12}$$

The phase shift $\Delta\Phi$ experienced by the X-rays in the sample is directly related to the refractive index decrement δ and can be expressed as:

$$\Delta\Phi(x, y, z) = k \int \delta(x, y, z)dz. \tag{2.13}$$

As only intensities can be measured by the detector, the phase shift needs to be encoded in an intensity variation. Several different techniques have been developed for this purpose in the last decades. Common X-ray phase-contrast imaging techniques give access to either the phase shift of the X-rays or its second or first derivative by exploiting free-space propagation or interferometric methods, as explained in more detail in Sect. 2.4.

The work of this Ph.D. project is mainly based on differential phase-contrast imaging methods, measuring the first derivative[2] of the phase $\partial\Phi/\partial x$ and/or $\partial\Phi/\partial y$. These typically rely on the phenomenon of refraction, the change in direction of the incoming X-ray beam. When passing through a sample of varying thickness, the X-rays entering the sample at different transverse positions experience different phase shifts. This effectively leads to a tilt of the outgoing wavefront, i.e. a change in direction by an angle α known as refraction angle. The refraction angle contains a component in the horizontal x and the vertical direction y, which can be expressed as (Paganin 2006):

$$\begin{aligned}\alpha_x(x, y) &= \frac{\lambda}{2\pi}\frac{\partial\Phi}{\partial x} = \frac{\partial}{\partial x}\int\delta(x, y, z)dz\\ \alpha_y(x, y) &= \frac{\lambda}{2\pi}\frac{\partial\Phi}{\partial y} = \frac{\partial}{\partial y}\int\delta(x, y, z)dz.\end{aligned} \tag{2.14}$$

2.3 X-ray Coherence

One of the key properties of an X-ray beam for many imaging applications that make use of the X-ray phase shift is its coherence in the observation plane. A sufficient degree of coherence is necessary for an interference pattern to form and be observed in the detector plane.[3] This is essential for many phase-contrast imaging methods, which rely on translating the phase modulations of the X-ray wavefront into intensity modulations that can be recorded by the detector.

The term coherence describes the correlation between wave fields in space and time and quantifies the length scale after which two waves are out of phase. One distinguishes between two types of coherence properties: transverse/spatial and longitudinal/temporal coherence.

2.3.1 *Transverse Coherence*

The spatial or transverse coherence is related to source size D and the propagation distance z. It can be quantified in terms of the transverse coherence length L_T, which is defined as the maximum separation between two points so that interference

[2]Some methods are only sensitive to the differential phase in one direction whereas others can measure the signal in both the horizontal and the vertical direction.

[3]It should be noted here that throughout the entire thesis the expressions "detector plane", "detector distance" etc. refer to the location of the part of the detector system where the X-rays are actually detected. For the measurements presented in the following chapters, this is the position of the scintillation screen that is used to convert X-rays to visible light photons. The latter are then recorded by a CCD or sCMOS camera after passing through an optical microscope.

effects can still be observed, i.e. the wave fields are correlated. It can be derived from geometrical considerations (Als-Nielsen and McMorrow 2011):

$$L_T = \frac{1}{2}\frac{\lambda}{D/z}, \tag{2.15}$$

where λ is the wavelength, D the source diameter and z the distance between source and observation plane.

It can immediately be concluded from Eq. (2.15) that for a high spatial coherence, i.e. a long transverse coherence length, it is beneficial to have a small source size and/or be far from the source. The latter is the reason why most high-coherence synchrotron beamlines are long beamlines with their experimental hutches up to hundreds of metres away from the X-ray source. It can also be understood from Eq. (2.15) that high-coherence imaging is more straightforward to perform with X-rays of longer wavelength (lower energy) and high-coherence imaging techniques are more challenging to be applied to high-energy applications.

2.3.2 *Longitudinal Coherence*

The temporal or longitudinal coherence is a measure for the correlation between wave fields of different wavelengths, which are travelling in the same direction. The longitudinal coherence length L_L is the distance after which the two wavefronts with a wavelength difference of $\Delta\lambda$ are out of phase (phase shift $\lambda/2$) (Als-Nielsen and McMorrow 2011):

$$L_L = \frac{1}{2}\frac{\lambda^2}{\Delta\lambda}. \tag{2.16}$$

In practice, Eq. (2.16) tells us that a long longitudinal coherence length requires a good energy resolution $\Delta\lambda/\lambda$. Therefore, monochromators are used for most coherent imaging applications. They select a narrow bandwidth (typically with an energy resolution of $\Delta\lambda/\lambda \approx 10^{-3} - 10^{-4}$) of the incoming X-ray beam spectrum and hence allow for a larger longitudinal coherence length compared to polychromatic illumination.

2.3.3 *Coherence Lengths at Synchrotron and Laboratory X-ray Sources*

To understand the limitations of X-ray imaging techniques that are based on some level of coherence, it is useful to get an idea about typical coherence lengths at

Table 2.1 Theoretical transverse coherence lengths calculated with Eq. (2.15) for different synchrotron beamlines and X-ray laboratory systems. Transverse coherence lengths $L_{T;\,h}$ in the horizontal and $L_{T;\,v}$ in the vertical for horizontal and vertical X-ray beam source sizes D_h and D_v (full width at half maximum) at an energy E and at a distance z from the source. It should be noted that the actual coherence lengths are shorter due to optics instabilities, amongst other factors. Source sizes were obtained from Malte Storm, Diamond Light Source, for Diamond I13 (simulated based on storage ring parameters, undulator and coupling); estimated from electron beam parameters (Diamond Light Source Ltd 2019) for Diamond I12; taken from European Synchrotron Radiation Facility (2019) for ESRF ID19; taken from Swiss Light Source, Paul Scherrer Institute (2019) for PSI (SLS) TOMCAT; as reported by Zanette et al. (2014) for the liquid-metal-jet laboratory source; as reported by Wang et al. (2016a) for the Nikon laboratory system

Source	D_h [μm]	D_v [μm]	E [keV]	z [m]	$L_{T;\,h}$ [μm]	$L_{T;\,v}$ [μm]
Diamond I13-2	553	13	20	220	12	525
Diamond I12	350	13	53	53	2	48
ESRF ID19	120	30	20	145	37	150
PSI-SLS TOMCAT	140	45	20	25	6	17
Excillum LMJ	8	9	16	1	5	4
Nikon XT H225	16	16	100	1	0.4	0.4

different types of X-ray sources.[4] Using the definition of the transverse coherence length in Eq. (2.15), examples are given in Table 2.1 for beamlines I13 and I12 at Diamond Light Source and ID19 at the ESRF under conditions similar to the ones given for the experiments presented in the following chapters of this thesis. For comparison, the transverse coherence lengths of some X-ray laboratory sources that have been used for X-ray speckle-based and grating-based imaging experiments by the author and other groups are listed in the last two lines of Table 2.1.

It can be seen that the transverse coherence lengths at the synchrotron are significantly larger and can reach hundreds of micrometres. However, it should be noted here that the values in Table 2.1 are based on theoretical calculations only including the source size, beam energy and distance to the source. In reality, the coherence length can be significantly shorter mainly due to optics instabilities, in particular of the monochromator, which increase the apparent source size (Wagner et al. 2017).

As can be seen from Table 2.1, the source size is typically much larger in the horizontal than in the vertical direction at synchrotron sources, leading to a shorter coherence length in this direction. Hence, for imaging methods that are sensitive in both transverse directions, such as X-ray speckle-based imaging and 2D grating interferometry, the visibility of the interference pattern is not the same along both axes and the signal sensitivity is not symmetric. For high-coherence applications at

[4]For an overview of the principles of different types of X-ray sources see Sect. 2.6.

synchrotrons, such as ptychography measurements, commonly primary slits are used to decrease the source size in the horizontal direction and achieve a larger transverse coherence length, albeit at the cost of a loss in flux.

In terms of the longitudinal coherence length at synchrotron sources, values in the hard X-ray regime are typically less than 1 μm. At 20 keV, the longitudinal coherence length when using a double-crystal monochromator with energy resolution $\Delta E/E \approx 10^{-4}$ is $L_L \approx 310\,\text{nm}$ and for a multilayer monochromator with energy resolution $\Delta E/E \approx 10^{-2}$ it is $L_L \approx 3\,\text{nm}$ (using Eq. (2.16)).

X-ray laboratory sources have a broad spectrum of X-ray energies typically spanning over several tens of kiloelectronvolts. While monochromatisation is in principle possible, it would lead to extremely low flux and is therefore not performed in practice. The spectral bandwidth can, however, be significantly reduced by the use of filters. In particular, filters can be chosen in a way to mainly retain one of the characteristic peaks, which allows to achieve some level of longitudinal coherence.

2.4 Phase-Sensitive X-ray Imaging Methods

In the last decades, a number of phase-sensitive X-ray imaging techniques have been developed including crystal interferometry (Bonse and Hart 1965; Momose 1995; Momose and Fukuda 1995; Momose et al. 1998), propagation-based phase-contrast imaging (Snigirev et al. 1995; Cloetens et al. 1996; Paganin et al. 2002; Cloetens et al. 1999), analyser-based imaging (Ingal and Beliaevskaya 1995; Davis et al. 1995; Chapman et al. 1997; Pagot et al. 2003), coded aperture phase-contrast imaging (Olivo and Speller 2007; Olivo et al. 2009; Diemoz et al. 2013; Hagen et al. 2014), grating interferometry (David et al. 2002; Momose et al. 2003; Pfeiffer et al. 2006; Weitkamp et al. 2005) and recently X-ray speckle-based imaging (Bérujon et al. 2012a, b; Morgan et al. 2012; Zanette et al. 2014). While crystal interferometry gives access to the phase shift, the other methods record either its first or second derivative. Phase integration can subsequently be applied to convert the differential phase signal into a phase shift signal.

Moreover, it should be noted that while in this thesis only full-field[5] X-ray imaging methods operating in the near field (Fresnel number $\geq$1) are considered, there are a number of far-field (Fresnel number $\ll$1) imaging methods, such as coherent diffractive imaging (CDI) and far-field ptychography (Miao et al. 1999; Nugent 2010; Rodenburg 2008; Thibault et al. 2008; Dierolf et al. 2010), that are not discussed here. These methods can achieve a very high spatial resolution—better than 10 nm in 2D and down to 15 nm in 3D tomography in the hard X-ray regime (Holler et al. 2017; Pfeiffer 2018; Shi et al. 2019)—and deliver quantitative information on the electron density distribution in the specimen. Ptychography, furthermore, allows for the reconstruction of the X-ray probe from the same data set. However,

[5] I.e. the full field of view is captured in all frames of the acquired data and scanning of the sample through the field of view is not required.

these techniques impose strict requirements on the temporal and spatial coherence of the X-ray beam and the data acquisition and reconstruction protocols are elaborate and computationally intensive. Moreover, the size of the objects that can be investigated with these techniques is limited.

In the following, the three full-field X-ray phase-contrast imaging techniques that have been employed and further developed in the scope of this Ph.D. project—propagation-based, grating-based and speckle-based phase-contrast imaging—are introduced. Particular emphasis is placed on X-ray grating interferometry and speckle-based imaging, which are the main focus of this Ph.D. project. More detailed explanations on the principles of these techniques are provided in Chaps. 4 and 5. In a last subsection, X-ray near-field ptychography is introduction as this technique is conceptually related to X-ray speckle-based imaging and is often mentioned along the same lines.

2.4.1 Propagation-Based Phase-Contrast Imaging

X-ray Propagation-based or inline phase-contrast imaging (Snigirev et al. 1995; Cloetens et al. 1996; Nugent et al. 1996) uses a simple setup, which does not require any specialised components apart from the conventional imaging equipment. The X-ray beam directly impinges onto the sample to be investigated and a detector is positioned at some distance downstream to record the signal, as shown in Fig. 2.3. The technique makes use of the phenomenon of Fresnel diffraction, which leads to edge-enhancement fringes arising at the interfaces of sample features upon propagation when illuminated by an X-ray beam under conditions of sufficient spatial coherence. Propagation-based phase-contrast imaging is sensitive to the second derivative of the phase and is hence best suited for detecting edges, i.e. sharp density gradients, but is

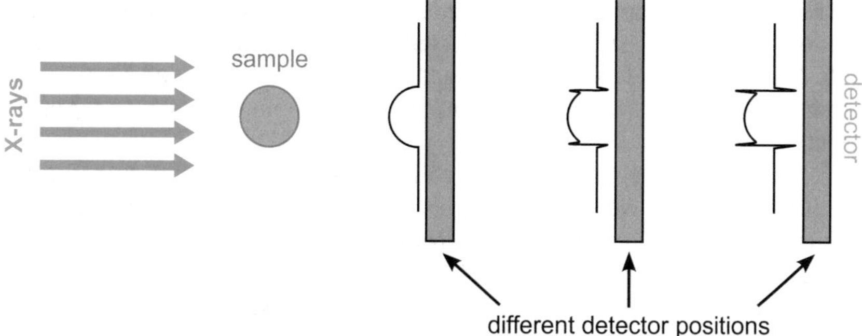

Fig. 2.3 Schematic drawing of a typical propagation-based phase-contrast imaging setup. With increasing detector distance the edge-enhancement fringes around the sample, which are caused by diffraction in the Fresnel regime, become more prominent

less sensitive to smooth density variations. The simple setup and the compatibility with polychromatic X-ray sources (Wilkins et al. 1996) make propagation-based phase-contrast imaging an easily implemented, fast and robust technique. However, it requires a relatively high degree of spatial coherence and a high-resolution detector system with a small effective pixel size that is able to resolve the fine propagation fringes.

As illustrated in Fig. 2.3, the prominence of the edge-enhancement fringes increases upon propagation. At propagation distances z, for which the relation $\sqrt{\lambda z} \ll a$ is fulfilled, features of size a are in the near-field regime and only one edge-enhancement fringe is visible. Increasing the distance z so that $\sqrt{\lambda z} \approx a$ leads to the holographic (Fresnel) regime where multiple fringes appear for the ideal case of a fully coherent source. In practice, partial coherence due to the finite source size plays a role and leads to a reduction in fringe visibility.

In the former case with $\sqrt{\lambda z} \ll a$, features with sharp edges are strongly enhanced in visibility compared to an image taken very close to the sample where propagation effects are small and mainly an absorption signal is observed. The edge-enhanced images are often used directly without any further reconstruction steps to reveal small features such as cracks and voids, in particular for materials science applications (Mayo et al. 2012). Alternatively, a phase retrieval step can be performed which leads to the removal of the edge-enhancement fringes. Phase retrieval significantly enhances image contrast and reduces noise, which enables the visualisation of very small density differences within the specimen.

A large number of approaches have been proposed for phase retrieval from propagation-based imaging data in the near field. A review of all of these would go beyond the scope of this thesis, but a few of the most commonly used ones are outlined in the following. Only direct phase retrieval methods are considered here, although it is important to note that iterative reconstruction methods exist. Most of the direct reconstruction approaches rely on a linearisation of the relationship between phase shift and recorded intensity, i.e. a linearisation of the Fourier transform of the Fresnel diffraction pattern (Langer et al. 2008). One solution is based on linearisation with respect to the propagation distance leading to the transport-of-intensity (TIE) equation (Teague 1983), which can be solved in different ways (Gureyev and Nugent 1996, 1997; Gureyev et al. 1999; Allen and Oxley 2001; Paganin and Nugent 1998; Paganin et al. 2002). Methods based on the TIE are limited to short propagation distances.

Another solution involves a linearisation of the object transmission function under assumption of a weakly absorbing sample and the use of the contrast-transfer-function (CTF) valid for the case of a slowly varying phase (Cloetens et al. 1999). The CTF describes how amplitude or phase information are transferred as a function of spatial frequency. The Fourier transforms of the intensity distribution at a given sample-detector distance and the phase can be considered linearly related under the approximation of a slowly varying phase. As the CTF shows an oscillating behaviour with spatial frequency, information at the minima is lost when using a single defocus distance. Therefore, images are taken at several longitudinal sample positions.

2.4.1.1 Single-Distance Phase Retrieval

The simplest way to perform phase retrieval of propagation-based imaging data is a single-distance approach, for which only one projection taken at a certain propagation distance is required. Paganin et al. (2002) have proposed a straightforward, robust algorithm to reconstruct the phase signal from the acquired image, which is based on the TIE concept:

$$\Phi(x, y) = \frac{\delta}{2\beta} \ln \left(\mathscr{F}^{-1} \left\{ \frac{\mathscr{F}\{I(x, y)/I_0(x, y)\}}{1 + [\lambda z \delta / (4\pi\beta)] (u^2 + v^2)} \right\} \right), \tag{2.17}$$

where δ/β is the ratio of the real and the imaginary part of the refractive index, $I(x, y)$ and $I_0(x, y)$ are the intensities measured by the detector placed at a distance z with and without the sample in the beam, respectively, λ is the X-ray wavelength and u and v are the coordinates in Fourier space. The reconstruction approach is based on solving the above-mentioned TIE (Teague 1983; Rytov et al. 1989) using Fourier methods including a number of assumptions. It is valid under conditions of monochromatic X-rays and for a sample of a single homogeneous material. A value for δ/β is provided for the reconstruction as a priori information. Furthermore, the sample needs to be placed in the near-field at a distance $z \ll a^2/\lambda$ with the smallest resolvable feature size a to avoid blurring of features (Weitkamp et al. 2011). However, in cases where the requirements of a homogeneous sample, a single X-ray wavelength and near-field conditions are not fully met or the δ/β-ratio is not known exactly, this algorithm can still deliver qualitative images with excellent contrast. The enhancement in contrast and reduction in noise compared to the raw images can be understood by looking at Eq. (2.17), which effectively acts as a low-pass filter.

For a practical implementation, one can estimate the maximum propagation distance that should be used with this technique from the width of the first Fresnel fringe $\sqrt{\lambda z}$. When assuming a minimum resolvable feature size a given by twice the effective detector pixel size p_{eff} (or the resolution of the detector system if known), the critical distance is given by (Weitkamp et al. 2011):

$$z_c = (2p_{\mathrm{eff}})^2/\lambda. \tag{2.18}$$

A few examples of critical propagation distances determined this way can be found in Table 2.2. In practice, the maximum propagation distance that should be used is typically reduced from the above theoretical value by the partial transverse coherence of the X-ray beam in the sample plane.

It is crucial to choose a sufficiently small propagation distance to preserve the best possible spatial resolution even for the smallest features present in the sample. On the other hand, high-frequency noise and artefacts, e.g. ring artefacts from dead detector pixels, scintillator defects and noise, might be more prominent for very short propagation distances as the inverse propagation distance $1/z$ contributes to the cutoff frequency of the low-pass filter in Eq. (2.17). Furthermore, care should be taken in selecting the optimal δ/β-ratio for reconstruction to avoid remaining edge-

Table 2.2 Examples of the maximum propagation distance that should be used for certain scan parameters when performing propagation-based phase-contrast imaging with a single-distance phase-retrieval algorithm

E [keV]	p_{eff} [μm]	z_c [mm]
10	0.45	6.5
	1.00	32.3
	5.00	806.6
20	0.45	13.1
	1.00	64.5
	5.00	1613.1

enhancement effects or blurring of the image. This is typically done empirically by testing a range of different parameters.

The single-distance phase-retrieval approach by Paganin et al. (2002) is straightforward to implement and robust and has hence become widely used for a large range of applications. However, it should be noted that it cannot provide quantitative phase information and often leads to low-frequency artefacts.

The method was employed for the results on propagation-based phase-contrast imaging of biomedical and palaeontology specimens performed in a number of collaborations during this Ph.D. project as outlined in Chap. 1.

2.4.1.2 Holographic Imaging and Holotomography

The Talbot effect (Talbot 1836; Cloetens et al. 1997), which we will encounter in more detail in Chap. 4, Sect. 4.2.1, tells us that when a phase object that is periodic with period a is illuminated by a partially coherent beam, no contrast is observed at certain distances na^2/λ ($n = 1, 2, 3, ...$). As a consequence, certain spatial frequencies of an object are lost at a given propagation distance, corresponding to zeros of the CTF. Hence, with a single-distance approach not all spatial frequencies of the specimen can be accessed. As mentioned above, another phase retrieval method for propagation-based phase-contrast data is based on the CTF approach and addresses this problem by recording Fresnel-diffraction images at several different sample-detector distances (Cloetens et al. 1999). The signal that is obtained can be estimated theoretically using a CTF-based derivation under assumptions of weak object absorption and a slowly varying phase. A regularised least-squares minimisation between measured and calculated signals delivers the phase-shift information.

This approach has been implemented in tomography to retrieve the 3D phase or density map of an object and is known as holotomography (Cloetens et al. 1999, 2006). Holotomography can provide high-contrast phase information at high spatial resolution (Cloetens et al. 2006). However, it is much more complicated to perform experimentally as well as from a reconstruction point of view compared to the single-distance approach in Sect. 2.4.1.1.

A mixed TIE- and CTF-based approach has been proposed for propagation-based imaging to overcome some of the limitations of both methods (Guigay et al. 2007). This algorithm including an additional regularisation step (Langer et al. 2010, 2012a) has been implemented for improved holotomography with reduced low-frequency artefacts (Langer et al. 2012b; Lang et al. 2012; Zanette et al. 2013; Lang et al. 2014).

2.4.1.3 Comparison of TIE- and CTF-Based Approaches

While the TIE-based approaches are simpler and more efficient in their experimental realisation, they are limited to short propagation distances and in the most common implementation (Paganin et al. 2002), see Sect. 2.4.1.1, rely on the assumption of a constant relationship between the object's refraction and absorption properties. The investigated object is typically considered to be comprised of a single material and a priori information on the sample is used in the reconstruction process.

The CTF-based approach, on the other hand, requires measurements at several propagation distances and a more elaborate reconstruction process and is only valid under conditions of weak absorption and slowly varying phase. It is, however, not limited to short distances, at which contrast is generally low, and relies on fewer assumptions on the object's properties. It has been shown that the two methods do not reach the same results for propagation distances approaching zero. To overcome the shortcomings and extend and link the validity of both methods, a mixed approach combining the CTF and TIE concepts has been proposed (Guigay et al. 2007; Langer et al. 2007). Assuming only a slowly varying object, one can arrive at an expression for the Fourier transform of the Fresnel diffraction image that includes a CTF-like term and an additional perturbation term. It approaches the TIE equation for small propagation distances. A solution is typically found after a small number of iterations.

Problems still arise at low spatial frequencies as the propagation-based diffraction images transfer contrast poorly in this case. It has been shown that a refinement of the phase-retrieval algorithm can help to mitigate this problem. One solution that has been proposed is based on incorporating a priori information on the absorption of the sample under a homogeneous-object assumption, which is applied at low frequencies to obtain a prior estimate of the phase in a regularisation step (Langer et al. 2010). The approach was later extended to multi-material objects (Langer et al. 2012a).

2.4.2 *X-ray Grating Interferometry*

X-ray grating interferometry is one of the younger phase-sensitive X-ray imaging methods. While long established for optical wavelengths (Lohmann and Silva 1971), it was first demonstrated in the X-ray regime just over 15 years ago (David et al. 2002; Momose et al. 2003; Pfeiffer et al. 2006).

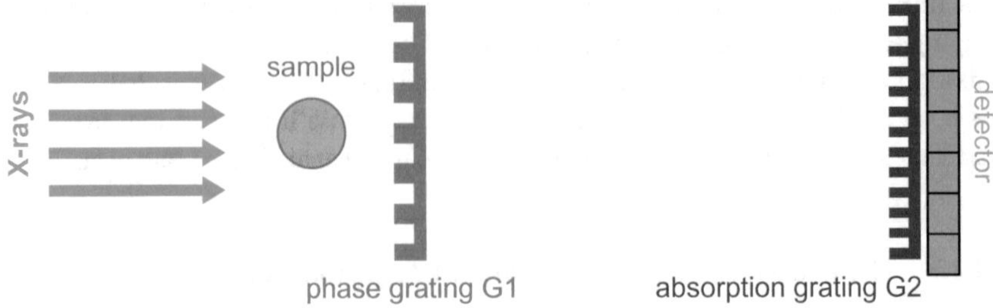

Fig. 2.4 Schematic drawing of a two-grating interferometry setup for X-ray phase-contrast imaging under coherent X-ray illumination. Under incoherent illumination, as given at X-ray laboratory tube sources, a third grating, the source grating G0, is added just behind the source. This creates a set of mutually incoherent but individually coherent X-ray sources

X-ray grating interferometry is a differential phase-contrast imaging method, which means that it is sensitive to the first derivative of the wavefront phase. In addition to the differential phase-contrast signal, grating interferometry also allows for the reconstruction of the transmission through the sample[6] and a small-angle scattering image, typically called dark-field signal (Pfeiffer et al. 2008; Yashiro et al. 2010). The latter is the result of scattering from small unresolved features in the sample, leading to a decrease in coherence of the illumination. While this signal provides an additional contrast channel and allows for visualising features beyond the resolving power of the imaging system, it is not straightforward to relate it to a particular physical property of the sample. Furthermore, other sources influencing the local visibility of the interference pattern, such as edges (Yashiro and Momose 2015), focussing effects (Wolf et al. 2015) and beam hardening (Chabior et al. 2011), can contribute to the dark-field signal. The multimodal image signals, which can be extracted from the same data set, give complementary information about the specimen, which can be valuable in particular for material characterisation applications.

The principle of X-ray grating interferometry is to encode the refraction, absorption and scattering in the specimen in a modulation of a reference interference pattern. For this purpose, a periodic interference pattern is created by placing a phase grating (G1) in the X-ray path. Diffraction leads to splitting of the beam, but as the separation angle between the first order diffracted beams is very small on the order of microradians, they interfere further downstream. The intensity and period of the periodic reference pattern is dependent on the distance from the beam-splitter and determined by the Talbot effect (Talbot 1836; Cloetens et al. 1997), see Sect. 4.2.1 in Chap. 4. For an interference pattern with maximum contrast, typically quantified in terms of the visibility (see Eq. (4.4) in Chap. 4, Sect. 4.2.1), the detector should be

[6]This modality is often simply referred to as absorption signal. However, it is not equivalent to a pure absorption image as it would be obtained from an optimised absorption scan in the contact regime. It typically also contains contributions from propagation effects. Hence, this signal is referred to as attenuation or propagation signal in parts of the thesis to distinguish it from an absorption-only signal.

placed at one of the fractional Talbot distances, the location of which is dependent on the X-ray wavelength, the grating period and the phase shift in the grating lines (see Eq. (4.5) in Chap. 4, Sect. 4.2.1).

When the specimen is inserted into the beam, the refraction, absorption and scattering of the X-rays in the sample leads to a displacement, an intensity reduction and a visibility reduction of the reference pattern, respectively. Commonly, data is acquired by performing a phase-stepping scan, for which the grating is stepped in several small, regular steps perpendicular to the direction of the grating lines (Weitkamp et al. 2005). The information about the sample can then be extracted from the phase-stepping curves with and without the specimen in the beam using a Fourier-based analysis, see Sect. 4.2.2 in Chap. 4.

For typical applications, the detector pixel size is not small enough to directly resolve the fine interference pattern of only a few micrometres across. Hence, an analyser or absorption grating G2 with the same period as the line interference pattern is added in front of the detector, see Fig. 2.4. G2 is made from a highly absorbing material like gold and translates the signal into intensity variations in the detector pixels.

Some major benefits of X-ray grating interferometry include its multimodal character and its high phase sensitivity reaching density resolutions better than 1 mg/cm^3, which is of particular importance for imaging biomedical specimens with minute density differences. Furthermore, unlike propagation-based phase-contrast imaging, for which phase retrieval relies on various approximations, X-ray grating interferometry, as well as speckle-based imaging, deliver real quantitative information on the phase shift in the sample. X-ray grating interferometry is, furthermore, widely accessible to users as it can be implemented at low-brilliance laboratory tube sources with the addition of a source grating G0 to the setup (Pfeiffer et al. 2006).

2.4.3 X-ray Speckle-Based Imaging

X-ray speckle-based imaging is the most recent addition to the group of X-ray phase-contrast imaging methods. Although X-ray far-field speckle have been used for almost 30 years for different applications such as X-ray photon correlation spectroscopy or CDI, see e.g. Sutton et al. (1991), Mochrie et al. (1997), Miao et al. (1999), Marchesini et al. (2003), Wochner et al. (2009), Hruszkewycz et al. (2012), Roseker et al. (2018), Rodenburg and Faulkner (2004), Rodenburg et al. (2007), Thibault et al. (2008), Thibault et al. (2009), amongst others, X-ray near-field speckle was reported for the first time only a decade ago (Cerbino et al. 2008). Four years later, the potential of using X-ray near-field speckle for phase-contrast imaging was realised and first experiments were conducted at synchrotron sources (Bérujon et al. 2012b, a; Morgan et al. 2012). Since then, X-ray speckle-based imaging has seen rapid development and increasing interest for a range of applications.

The technique relies on a similar principle as X-ray grating interferometry and also delivers multimodal image signals. A reference interference pattern is used as

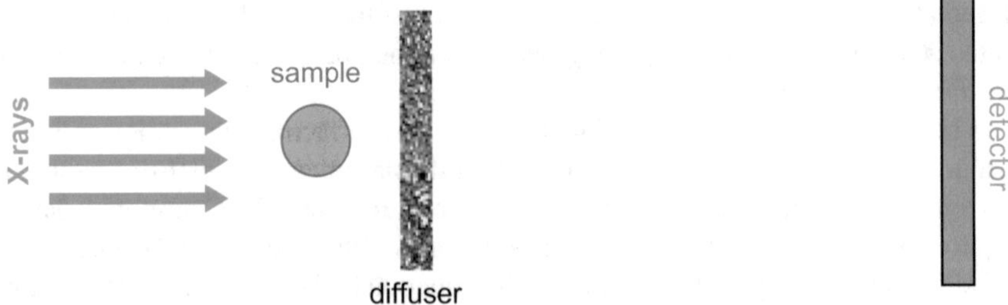

Fig. 2.5 Schematic drawing of a setup for X-ray speckle-based imaging. Scattering and subsequent interference of X-rays from the diffuser particles lead to the formation of a near-field speckle pattern that is used as a wavefront marker. The multimodal image signals are retrieved from the sample-induced distortions of the speckle pattern

a wavefront marker and sample-induced modulations of the pattern with respect to the reference are analysed to obtain the differential phase, transmission and dark-field images of the specimen. Instead of a periodic grating pattern, a random speckle pattern is used, which makes the setup significantly more flexible (Fig. 2.5).

The speckle pattern is usually created by scattering and subsequent interference of X-rays from a diffuser, which is comprised of randomly distributed small particles. Commonly, a piece of commercially available sandpaper or biological filter membrane is used as a diffuser. Refraction, absorption and small-angle scattering in the sample illuminated by a speckle-modulated X-ray beam result in a displacement, change in intensity and change in visibility of the pattern. As the speckle pattern is a two-dimensional (2D) structure, as opposed to the one-dimenisonal (1D) nature of e.g. a line pattern created by a phase grating, the displacement in both the horizontal and the vertical directions can be determined from the data set, which improves the quality of the integrated phase image.

As explained in more detail in Chap. 5, the data acquisition can be carried out in different operational modes, such as the single-shot speckle-tracking mode (Bérujon et al. 2012b; Morgan et al. 2012) and the multi-frame 2D scanning (Bérujon et al. 2012a) or 1D (Wang et al. 2016b) scanning modes. The latter require high-precision scanning stages for stepping of the diffuser in steps smaller than the speckle size. Following these first data acquisition schemes, advanced mixed approaches were developed that allow for a simpler and more flexible scanning and reconstruction routine (Bérujon and Ziegler 2015, 2016, 2017; Zdora et al. 2017). One of them, coined unified modulated pattern analysis (UMPA), was developed during this Ph.D. project and will be discussed in detail in Chap. 6. The analysis of speckle-imaging data is performed in real space via windowed cross-correlation (Bérujon et al. 2012b, a; Morgan et al. 2012) or least-squares minimisation algorithms (Zanette et al. 2014; Zdora et al. 2017). Just recently, another approach has been proposed that is based on the concept of geometric flow and only requires a single image with the sample in the beam (Paganin et al. 2018).

It should be noted that, in contrast to the case of X-ray grating interferometry, speckle-based imaging requires the interference pattern to be resolved directly by the detector and hence smaller pixel sizes are necessary than for a two-grating interferometer.

Over the last years, X-ray speckle-based imaging has raised interest in the imaging community thanks to its high phase sensitivity, quantitative character and its simple, cost-effective experimental setup that only requires a diffuser as an additional optical element. The diffuser material can be chosen according to the desired speckle size and the grain size should be on the order of several pixels in the detector plane. Speckle-based imaging, furthermore, has relatively relaxed requirements on the temporal and spatial coherence of the X-ray beam and is hence compatible with polychromatic laboratory sources (Zanette et al. 2014, 2015). The work performed during this Ph.D. project contributed to this development, see Chap. 1 and Sect. 9.4 in Chap. 9. An in-depth treatment of the working principles, characteristics and developments of the X-ray speckle-based technique can be found in the author's recent review article (Zdora 2018) and Chap. 5 of this thesis.

2.4.4 X-ray Near-Field Ptychography

Although not explored during this Ph.D. project, another technique worth mentioning here is near-field ptychography (Stockmar et al. 2013, 2015b, a; Clare et al. 2015), as it has strong parallels to X-ray speckle-based imaging and could in principle be operated with the same experimental setup.

Near-field ptychography is based on shining a structured illumination, often realised by a diffuser in the beam, onto the specimen and acquiring images in the holographic regime at different lateral positions of the specimen with respect to the beam. Signal reconstruction is performed with algorithms that are analogous to the far-field ptychography case and the phase information about the sample as well as the illuminating probe can be recovered from the same data set.

The key difference to far-field ptychography is that the specimen is placed downstream of the focus, rather than in the focal plane, and the detector is typically located at a distance of less than a metre from the sample, which effectively leads to near-field conditions according to the Fresnel scaling theorem (Paganin 2006). Only information encoded in the Fresnel fringes arising in the holographic near-field regime is used in the reconstruction, no far-field information is required.

Compared to far-field ptychography, near-field ptychography requires significantly less scanning steps and allows imaging of a much larger field of view. As the technique uses an approach that illuminates a large part of the sample, it is suitable for use at existing propagation-based phase-contrast imaging setups. It, furthermore, does not impose as stringent coherence requirements as far-field methods. On the other hand, the achievable spatial resolution of near-field ptychography (typically in the range of several tens of nanometers to 100 nm) cannot reach the level of far-field ptychography. The spatial resolution of near-field ptychography is limited mainly by the effective pixel size but is also influenced by other factors such as the size of the focal spot (i.e. the source size), the width of the Fresnel fringes and the stepping scheme. For thick specimens, it, moreover, needs to be considered that the projection approximation does not hold, which can further reduce the resolution when not properly taken into account (Stockmar et al. 2015a).

Another implementation of near-field ptychography has been proposed where scanning of the sample is not performed laterally, but longitudinally along the beam direction, i.e. diffraction images are recorded at different sample-detector distances (Robisch and Salditt 2013). It has also been demonstrated that both lateral and longitudinal shifts can be used (Robisch et al. 2015). This approach has been shown to be suitable for successful reconstruction of probe and object under illumination with extended beams and without the use of a diffuser or other wavefront modulators that are commonly used to increase diversity in near-field ptychography. These implementations can be seen as hybrids between holographic inline imaging and near-field ptychography. They could be used for probe retrieval in holotomography measurements to eliminate artefacts and are promising for wavefront characterisation.

Although the reconstruction approach for near-field ptychography differs substantially from the one used for X-ray speckle-based imaging, the two techniques share some fundamental concepts. Both use a structured illumination and image reconstruction is typically performed by combining the information from several relative lateral positions of diffuser with respect to the sample. In principle, data from a near-field ptychography setup could be analysed with both a ptychographic and a speckle-imaging algorithm. While speckle-based imaging does not deliver the X-ray probe, it can be more flexible in its experimental setup and the required number of diffuser positions. Furthermore, the constraints in the lateral and temporal coherence of the X-ray beam are more relaxed compared to near-field ptychography and it can be operated also at shorter propagation distances. Fresnel fringes, as arising in the holographic regime, are not required—and in fact not desired—for X-ray speckle-based imaging.

2.4.5 Implications of Partial Coherence for X-ray Grating- and Speckle-Based Phase-Contrast Imaging

The X-ray phase-sensitive imaging methods outlined in the previous subsections all rely on some level of partial coherence. However, some of them do not impose too strict requirements on the coherence of the beam and can still be operated at laboratory sources, including conventional micro-focus sources, with suitable modifications of the setup. Although a higher degree of coherence is beneficial and will increase image quality, both of the main methods explored in this thesis, X-ray grating interferometry and X-ray speckle-based phase-contrast imaging, can be operated under conditions of limited spatial and temporal coherence.

The main effect of a reduced coherence of the illumination is a reduction in the visibility of the interference pattern. For both techniques, the level of transverse coherence has a more significant impact on the results than the level of longitudinal coherence. However, it should be noted that artefacts can arise in the reconstructed images due to beam hardening in case of a low longitudinal coherence.

It has been shown for X-ray grating interferometry that the lateral fringe contrast remains high even when the broad X-ray spectrum is used as long as a sufficient transverse coherence is given, in particular for low fractional Talbot distances and π-shifting gratings (Engelhardt et al. 2008; Thüring and Stampanoni 2014; Wang et al. 2010). The contrast modulations along the propagation direction, however, become smooth as a result of the superposition of the patterns created by X-rays of different energies. This in turn lifts the restriction of using inter-grating distances corresponding to the exact fractional Talbot distances determined by the design energy of the grating (Engelhardt et al. 2008; Hipp et al. 2014). In general, for an acceptable contrast of the interference pattern, the ratio of design X-ray wavelength λ_0 and wavelength spread $\Delta\lambda$ of the beam should be larger than the Talbot order n (see Eq. (4.5) in Chap. 4, Sect. 4.2.1) (Weitkamp et al. 2006, 2005):

$$\frac{\lambda_0}{\Delta\lambda} \gtrsim n, \tag{2.19}$$

which is a relaxed condition on the polychromaticity of the beam.

On the other hand, an extended source, resulting in a low level of transverse coherence, can have a dramatic impact on the visibility of the interference pattern. The intensity of the pattern in the observation plane is in this case a convolution of the intensity achieved with a point source and the projected intensity distribution of the source (Zanette 2011; Bech 2009). The latter can be assumed to be a Gaussian distribution with the standard deviation of the Gaussian given by the demagnified source size $w = D \times (d/z)$ in the observation plane, where D is the actual source size, d the grating-detector distance and z the source-detector distance. As a consequence, the interference pattern becomes blurred and this effect increases rapidly with the distance from the phase grating. As derived by Bech (2009) and Zanette (2011), one can estimate the maximum acceptable projected source size w required to achieve a certain visibility v (see Eq. (4.4) in Chap. 4, Sect. 4.2.1) of the interference pattern:

$$w < \frac{p_1}{\pi}\sqrt{-\frac{1}{2}\ln\left(\frac{\pi^2}{8}v\right)}, \tag{2.20}$$

where p_1 is the period of the phase grating. A similar expression was derived by Weitkamp et al. (Weitkamp et al. 2006):

$$w < 0.53 p_2 \sqrt{-\ln v}, \tag{2.21}$$

where p_2 is the period of the analyser grating. It can immediately be understood from Eqs. (2.20) and (2.21) that for the common parameters used in grating interferometry, the required maximum source size cannot by reached with conventional laboratory sources. For example, for a visibility of $v = 0.2$, a phase grating period of $p_1 = 4.8\,\mu\text{m}$ and an analyser grating period of $p_2 = 2.4\,\mu\text{m}$, a source size of $w < 1.3\,\mu\text{m}$ or $w < 1.6\,\mu\text{m}$ for Eqs. (2.20) and (2.21), respectively, would be required. This can be achieved at synchrotron sources by exploiting a long distance z from the source, but typically not at laboratory setups, for which space is limited and the photon flux is low. However, this problem has been addressed at conventional laboratory sources by introducing a source grating G0 (Pfeiffer et al. 2006), as described in Sect. 4.2.4, Chap. 4.

For X-ray speckle-based imaging the requirements on the spatial and temporal coherence of the X-ray beam are similarly relaxed. The influence of a polychromatic X-ray beam on the visibility of the speckle pattern was investigated in a simulation study and found not to be a limiting factor (Zdora et al. 2015). Under otherwise ideal conditions, simulations showed a drop in visibility from 43% for a monochromatic beam of energy 20 keV down to 29% for a polychromatic spectrum of a width of 40 keV with the same mean energy (Zdora et al. 2015). This is still easily sufficient for successful image reconstruction and shows that speckle-based imaging is relatively robust to the polychromaticity of the illuminating beam. On the other hand, it was shown that a broad X-ray spectrum leads to beam hardening artefacts in the sample, which can have a major impact on the reconstructed signals. This is, however, not an intrinsic problem of speckle-based imaging and occurs with any kind of X-ray imaging method.

It has been demonstrated that the partial spatial coherence of the X-ray beam can be directly determined from the normalised intensity of the speckle pattern by means of a Fourier power spectrum analysis (Cerbino et al. 2008; Kashyap et al. 2015). The latter can be modelled as a product of the scattered intensity distribution of the diffuser particles and a transfer function that contains contributions from the detector response, the Talbot effect and the partial coherence of the beam (Cerbino et al. 2008). As the first two contributions are known, the coherence lengths of the illuminating beam can be determined from the measured speckle pattern.

However, a detailed study on the influence of the spatial coherence on the visibility of the speckle pattern under different setup conditions has not been reported to date. Generally, the lateral coherence length of the beam should be larger than the size of the scattering grains in the diffuser, in order to obtain a speckle pattern created by scattering and interference effects. While this condition is easily given at synchrotrons, it is more challenging to achieve at laboratory sources. As will be explained in later chapters, speckle-based imaging has, however, been successfully implemented at high-end laboratory sources with small spot sizes such as the liquid-metal-jet source by Excillum (Zanette et al. 2014), at which a sufficient transverse coherence length could be realised, see Sect. 2.3.3. For micro-focus conventional solid-target X-ray sources, it has been proposed to use a random absorption mask instead of a phase modulator, which leads to the creation of a random pattern based mainly on absorption effects that can also be used as a wavefront marker, albeit at the cost of reduced flux (Wang et al. 2016a). Here, it should be mentioned that a related idea was proposed more than a decade ago using a set of periodic absorption masks, called coded apertures, for differential phase-contrast imaging at laboratory sources (Olivo and Speller 2007; Olivo et al. 2009). An array of individual beams is created by a mask upstream of the sample and a second mask in front of the detector is used to measure the deflection of X-rays due to refraction in the sample. As this approach is independent of interference effects and simply based on X-ray absorption in the masks, it is compatible with laboratory sources under conditions of very low spatial and temporal coherence.

Systematic theoretical and experimental studies on the spatial and temporal coherence requirements of the speckle-based technique have not been reported to date, but should be performed in the future to allow for better planning and optimisation of experiments in particular at laboratory sources.

2.4.6 Choosing an Imaging Method: X-ray Speckle-Based Imaging, Grating Interferometry and Propagation-Based Imaging

It is important to note that one X-ray phase-contrast imaging method cannot meet the requirements for every application. The best suitable technique should be chosen depending on the type of sample, the desired properties of the retrieved images and the given experimental conditions. Here, a short summary is provided of the properties and setup conditions for X-ray speckle-based imaging (SBI) and grating interferometry (XGI) in comparison with propagation-based imaging (PBI), the third most common full-field imaging method, as explained above. This can serve as a guide for users on which imaging method to choose for a particular study under given experimental conditions. A detailed comparison of SBI and XGI, the two methods that have been investigated in this Ph.D. project and are discussed in more detail in the following chapters, can be found in Appendix A.

The experimental arrangement for PBI is the simplest as it relies on free-space propagation and does not require additional optical elements. SBI is also operated with a relatively simple setup and only uses a cost-effective diffuser in addition to the standard X-ray imaging equipment. XGI relies on the use of custom-made, relatively costly gratings, which need to be moved in small and equidistant steps demanding high-precision scanning motors. While these are also required for some implementations of SBI, UMPA can be performed with less precise stages as the step sizes can be large and irregular.

All three methods do not impose strong demands on the longitudinal (temporal) coherence of the X-rays and have been demonstrated with polychromatic X-ray beams. The requirements on the transverse (spatial) coherence conditions for PBI and SBI are comparable and moderate. While for the first it is necessary to observe Fresnel fringes, the latter relies on the formation of near-field speckle via interference effects.

The coherence requirements for SBI are further relaxed when mainly absorption effects are exploited for creating a random reference pattern. XGI can be operated at low-brilliance sources with very limited transverse coherence by using a source grating.

The optimal sample-detector distance for single-distance PBI is determined by the width of the first Fresnel zone and is typically in the range of a few to several tens of centimetres. For differential phase-contrast methods, such as SBI and XGI, the signal sensitivity generally improves with increasing propagation distance. However, with the current reconstruction approaches strong edge-enhancement fringes that arise at long propagation distances due to Fresnel diffraction can lead to artefacts in the reconstructed images. While XGI performs best when the grating-detector (or phase grating-absorption grating) distance is a fractional Talbot distance, SBI does not require specific distances and allows for more freedom in the experimental arrangement. Propagation distances for the two methods are typically on the order of tens of centimetres up to a few metres.

Another experimental factor is the total acquisition time. PBI only requires a single sample image when used in single-distance implementation. XGI is most commonly performed in phase-stepping mode, for which sample images at four or more grating positions are acquired. Moreover, the use of an absorption grating and/or a source grating can lead to longer scan times due to the higher X-ray absorption. For SBI, the acquisition time is dependent on the operational mode that is chosen. For single-shot speckle tracking, only one image is required and scan times can be comparable to PBI. For scanning-based techniques, the larger number of steps leads to an increased time effort. Among these, the unified modulated pattern analysis (UMPA) allows for a significant reduction in the number of steps compared to other implementations of SBI and scan times can be similar to XGI measurements. It should be considered that generally the acquisition time per frame strongly depends on the experimental setup, which is different for the three types of phase-contrast imaging techniques.

An essential point for choosing the most suitable imaging method for a particular application is the quality and information content of the images. Generally, the sensitivity of differential phase-contrast methods, XGI and SBI, is higher than for methods that carry the information in the second derivative of the wavefront phase such as PBI. Furthermore, XGI and SBI are able to deliver quantitative information on the X-ray phase shift in the sample and on the 3D electron distribution when employed in tomography, see Sect. 2.5.4. Single-distance PBI, on the other hand, requires a priori input, which makes it mainly suitable for qualitative imaging applications. Other PBI approaches like holotomography can deliver quantitative information, but data analysis is much more elaborate. Moreover, SBI and XGI provide additional complementary absorption and small-angle scattering signals, not accessible from PBI data.

As a last point, the effective pixel size, field of view and spatial resolution are considered. PBI is the method of choice for small effective pixels (on the order of sub-micrometre up to a few micrometres in size), which are required to resolve the fine Fresnel propagation fringes. The small pixels allow for a high spatial resolution, but also limit the field of view. Hence, PBI is most suitable for high-resolution qualitative imaging applications. XGI, on the other hand, can be operated with large pixels of tens to hundreds of micrometres when using an absorption grating in front of the detector, which allows for imaging of large samples of tens of centimetres in size. A drawback is here often the limited size of the active area of the gratings that are elaborate to fabricate. The spatial resolution of XGI is lower than for PBI and limited to typically tens of micrometres. SBI provides an intermediate solution. The pixel size needs to be small enough (micrometres to tens of micrometres) to resolve the speckles, but can be larger than for PBI. Therefore, the available field of view is typically larger than for PBI, but smaller than for XGI. In contrast to XGI, diffusers for SBI are commonly available with large areas and are hence not limiting the field of view. The spatial resolution of the images obtained with SBI depends on the operational mode and is typically tens of pixels for single-shot imaging, a few pixels for UMPA and down to the detector system resolution for 2D speckle scanning.

In summary, single-distance PBI is the simplest and fastest of the three techniques and is most suitable for setups with small pixel sizes for high-resolution applications that do not require quantitative or multimodal information. XGI is well adapted for use at low-brilliance sources and with large-pixel detectors, enabling multimodal imaging of large samples at lower resolution but with excellent sensitivity. SBI in its current implementations combines benefits of both of the other two methods. It delivers quantitative and multimodal images at medium spatial resolution and field of view. While its angular sensitivity can reach nanoradians like for XGI, its practical implementation and experimental setup are significantly simpler and more robust.

2.5 X-ray Computed Tomography

The previous sections covered the use of X-rays to obtain 2D images of an object exploiting either absorption, phase-shift or scattering in the specimen. However, for most applications it is of interest to retrieve 3D volumetric information. The principles of X-ray computed tomography (CT) are outlined in this section. It is, furthermore, explained how the 3D maps of the refractive index decrement δ, electron density ρ_{el} and mass density ρ_m, as well as the distributions of the attenuation coefficient μ and linear diffusion coefficient ε can be obtained from the phase, absorption and dark-field signals, respectively, which can be retrieved with multimodal X-ray imaging methods, such as speckle-based imaging and grating interferometry.

2.5.1 *Background and Basic Principle*

In the early days, X-ray imaging was performed solely in 2D projection mode, for example for medical radiography (Codman 1896; Editorial 1896b, a; Spiegel 1995). Medical imaging was in fact one of the first applications of X-rays and hence early developments started with the improvement of medical diagnostics in mind. A major milestone for X-ray imaging was the extension from 2D projection to 3D volumetric imaging. This was first proposed by Cormack (1963) and Hounsfield (1973) who came up with an idea for visualising the inner structure of an object in three dimensions from combining projections at different viewing angles, later called X-ray computed tomography (CT). Since its invention in 1972, CT has become an invaluable tool in medicine and was later also applied to other areas of research for non-destructive 3D visualisation.

The principle of a tomography scan is to rotate the sample and record images at regularly spaced angular intervals.[7] Ideally, as it is the case for the measurements in this thesis, the specimen is rotated by a minimum of 180° and several hundreds or thousands of projections are acquired. However, missing wedges and a reduced number of projections due to experimental constraints can be tolerated albeit at a loss of spatial resolution. For the case of non-parallel beam geometries, an extended angular range is required to account for the beam divergence. For a fan or cone beam setup, the full field of view can be reconstructed when a minimum rotation range of 180° plus the fan or cone angle, respectively, is covered. The recorded projections are subsequently combined using mathematical algorithms to reconstruct the 3D volume of the object. Appropriate weighting of the data has to be applied for the fan- and cone-beam cases.

Tomographic imaging can be performed with conventional X-ray absorption imaging, but also phase-sensitive imaging techniques. For multimodal imaging methods such as X-ray grating interferometry and speckle-based imaging, it is possible to obtain three separate tomograms, the absorption, the phase and the dark-field tomogram, from a single data set. The first quantifies the absorption properties, the second the phase-shifting properties, and the last the small-angle scattering power within the sample, which gives information about features in the specimen that are beyond the spatial resolution limit of the system (Yashiro et al. 2010).

There are several different algorithms for tomographic reconstruction from the projections including Fourier-domain methods (Mersereau and Oppenheim 1974), iterative methods (Dudgeon and Mersereau 1984) and filtered back-projection (FBP) (Kak and Slaney 2001). The latter is the most commonly employed method as it is fast and easy to implement. It shows excellent performance for data recorded at equally spaced projection angles over at least 180°. However, in cases of limited angle tomography, missing angles or noisy data, iterative methods can yield superior results, see e.g. Nassi et al. (1982), Sidky et al. (2006), Duan et al. (2009), Delaney and Bresler (1998), Hanson and Wecksung (1983). For cone-beam tomography, the algorithm proposed by Feldkamp et al. is often used, which is an extension of the common filtered back-projection approach for direct reconstruction of the 3D volume from a set of 2D projections (Feldkamp et al. 1984).

---

[7] In clinical CT measurements the detector is rotated around the patient for practical reasons, which is equivalent as only the relative movement between object and detector matters.

2.5.2 Filtered Back-Projection

As mentioned, the filtered back-projection algorithm is the most common approach for tomographic reconstruction of the three-dimensional volume from the acquired projections at different viewing angles of the sample. Filtered back-projection was used for all the tomography reconstructions presented in this thesis.

The principle of the basic back-projection algorithm is simple. From the projections recorded at different rotational angles, an estimation of the sample can be obtained by projecting them back along the original acquisition path. For the case of a parallel beam, this can be done individually for each row of the image. The whole slice can then be reconstructed by summation of the back-projections from all angles. However, simple back-projection leads to blurring at the edges of the reconstructed object. This can be corrected for by applying a high-pass filter before or after back-projection. Mathematically, a simple ramp filter is the correct choice of filter and is most commonly used for conventional absorption tomography, but in the presence of significant noise other filters giving less weight to the highest frequencies, such as Hamming, Hann, Cosine or Shepp-Logan filters, can achieve better results. The combination of back-projection with a high-pass filter is called filtered back-projection.

In the following, we will only consider a single slice of the sample in the plane perpendicular to the rotation axis (x-z plane). For the case of parallel beam geometry, the whole sample volume can be reconstructed as a set of the individual slices. Figure 2.6a shows one slice of the object described by the 2D object function $f(x, z)$. The 1D projection of the 2D sample function $f(x, z)$ at an angle θ, also called the Radon transform, can be described in the rotated coordinate system (x', z'):

$$\mathscr{R}(x', \theta) = \int f(x' \cos\theta - z' \sin\theta, x' \sin\theta + z' \cos\theta) dz'. \tag{2.22}$$

The rotated coordinate system (x', z') of the sample is related to the coordinate system (x, z) of the source/detector through the following equations:

$$\begin{aligned} x &= x' \cos\theta - z' \sin\theta \qquad & x' &= x \cos\theta + z \sin\theta \\ z &= x' \sin\theta + z' \cos\theta \qquad & z' &= -x \sin\theta + z \cos\theta \end{aligned} \tag{2.23}$$

The reconstruction of the slice $f(x, z)$ can be performed via back-projection of $\mathscr{R}(x', \theta)$ for all angles θ. The back-projection operator $\mathscr{B}$ is defined as:

$$\mathscr{B}g(x', \theta) = \int_0^{\pi} f(x \cos\theta + z \sin\theta, \theta) d\theta. \tag{2.24}$$

However, as mentioned above, simple back-projection leads to blurring of the sample in the reconstructed image. To correct for this, the projections are filtered before the back-projection step. The filtering step is performed in Fourier space.

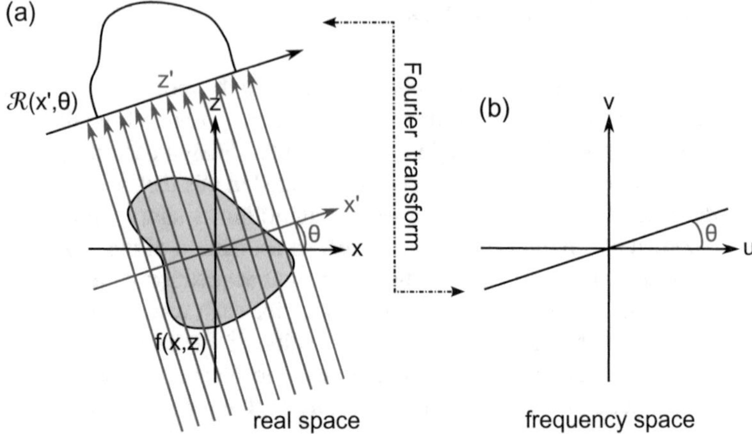

Fig. 2.6 Principle of the Fourier slice theorem. The projection of the two-dimensional object function $f(x, z)$ under the angle θ is the Radon transform $\mathscr{R}(x', \theta)$. The Fourier slice theorem states that **a** the 1D Fourier transform of $\mathscr{R}(x', \theta)$ is equivalent to **b** a slice of the 2D Fourier transform of $f(x, z)$ under the angle θ. Figure after Kak and Slaney (2001)

One can now exploit the Fourier slice theorem (also known as central slice theorem), which states that taking the 1D Fourier transform of a projection (i.e. the Radon transform) of the function $f(x, z)$ at viewing angle θ is equivalent to the central slice at angle θ through the 2D Fourier transform of $f(x, z)$, as illustrated in Fig. 2.6. The Fourier slice theorem implies that the Radon transform $\mathscr{R}(x', \theta)$ describes completely any object $f(x, z)$ that is Fourier transformable and that $f(x, z)$ can be uniquely retrieved from its Radon transform, which is directly related to its 2D Fourier transform. We can now apply a filter in Fourier space to the Fourier transform of the Radon transform, apply an inverse Fourier transform and subsequently back-project the filtered Radon transform to reconstruct the object function. The shape of the filter needs to account for the different sampling in real and Fourier space. As the data spacing is radial in Fourier space, the sampling is denser for lower spatial frequencies and sparser for higher spatial frequencies. To account for this, a high-pass ramp filter is applied.

In summary, we can express the reconstruction of the slice $f(x, z)$ as the back-projection of the filtered Radon transform:

$$f(x, z) = \int_0^{\pi} \mathscr{R}_f(x \cos\theta + z \sin\theta, \theta) d\theta, \tag{2.25}$$

where $\mathscr{R}_f$ is the filtered Radon transform defined as:

$$\mathscr{R}_f(x', \theta) = \mathscr{F}^{-1} \left\{ h(\nu) \left[\mathscr{F} \left\{ \mathscr{R}(x', \theta) \right\} (\nu, \theta) \right] \right\}. \tag{2.26}$$

Here, $\mathscr{F}$ and $\mathscr{F}^{-1}$ denote a Fourier transform and inverse Fourier transform, respectively, and $h(v)$ is the filter function in Fourier space (Fourier coordinate v), most commonly a ramp filter.

2.5.3 *Absorption Tomogram*

The absorption tomogram is reconstructed from the absorption projections and gives 3D information about the absorption properties of the sample in form of a 3D map of the attenuation coefficient μ. As derived in Eq. (2.11), the transmission T through the sample is given by the following relation know as Beer-Lambert Law:

$$T = \exp\left(-\int \mu(x, y, z)dz\right), \tag{2.27}$$

where μ is the linear attenuation coefficient, which is dependent on the material and the X-ray energy. The first step in the tomographic reconstruction of μ is to apply the logarithm to the absorption projections and multiply by -1. Subsequently, filtered back-projection can be employed to obtain the absorption tomogram. It should be noted that for the case that data acquisition is not performed in the contact regime, but with the detector at some propagation distance away from the sample, as it is commonly the case for X-ray grating interferometry and speckle-based imaging, the tomogram reconstructed from the attenuation projections will also contain contributions from edge-enhancement effects if a sufficient degree of spatial coherence is given. The fringes arise upon propagation due to Fresnel diffraction under conditions of sufficient coherence. The resulting tomogram hence does not provide a true map of μ and we will refer to it as attenuation or propagation tomogram in most parts of this thesis.

2.5.4 *Phase Tomogram*

The phase tomogram delivers the 3D distribution of the refractive index decrement δ. The latter is directly proportional to the electron density ρ_{el} for X-ray energies far from the absorption edges of the sample materials and we can hence retrieve a 3D electron density map of the specimen.

The refractive index decrement δ is directly related to the phase shift $\Delta\Phi$ of the X-rays, see Eq. (2.13):

$$\Delta\Phi = \frac{2\pi}{\lambda} \int \delta(x, y, z)dz. \tag{2.28}$$

For the different phase-sensitive imaging techniques presented in this thesis, the phase signal $\Delta\Phi$ is either directly recovered computationally from the recorded signal (propagation-based imaging) or the first derivative $\partial\Phi/\partial\mathbf{r}$ is obtained as a reconstruction output (grating- and speckle-based imaging). In the first case, the tomographic reconstruction is performed from the phase projections the same way as for the absorption volume discussed in the previous section. However, no logarithm is applied to the projections and the reconstructed phase volume is multiplied by $\lambda/2\pi$ to obtain the 3D distribution of δ, which can then be converted to a map of ρ_{el}.

In the case of using the differential phase signal (here along the horizontal axis x), the following relation holds:

$$\frac{\partial\Phi}{\partial x} = \frac{2\pi}{\lambda}\frac{\partial}{\partial x}\int \delta(x, y, z)dz. \tag{2.29}$$

Integration of the signal needs to be performed to obtain the δ map. This can be done before the tomographic reconstruction step by simple 1D (for methods sensitive to the differential phase in only one direction) or 2D (for methods sensitive to the differential phase in both directions) integration of the differential phase projections. Subsequently, conventional filtered back-projection, as for the case of propagation-based phase tomography, is performed. Alternatively, the 1D integration can be included in the filtered back-projection step by applying the following imaginary filter $\tilde{h}(\nu)$ in Fourier space (Pfeiffer et al. 2007):

$$\tilde{h}(\nu) = \frac{|\nu|}{2\pi i\ \nu} = \frac{1}{2\pi i\ \text{sgn}(\nu)}, \tag{2.30}$$

where sgn is the sign function and ν the coordinate in Fourier space. In real space, this corresponds to the Hilbert transform. The approach is equivalent to integration and subsequent filtering with a ramp filter thanks to the Fourier derivative theorem, which states that the Fourier transform of the derivative of a function is equal to the Fourier transform of the function multiplied by $2\pi i\ \nu$. Advantages of using the Hilbert filter include the simpler reconstruction process without additional integration step and the reduction of artefacts in the phase images.

It should be noted that the method based on the Hilbert filter can only be used when the differential phase signal is given along the direction perpendicular to the sample's rotation axis. Hilbert filtering was employed for the tomography reconstructions from the grating interferometry data presented in this thesis. For speckle-based imaging, 2D phase integration of the differential phase projections was performed and conventional filtered back-projection with a ramp filter was applied to the phase projections.

As mentioned above, for energies far from the absorption edges of the sample material, the refractive index decrement δ is directly proportional to the electron density ρ_{el} and hence the 3D map of δ can be converted to an electron density map of the specimen (Als-Nielsen and McMorrow 2011):

$$\rho_{el} = \delta\frac{2\pi}{r_0\lambda^2}, \tag{2.31}$$

where r_0 is the classical electron radius. The electron density map can then be converted to a mass density map via the relation:

$$\rho_m = \rho_{el} \frac{A}{N_A Z}, \tag{2.32}$$

where A is the molecular mass, N_A the Avogadro number and Z the atomic number. For materials made from light elements, the approximation $A/Z = 2$ is valid (Guinier 1994).

Combining the complementary information from the absorption and density maps can give significantly aid the discrimination of different materials in a specimen (Zanette et al. 2015).

2.5.5 Dark-Field Tomogram

For multimodal imaging methods that in addition to the absorption and phase images allow for the reconstruction of the dark-field signal, a dark-field tomogram can be obtained from the same data set.

As demonstrated first for analyser-based dark-field imaging and later validated for X-ray grating interferometry, the second moment of the scattering angle distribution can be expressed as a line integral of a general scattering parameter (Khelashvili et al. 2006; Wang et al. 2009) and hence conventional filtered back-projection can be used to obtain a tomographic volume, analogous to the absorption tomogram case. A linear diffusion coefficient

$$\varepsilon = \frac{\sigma^2}{\Delta z} \tag{2.33}$$

can be defined with the total scattering width σ and the slice thickness Δz (Bech et al. 2010). Hence the scattering width σ along the beam path z is given by $\sigma^2 = \int \varepsilon(z) dz$.

Often, approximate expressions can be derived for σ^2 through the specimen for the various X-ray imaging methods that are capable of delivering dark-field information. For the case of grating interferometry, ε can be directly related to the visibility v of the interference pattern as:

$$\int \varepsilon(z) dz = -\frac{p_2^2}{2\pi^2 d^2} \ln v, \tag{2.34}$$

where d is the inter-grating distance and p_2 the period of the absorption grating (Bech et al. 2010). A similar expression has been derived for speckle-based imaging (Wang et al. 2016c):

$$\int \varepsilon(z) dz = -\frac{\bar{\xi}^2}{8\pi^4 d^2} \ln \gamma^{\max}, \tag{2.35}$$

where $\bar{\zeta}$ is average speckle size, d the propagation distance and $\gamma^{\max}$ the maximum cross-correlation coefficient, which quantifies the loss in the visibility of the speckle pattern, see Chap. 5, Sect. 5.5.

Using the above relations, a dark-field tomogram, giving the 3D distribution of the linear diffusion coefficient ε, is obtained via tomographic reconstruction, e.g. using a filtered back-projection algorithm.

2.6 X-ray Sources

In the previous sections X-rays were introduced as electromagnetic radiation and their fundamental interactions with matter were explained and related to the different image formation processes. The mechanisms of how the X-rays are generated in the first place are the subject of this section. It contains an overview of the historic and recent developments of different types of X-ray sources commonly used for X-ray imaging purposes and explains how X-ray radiation is generated by them.

Before we go into details of the processes of X-ray generation and the developments of X-ray sources, a quantity to assess and compare the quality of X-ray sources needs to be defined. The quality of the X-rays in the observation plane crucially depends on the properties of the source, in particular the flux, i.e. the number of photons emitted per second, the divergence of the X-ray beam, the cross-sectional area of the source and the number of photons that are within 0.1% bandwidth of the central wavelength. These quantities are combined in a factor called brilliance B (Als-Nielsen and McMorrow 2011):

$$B = \frac{\text{flux}\,[\text{s}^{-1}]}{(\text{beam divergence}\,[\text{mrad}^2]) \times (\text{source cross section}\,[\text{mm}^2]) \times (0.1\%\ \text{bandwidth})}. \tag{2.36}$$

The product of the first two factors in the denominator of Eq. (2.36) are the product of the vertical and horizontal beam emittances, where the emittance is defined as the product of source size and divergence.

Generally, a high brilliance of the source is desirable, which means that it delivers a large number of X-ray photons per unit time of a certain wavelength in a certain direction focussed in a small spot. Practically, a high-brilliance X-ray source delivers an X-ray beam with high coherent flux and allows for using short scan times.

The increase in beam brilliance is an ongoing pursuit in the development of new-generation synchrotron sources and X-ray free electron lasers, see Sect. 2.6.3.

2.6.1 X-ray Tube Sources

When Röntgen discovered X-rays in 1895, he was working with a Crookes tube source, a discharge tube consisting of an anode and a cathode in a glass bulb (Crookes 1879). It was originally designed for observing electrical discharge when a high voltage was applied to the electrodes, which later led to the discovery of electrons, first called cathode rays (Thomson 1897, a). Röntgen, however, immediately realised that the radiation he just discovered was not the same as the previously described cathode rays (Röntgen 1895, 1898). After the initial observations of X-rays on fluorescent screens, Röntgen very soon noticed their high penetration power and took first images of several objects as well as a human hand. The early communications by Röntgen were followed by rising interest from the community and existing tube sources were developed, improved and optimised for the purpose of producing X-ray radiation.

From a technical point of view, the principle of an X-ray tube is simple and has remained the same as for the early discharge tubes used by Röntgen: electrons are emitted from the cathode and accelerated by a high voltage between cathode and anode. When the electrons hit the anode target, the interaction with the target material leads to the emission of X-rays. This happens mainly via two processes: the generation of bremsstrahlung and the emission of characteristic fluorescence X-rays. Bremsstrahlung is caused by the deceleration of electrons in the target material and gives a broad continuous spectrum of X-rays, as shown in Fig. 2.7c. The maximum photon energy of the distribution is determined by the accelerating voltage and the average photon energy lies at approximately 1/3 of the maximum energy. Characteristic X-rays, on the other hand, are the result of incoming electrons ejecting an inner-shell electron from the target atoms, which is followed by outer-shell electrons filling the core hole. This is accompanied by the emission of X-rays with a wavelength characteristic to the energy levels of the target material. In the X-ray tube spectrum, this effect can be observed as sharp peaks sitting on top of the continuous bremsstrahlung spectrum, see Fig. 2.7c.

One of the first major developments in X-ray tube sources was the introduction of the Coolidge tube in 1913, which uses a heated tungsten filament as a cathode (Coolidge 1913), see Fig. 2.7a. The anode can be either a transmission or a reflection target. For the first, the electrons pass through the material and photons are emitted in the same direction as the electron beam. A reflection target, on the other hand, is arranged at an angle so that X-rays produced at 90° to the beam direction can escape. This was a significant step forward for X-ray science, as operation of the new Coolidge tube was much more flexible and reliable than previously available tubes and allowed for optimising the voltage and current independently from each other.

During operation of the X-ray tube, a significant amount of heat is generated at the anode, which limits the power that can be applied and hence the X-ray flux. In the first implementations of the Coolidge tube, the anode was cooled with a water circuit, see Fig. 2.7a. Half a decade later, the limitation set by the cooling efficiency was partly overcome by the introduction of a rotating anode, see Fig. 2.7b, improving

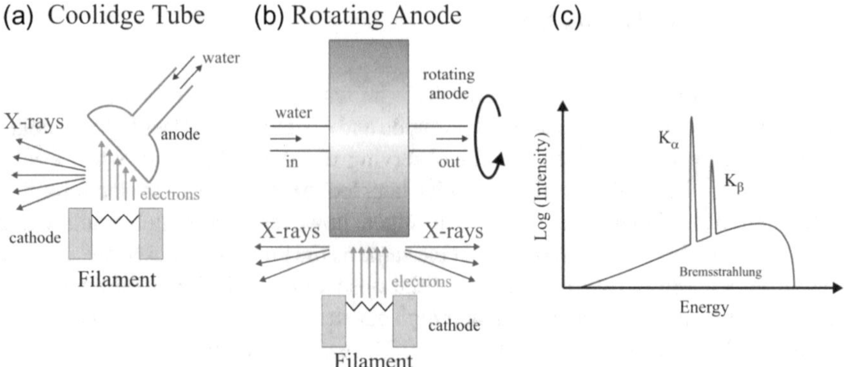

Fig. 2.7 X-ray tube sources. A heated filament emits electrons that are accelerated towards the anode by a high voltage. Interaction with the anode material leads to the production of X-rays. **a** Coolidge tube with reflection target. **b** Rotating anode tube for improved heat dissipation, allowing for a higher tube voltage. **c** Typical X-ray spectrum emitted by a tube source. The continuous bremsstrahlung spectrum is overlaid by characteristic X-ray peaks. The tube voltage determines the maximum X-ray energy of the spectrum. Figures reprinted with permission from Als-Nielsen and McMorrow (2011), copyright (2011) by John Wiley & Sons, Ltd

heat dissipation and hence allowing for a much higher tube power. Rotating anode tubes are nowadays still the workhorse for laboratory X-ray imaging setups.

Another step towards higher-flux sources was taken about 16 years ago with the development of liquid-metal-jet anodes (Hemberg et al. 2003). The principle of this novel X-ray source is to continuously deliver fresh anode material in form of a liquid-metal stream, which lifts the limitation set by the melting point of the target material. This has the advantage of allowing for a much higher electron-beam power and hence photon flux, significantly reducing scan times. Typically, a gallium(Ga)/indium(In)/tin(Sn) alloy is used as target material. The fluorescence emission lines for gallium are similar to the ones of copper and for indium close to the ones for silver. The high photon flux at small source size in the range of a few microns to a few tens of microns (limited by the electron beam power) lead to a significantly higher brilliance (see Eq. (2.36)) of the liquid-metal-jet source compared to other laboratory sources. Furthermore, the small source diameter enables a higher spatial coherence (see Sect. 2.3) than common tube systems, which typically have much larger spot sizes at the same tube power. These properties make the source ideal for imaging techniques that require some level of coherence of the X-ray beam but do not impose too stringent requirements, such as X-ray speckle-based imaging. On the other hand, for the operation of the machine with a small source size the maximum power that can be applied is in the range of tens of Watts, which limits the total photon flux. Furthermore, they are more costly than conventional sources and hence not as easily accessible to users. The implementation of X-ray speckle-based imaging at a liquid-metal-jet source was part of this Ph.D. project, see Chap. 9, Sect. 9.4.

One of the most recent developments in X-ray laboratory sources is the design of an anode comprised of small micron-sized metal emitters that are embedded in a diamond substrate and allow for rapid cooling of the anode material due to large thermal gradients (Yun et al. 2015). This type of source provides another solution to operate at much higher power than conventional sources and to achieve a small spot size of a few micrometres, resulting in increased brilliance. An additional advantage is the possibility to implement multiple materials in a single target for fast energy switching.

2.6.2 Synchrotron Sources

X-ray tube sources are still used and further developed nowadays and are an important workhorse for routine, high-throughput X-ray imaging applications. However, the major advances in X-ray science in the 20th century would not have been possible without synchrotron X-ray sources. X-ray synchrotron radiation was first discovered in 1947 and, while it was initially regarded as an undesired effect in particle physics experiments, the remarkable properties of synchrotron X-rays were realised a decade later, followed soon by the construction of dedicated synchrotron radiation facilities (Pollock 1983; Bilderback et al. 2005). Synchrotrons deliver X-rays of significantly higher brilliance than conventional tube sources. Their increasing availability and the development of third-generation synchrotron sources (Kunz 2001) in the late 20th century have proven as a major milestone for X-ray science and enabled imaging with unprecedented resolution and image quality. Among the developments in X-ray imaging techniques pushed by the use of synchrotron sources was the implementation of X-ray phase-contrast imaging. The principle of phase contrast was discovered for visible light in the 1930s (Zernike 1942) and was successfully applied also to X-rays in 1965 (Bonse and Hart 1965). However, only with the availability of highly brilliant X-rays from third-generation synchrotrons X-ray phase-contrast imaging was further developed and its huge potential was realised.

Most of the work on technique development for X-ray imaging is still conducted at synchrotrons to date, including the majority of the experiments performed during this Ph.D. project. These were mainly carried out at the beamline I13 at Diamond Light Source in the UK and the beamline ID19 at the European Synchrotron Radiation Facility (ESRF) in France.

Nowadays, there are more than 50 synchrotrons around the world conducting research in different areas from biological imaging to materials sciences, chemical mapping and crystallography. The core part of a synchrotron is the electron storage ring, which typically has a doughnut-shaped appearance, as seen in the photograph of Diamond Light Source in Fig. 2.8a.

The principle of synchrotron facilities is to create X-rays by acceleration of charged particles, typically electrons, travelling at the speed of light. This is done in several steps. The electrons are generated by heating of a cathode and are subsequently accelerated towards an anode, which gives them an energy of typically

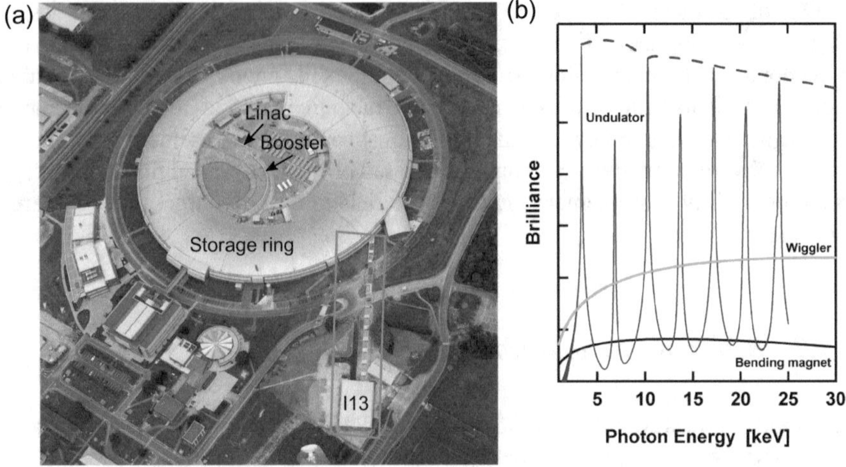

Fig. 2.8 Typical layout of a synchrotron facility and X-ray beam spectra produced by bending magnets and insertion devices. **a** Aerial photograph of Diamond Light Source, UK. The linear accelerator, the booster ring and the storage ring are labelled. Beamline I13, where some of the experiments in this thesis were conducted, is highlighted by an orange box. Image courtesy of Diamond Light Source Ltd. **b** Qualitative plots of the X-ray spectra produced by a bending magnet, a wiggler and an undulator insertion device. *Source* http://photon-science.desy.de/research/studentsteaching/primers/synchrotron_radiation/index_eng.html

several tens of kiloelectronvolts (90 keV at Diamond Light Source). A linear accelerator (linac) is then used to increase the energy of the electrons (to about 100 MeV at Diamond Light Source; 170 MeV at the ESRF), which is achieved using radio-frequency cavities. The next step is the booster ring, where the electrons are gradually accelerated further to reach the designated storage ring energy (3 GeV for Diamond Light Source, 6 GeV for ESRF). During each lap of the electrons around the ring, they are accelerated by radio-frequency cavities in straight sections and dipole bending magnets in the curved sections to keep them on their circular track. To account for the increasing energy of the electrons, the magnetic field of the bending magnets is increased with every round. When the electrons reach their final energy, they are injected into the storage ring, which consists of several tens of straight sections and bending magnets that keep the electrons on their defined path. X-rays are generated at the bending magnets due to the fact that charged particles, here electrons, travelling near the speed of light emit radiation when accelerated/decelerated. In addition, insertion devices (IDs) are placed in the straight section of the ring, which emit X-rays of higher brilliance than the bending magnets.

There are two types of insertion devices: undulators and wigglers. Both consist of a series of dipole magnets arranged periodically to produce an alternating magnetic field, which induces oscillations of the electrons that are passing through. In wigglers, the electrons undergo sinusoidal oscillations of very large amplitude and the radiation cones emitted from each oscillation overlap incoherently and add up creating a continuous spectrum over a wide range of wavelengths, but with an intensity much higher than the radiation from a bending magnet, see green curve in Fig. 2.8b. In undulators, the amplitude of the oscillations is smaller and the radiation cones add up coherently leading to a spectrum with distinct peaks of extremely high intensities, as illustrated by the blue curve in Fig. 2.8b.

At the bending magnets and insertion devices, the X-rays are extracted from the ring and guided to the beamlines, where the experimental endstations are located. Typically, the beam first passes through an optics hutch, which houses optical components, such as slits for collimation and defining the beam size, mirrors for focussing and monochromators for selecting a narrow wavelength range from the X-ray spectrum. The latter can be located in a second optics hutch further downstream. The beam is then guided into the experimental hutch, where the experimental equipment, such as sample stage and detector, is located and the measurements are performed.

As the electrons lose a significant part of their energy through the emission of synchrotron radiation, radio-frequency cavities are installed in the storage ring to bring the electrons back to their design energy and keep them on their fixed track around the ring.

X-ray radiation from a synchrotron source, in particular from undulator insertion devices, typically has very high brilliance, many orders of magnitude higher than conventional laboratory sources. Apart from the high flux, a small emittance, typical for X-ray synchrotron radiation sources, contributes to its high brilliance. The photon beam emittance is determined both by the emittance of the electron beam and the emittance of the photon beam produced by a single electron travelling along the source path (Als-Nielsen and McMorrow 2011). The latter is driven by relativistic forward-focussing effects. The emittance of the electron beam is the results of a combination of radiation damping and quantum diffusion effects. In the horizontal direction it is caused by dispersion in the bending plane due to the Lorentz forces from the magnets acting in this direction, while along the vertical axis there is no intrinsic diffusion and hence the emittance is significantly smaller.[8] As a consequence, the sources at synchrotrons are significantly smaller in the vertical than in the horizontal direction.

Thanks to the small source size and long distance from the source, the X-rays typically have long transverse (spatial) coherence lengths (see Eq. (2.15)) at the experimental endstations, which is essential for coherent imaging techniques such as CDI and ptychography, but also beneficial for near-field imaging methods like X-ray grating interferometry and speckle-based imaging. The high spatial coherence allows for the operation of X-ray grating interferometry without a source grating G0 (as explained in Chap. 4) and leads to a high visibility of the speckle pattern for speckle-based imaging. While the temporal coherence requirements are quite high for coherent imaging techniques, X-ray grating- and speckle-based imaging do not impose strong restrictions, but do deliver better results for higher temporal

[8]Typically, nevertheless a finite vertical emittance can be observed due to coupling effects. Coupling describes the case when a horizontal component of the magnetic fields exists due to misalignments such as the rotation of quadrupole magnets and the off-centre position of sextupole magnets and insertion devices.

coherence. Double-crystal monochromators (DCM), which are commonly installed at synchrotron beamlines, achieve very narrow wavelength bandwidths $\Delta\lambda$ and hence give a large temporal coherence length, even for small wavelengths. Often also multilayer monochromators (MLM) are available, which provide a higher flux at the cost of a broader bandwidth of photon wavelengths. This can be useful to reduce scan times when using methods that do not crucially rely on a high temporal coherence of the X-ray beam.

For coherent imaging applications, it is often beneficial to make use of radiation from an undulator insertion device rather than a bending magnet or wiggler due to the higher coherent flux. Generally, the emittance of undulators is smaller and the concentration of the radiation in distinct harmonics allows for achieving a high temporal coherence at high flux. On the other hand, wigglers are useful for applications at higher X-ray energies and in particular superconducting wigglers can deliver a very high photon flux. The total integrated power output by undulators is significantly less, but they provide a higher flux per unit solid angle.

In the scope of this Ph.D. project, X-ray grating interferometry was implemented for the first time at beamline I13 at Diamond Light Source using a DCM and an MLM, see Chap. 4, Sects. 4.3 and 4.4. X-ray speckle-based imaging was conducted with monochromatic X-rays from a DCM, see Chaps. 6 and 7, but also with a filtered white beam, i.e. using the full bandwidth of the X-ray spectrum of the beam coming from the undulator with some shaping of the spectrum by filters, see Chap. 8. The latter is possible thanks to the relatively relaxed temporal coherence requirements of the technique.

2.6.3 *Novel X-ray Sources*

In the recent years, a number of novel X-ray sources have been developed, opening up a new era for X-ray science. One of the most important developments in the last decade was the realisation of hard X-ray free electron lasers (XFELs) (Feldhaus et al. 2005; Pellegrini 2012, 2016). The Linac Coherent Light Source (LCLS) at SLAC National Accelerator Laboratory, USA, was the first hard XFEL to start operation in 2009 (Emma et al. 2010; Bostedt et al. 2013). Since then, five more hard XFELs have become operational, which are located in different countries across Europe, the USA and Asia.

XFELs can produce X-rays with a 10^9 to 10^{10} times higher peak brilliance, i.e. the brilliance for a single pulse, than third generation synchrotrons (Pellegrini 2016), as illustrated in Fig. 2.9. Furthermore, while third generation synchrotrons can generate pulses of several tens of picoseconds, XFELs can achieve ultra-short pulses of tens of femtoseconds duration, making it possible to study the structure and dynamics of atomic and molecular systems at extremely small spatial and temporal length scales, see e.g. Chapman (2019); Roseker et al. (2018); Wiedorn et al. (2018), amongst others.

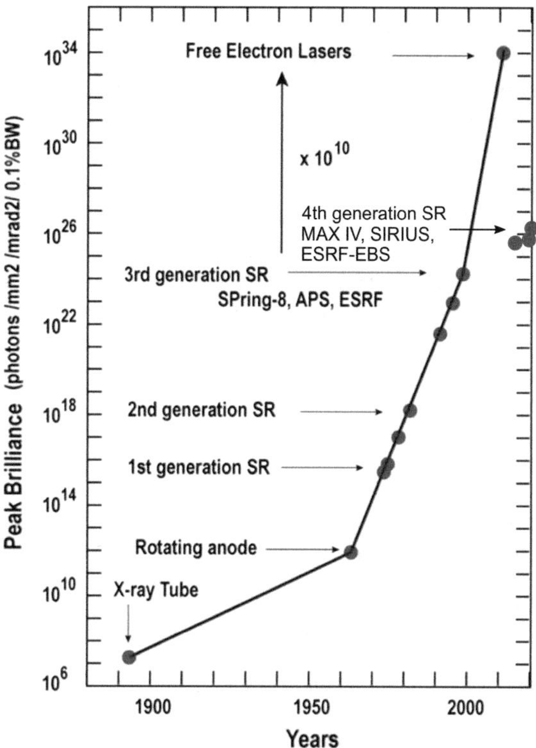

Fig. 2.9 Evolution of the peak brilliance of X-ray sources from tube sources up to X-ray free electron lasers over the last century. Figure adapted and reprinted with permission from Shintake (2007), copyright [2007] IEEE

At around the same time, fourth-generation synchrotron sources[9] were being developed, making use of the multi-bend achromat technology (Einfeld et al. 1995, 2014; Einfeld 2014) to decrease emittance. This leads to a significant increase in brilliance of one to two orders of magnitude of these sources compared to third generation synchrotrons, see Fig. 2.9. The first fourth-generation synchrotron, MAX IV Laboratory in Lund (Sweden) started operation in 2016 and Sirius in Campinas (Brazil) is expected to open in 2020. Furthermore, upgrades of several third generation synchrotron sources are currently planned or underway. The first one is the ESRF, which is scheduled to start operation as ESRF-EBS (extremely brilliant source) (Einfeld 2018) in 2020. It will be the first high-energy fourth-generation storage ring and is designed to deliver an X-ray beam of 100 times higher brilliance and coherent fraction than before the upgrade.

[9]Often XFELs are referred to as fourth-generation X-ray sources. At the same time, the term "fourth-generation synchrotron sources" is used for the newly developed high-brilliance storage rings.

In the field of high-brilliance compact and cost-efficient X-ray sources, some of the recent developments include the Munich compact light source (MuCLS) based on inverse Compton scattering (Eggl et al. 2016; Dierolf et al. 2018), which was installed a few years ago at TU Munich (Germany), and laser-wakefield-driven compact sources as installed at several institutions in the USA (Kneip et al. 2010; Wang et al. 2013; Albert et al. 2017), Germany (Wenz et al. 2015) and the UK (Cole et al. 2018). While these types of sources do not reach the brilliance, flux and coherence properties of synchrotron sources, they significantly outperform conventional laboratory sources and are yet of much smaller dimensions and cost than synchrotron sources, which makes them more accessible to users.

References

Albert F, Lemos N, Shaw JL, Pollock BB, Goyon C, Schumaker W, Saunders AM, Marsh KA, Pak A, Ralph JE, Martins JL, Amorim LD, Falcone RW, Glenzer SH, Moody JD, Joshi C (2017) Observation of Betatron X-ray radiation in a self-modulated laser wakefield accelerator driven with picosecond laser pulses. Phys. Rev. Lett. 118(13):134801

Allen L, Oxley M (2001) Phase retrieval from series of images obtained by defocus variation. Opt. Commun. 199(1):65–75

Als-Nielsen J, McMorrow D (2011) Elements of modern X-ray physics, 2nd edn. Wiley, Chichester, UK

Bech M (2009) X-ray imaging with a grating interferometer. University of Copenhagen, Copenhagen, Denmark Ph.D. thesis

Bech M, Bunk O, Donath T, Feidenhans'l R, David C, Pfeiffer F (2010) Quantitative x-ray dark-field computed tomography. Phys Med Biol 55(18):5529–5539

Bérujon S, Wang H, Sawhney K (2012a) X-ray multimodal imaging using a random-phase object. Phys Rev A 86(6):063813

Bérujon S, Ziegler E (2015) Near-field speckle-scanning-based x-ray imaging. Phys Rev A 92(1):013837

Bérujon S, Ziegler E (2016) X-ray multimodal tomography using speckle-vector tracking. Phys Rev Appl 5(4):044014

Bérujon S, Ziegler E (2017) Near-field speckle-scanning-based x-ray tomography. Phys Rev A 95(6):063822

Bérujon S, Ziegler E, Cerbino R, Peverini L (2012b) Two-Dimensional X-ray beam phase sensing. Phys Rev Lett 108(15):158102

Bilderback DH, Elleaume P, Weckert E (2005) Review of third and next generation synchrotron light sources. J Phys B 38(9):S773–S797

Bonse U, Hart M (1965) An X-ray interferometer. Appl Phys Lett 6:155–156

Bostedt C, Bozek JD, Bucksbaum PH, Coffee RN, Hastings JB, Huang Z, Lee RW, Schorb S, Corlett JN, Denes P, Emma P, Falcone RW, Schoenlein RW, Doumy G, Kanter EP, Kraessig B, Southworth S, Young L, Fang L, Hoener M et al (2013) Ultra-fast and ultra-intense x-ray sciences: first results from the Linac Coherent light source free-electron laser. J Phys B 46(16):164003

Cerbino R, Peverini L, Potenza MAC, Robert A, Bösecke P, Giglio M (2008) X-ray-scattering information obtained from near-field speckle. Nat Phys 4(3):238–243

Chabior M, Donath T, David C, Bunk O, Schuster M, Schroer C, Pfeiffer F (2011) Beam hardening effects in grating-based x-ray phase-contrast imaging. Med Phys 38(3):1189–1195

Chapman D, Thomlinson W, Johnston RE, Washburn D, Pisano E, Gmür N, Zhong Z, Menk R, Arfelli F, Sayers D (1997) Diffraction enhanced x-ray imaging. Phys Med Biol 42(11):2015
Chapman HN (2019) X-ray free-electron lasers for the structure and dynamics of macromolecules. Annu Rev Biochem 88(1):35–58
Clare RM, Stockmar M, Dierolf M, Zanette I, Pfeiffer F (2015) Characterization of near-field ptychography. Opt Express 23(15):19728–19742
Cloetens P, Barrett R, Baruchel J, Guigay J-P, Schlenker M (1996) Phase objects in synchrotron radiation hard x-ray imaging. J Phys D: Appl Phys 29(1):133–146
Cloetens P, Guigay JP, De Martino C, Baruchel J, Schlenker M (1997) Fractional Talbot imaging of phase gratings with hard x rays. Opt Lett 22(14):1059–1061
Cloetens P, Ludwig W, Baruchel J, Dyck DV, Landuyt JV, Guigay JP, Schlenker M (1999) Holotomography: quantitative phase tomography with micrometer resolution using hard synchrotron radiation x-rays. Appl Phys Lett 75(19):2912–2914
Cloetens P, Mache R, Schlenker M, Lerbs-Mache S (2006) Quantitative phase tomography of Arabidopsis seeds reveals intercellular void network. Proc Natl Acad Sci 103(39):14626–14630
Codman EA (1896) Radiograph of fetal arm. Boston Med Surg J 134(13):327
Cole JM, Symes DR, Lopes NC, Wood JC, Poder K, Alatabi S, Botchway SW, Foster PS, Gratton S, Johnson S, Kamperidis C, Kononenko O, De Lazzari M, Palmer CAJ, Rusby D, Sanderson J, Sandholzer M, Sarri G, Szoke-Kovacs Z, Teboul L et al (2018) High-resolution μCT of a mouse embryo using a compact laser-driven X-ray betatron source. Proc Natl Acad Sci 115(25):6335–6340
Coolidge WD (1913) A powerful Röntgen ray tube with a pure electron discharge. Phys Rev 2(6):409–430
Cormack AM (1963) Representation of a function by its line integrals, with some radiological applications. J Appl Phys 34(9):2722–2727
Crookes W (1879) V the Bakerian Lecture—on the illumination of lines of molecular pressure, and the trajectory of molecules. Philos Trans R Soc 170:135–164
David C, Nöhammer B, Solak HH, Ziegler E (2002) Differential x-ray phase contrast imaging using a shearing interferometer. Appl Phys Lett 81(17):3287–3289
Davis TJ, Gao D, Gureyev TE, Stevenson AW, Wilkins SW (1995) Phase-contrast imaging of weakly absorbing materials using hard X-rays. Nature 373(6515):595–598
Delaney AH, Bresler Y (1998) Globally convergent edge-preserving regularized reconstruction: an application to limited-angle tomography. IEEE Trans Image Process 7(2):204–221
Diamond Light Source Ltd (2019) Machine. https://www.diamond.ac.uk/Science/Machine.html. Accessed 10 July 2019
Diemoz PC, Endrizzi M, Zapata CE, Pešić ZD, Rau C, Bravin A, Robinson IK, Olivo A (2013) X-ray phase-contrast imaging with nanoradian angular resolution. Phys Rev Lett 110(13):138105
Dierolf M, Günther B, Gradl R, Jud C, Eggl E, Gleich B, Achterhold K, Pfeiffer F (2018) The Munich compact light source—operating an inverse compton source in user mode. In: High-Brightness sources and light-driven interactions. Optical Society of America. paper EM2B.3, OSA Technical Digest (online)
Dierolf M, Menzel A, Thibault P, Schneider P, Kewish CM, Wepf R, Bunk O, Pfeiffer F (2010) Ptychographic X-ray computed tomography at the nanoscale. Nature 467:436
Duan X, Zhang L, Xing Y, Chen Z, Cheng J (2009) Few-View projection reconstruction with an iterative reconstruction-reprojection algorithm and TV constraint. IEEE Trans Nucl Sci 56(3):1377–1382
Dudgeon DE, Mersereau RM (1984) Multidimensional digital signal processing. Prentice-Hall, Englewood Cliffs, NJ, US
Editorial (1896a) On the application of the Röntgen rays to the diagnosis of arterio-sclerosis. Boston Med Surg J 134(22):550–551
Editorial (1896b) Rare anomalies of the phalanges shown by the Röntgen process. Boston Med Surg J 134(8):198–199

Eggl E, Dierolf M, Achterhold K, Jud C, Günther B, Braig E, Gleich B, Pfeiffer F (2016) The Munich compact light source: initial performance measures. J Synchrotron Rad 23(5):1137–1142

Einfeld D (2014) Multi-bend achromat lattices for storage ring light sources. Synchrotron Rad News 27(6):4–7

Einfeld D (2018) EBS storage ring technical report, Technical report, the European Synchrotron—ESRF (2018). http://www.esrf.eu/files/live/sites/www/files/about/upgrade/documentation/Design%20Report-reduced-jan19.pdf. Accessed 22 Mar 2019

Einfeld D, Plesko M, Schaper J (2014) First multi-bend achromat lattice consideration. J Synchrotron Rad 21(5):856–861

Einfeld D, Schaper J, Plesko M (1995) Design of a diffraction limited light source (DIFL). Proc PAC 1:177–179

Emma P, Akre R, Arthur J, Bionta R, Bostedt C, Bozek J, Brachmann A, Bucksbaum P, Coffee R, Decker F-J, Ding Y, Dowell D, Edstrom S, Fisher A, Frisch J, Gilevich S, Hastings J, Hays G, Hering P, Huang Z et al (2010) First lasing and operation of an ångstrom-wavelength free-electron laser. Nat Photonics 4:641–647

Engelhardt M, Kottler C, Bunk O, David C, Schroer C, Baumann J, Schuster M, Pfeiffer F (2008) The fractional Talbot effect in differential x-ray phase-contrast imaging for extended and polychromatic x-ray sources. J Microsc 232(1):145–157

European Synchrotron Radiation Facility (2019) Beamline ID19. https://www.esrf.eu/home/UsersAndScience/Experiments/StructMaterials/ID19/over.html. Accessed 10 July 2019

Feldhaus J, Arthur J, Hastings JB (2005) X-ray free-electron lasers. J Phys B 38(9):S799–S819

Feldkamp LA, Davis LC, Kress JW (1984) Practical cone-beam algorithm. J Opt Soc Am A 1(6):612–619

Guigay JP, Langer M, Boistel R, Cloetens P (2007) Mixed transfer function and transport of intensity approach for phase retrieval in the Fresnel region. Opt Lett 32(12):1617–1619

Guinier A (1994) X-ray diffraction in crystals, imperfect crystals, and amorphous bodies. Dover Books on Physics Series, Dover, New York, NY, US

Gureyev T, Nugent K (1997) Rapid quantitative phase imaging using the transport of intensity equation. Opt Commun 133(1):339–346

Gureyev TE, Nugent KA (1996) Phase retrieval with the transport-of-intensity equation. II. Orthogonal series solution for nonuniform illumination. J Opt Soc Am A 13(8):1670–1682

Gureyev TE, Raven C, Snigirev A, Snigireva I, Wilkins SW (1999) Hard x-ray quantitative non-interferometric phase-contrast microscopy. J Phys D: Appl Phys 32(5):563–567

Hagen CK, Munro PRT, Endrizzi M, Diemoz PC, Olivo A (2014) Low-dose phase contrast tomography with conventional x-ray sources. Med Phys 41(7):070701

Hanson KM, Wecksung GW (1983) Bayesian approach to limited-angle reconstruction in computed tomography. J Opt Soc Am 73(11):1501–1509

Hemberg O, Otendal M, Hertz HM (2003) Liquid-metal-jet anode electron-impact x-ray source. Appl Phys Lett 83(7):1483–1485

Henke BL, Gullikson EM, Davis JC (1993) X-ray interactions: photoabsorption, scattering, transmission, and reflection at E = 50–30,000 eV, Z = 1–92. Atomic Data Nucl Data Tables 54(2):181–342

Hipp A, Willner M, Herzen J, Auweter S, Chabior M, Meiser J, Achterhold K, Mohr J, Pfeiffer F (2014) Energy-resolved visibility analysis of grating interferometers operated at polychromatic X-ray sources. Opt Express 22(25):30394–30409

Holler M, Guizar-Sicairos M, Tsai EHR, Dinapoli R, Müller E, Bunk O, Raabe J, Aeppli G (2017) High-resolution non-destructive three-dimensional imaging of integrated circuits. Nature 543:402

Hounsfield GN (1973) Computerized transverse axial scanning (tomography): Part 1. Description of system. Br J Radiol 46(552):1016–1022

Hruszkewycz SO, Sutton M, Fuoss PH, Adams B, Rosenkranz S, Ludwig KF, Roseker W, Fritz D, Cammarata M, Zhu D, Lee S, Lemke H, Gutt C, Robert A, Grübel G, Stephenson GB (2012) High contrast X-ray speckle from atomic-scale order in liquids and glasses. Phys Rev Lett 109(18):185502

Ingal VN, Beliaevskaya EA (1995) X-ray plane-wave topography observation of the phase contrast from a non-crystalline object. J Phys D: Appl Phys 28(11):2314–2317

Kak AC, Slaney M (2001) Principles of computerized tomographic imaging. Society for Industrial and Applied Mathematics, Philadelphia, PA, US

Kashyap Y, Wang H, Sawhney K (2015) Two-dimensional transverse coherence measurement of hard-x-ray beams using near-field speckle. Phys Rev A 92(3):033842

Khelashvili G, Brankov JG, Chapman D, Anastasio MA, Yang Y, Zhong Z, Wernick MN (2006) A physical model of multiple-image radiography. Phys Med Biol 51(2):221

Kneip S, McGuffey C, Martins JL, Martins SF, Bellei C, Chvykov V, Dollar F, Fonseca R, Huntington C, Kalintchenko G, Maksimchuk A, Mangles SPD, Matsuoka T, Nagel SR, Palmer CAJ, Schreiber J, Phuoc KT, Thomas AGR, Yanovsky V, Silva LO et al (2010) Bright spatially coherent synchrotron X-rays from a table-top source. Nat Phys 6:980

Kunz C (2001) Synchrotron radiation: third generation sources. J Phys Condens Matter 13(34):7499–7510

Lang S, Müller B, Dominietto MD, Cattin PC, Zanette I, Weitkamp T, Hieber SE (2012) Three-dimensional quantification of capillary networks in healthy and cancerous tissues of two mice. Microvasc Res 84(3):314–322

Lang S, Zanette I, Dominietto M, Langer M, Rack A, Schulz G, Le Duc G, David C, Mohr J, Pfeiffer F, Müller B, Weitkamp T (2014) Experimental comparison of grating- and propagation-based hard X-ray phase tomography of soft tissue. J Appl Phys 116(15):154903

Langer M, Cloetens P, Guigay J-P, Peyrin F (2008) Quantitative comparison of direct phase retrieval algorithms in in-line phase tomography. Med Phys 35(10):4556–4566

Langer M, Cloetens P, Guigay JP, Valton S, Peyrin F (2007) Quantitative evaluation of phase retrieval algorithms in propagation based phase tomography. In: 2007 4th IEEE international symposium on biomedical imaging: from Nano to Macro, pp 552–555

Langer M, Cloetens P, Pacureanu A, Peyrin F (2012a) X-ray in-line phase tomography of multimaterial objects. Opt Lett 37(11):2151–2153

Langer M, Cloetens P, Peyrin F (2010) Regularization of phase retrieval with phase-attenuation duality prior for 3-D holotomography. IEEE T Image Process 19(9):2428–2436

Langer M, Pacureanu A, Suhonen H, Grimal Q, Cloetens P, Peyrin F (2012b) X-ray phase nanotomography resolves the 3D human bone ultrastructure. PLoS One 7(8):1–7

Lohmann A, Silva D (1971) An interferometer based on the Talbot effect. Opt Commun 2(9):413–415

Marchesini S, Chapman HN, Hau-Riege SP, London RA, Szoke A, He H, Howells MR, Padmore H, Rosen R, Spence JCH, Weierstall U (2003) Coherent X-ray diffractive imaging: applications and limitations. Opt Express 11(19):2344–2353

Mayo S, Stevenson A, Wilkins S (2012) In-Line phase-contrast X-ray imaging and tomography for materials science. Materials 5(5):937–965

Mersereau RM, Oppenheim AV (1974) Digital reconstruction of multidimensional signals from their projections. Proc IEEE 62(10):1319–1338

Miao J, Charalambous P, Kirz J, Sayre D (1999) Extending the methodology of X-ray crystallography to allow imaging of micrometre-sized non-crystalline specimens. Nature 400:342

Mochrie SGJ, Mayes AM, Sandy AR, Sutton M, Brauer S, Stephenson GB, Abernathy DL, Grübel G (1997) Dynamics of block Copolymer Micelles Revealed by X-ray intensity fluctuation spectroscopy. Phys Rev Lett 78(7):1275–1278

Momose A (1995) Demonstration of phase-contrast X-ray computed tomography using an X-ray interferometer. Nucl Instrum Methods Phys Res A 352(3):622–628

Momose A, Fukuda J (1995) Phase-contrast radiographs of nonstained rat cerebellar specimen. Med Phys 22(4):375–379

Momose A, Kawamoto S, Koyama I, Hamaishi Y, Takai K, Suzuki Y (2003) Demonstration of X-ray talbot interferometry. Jpn J Appl Phys 42:L866–L868

Momose A, Takeda T, Itai Y, Yoneyama A, Hirano K (1998) Phase-Contrast tomographic imaging using an X-ray interferometer. J Synchrotron Rad 5(3):309–314

Morgan KS, Paganin DM, Siu KKW (2012) X-ray phase imaging with a paper analyzer. Appl Phys Lett 100(12):124102

Nassi M, Brody WR, Medoff BP, Macovski A (1982) Iterative reconstruction-reprojection: an algorithm for limited data cardiac-computed tomography. IEEE Trans Biomed Eng 29(5):333–341

Nesterets YI (2008) On the origins of decoherence and extinction contrast in phase-contrast imaging. Opt Commun 281(4):533–542

Nugent KA (2010) Coherent methods in the X-ray sciences. Adv Phys 59(1):1–99

Nugent KA, Gureyev TE, Cookson DF, Paganin D, Barnea Z (1996) Quantitative phase imaging using hard X-rays. Phys Rev Lett 77(14):2961–2964

Olivo A, Bohndiek SE, Griffiths JA, Konstantinidis A, Speller RD (2009) A non-free-space propagation x-ray phase contrast imaging method sensitive to phase effects in two directions simultaneously. Appl Phys Lett 94(4):044108

Olivo A, Speller R (2007) A coded-aperture technique allowing x-ray phase contrast imaging with conventional sources. Appl Phys Lett 91(7):074106

Paganin D, Mayo SC, Gureyev TE, Miller PR, Wilkins SW (2002) Simultaneous phase and amplitude extraction from a single defocused image of a homogeneous object. J Microsc 206(1):33–40

Paganin D, Nugent KA (1998) Noninterferometric phase imaging with partially coherent light. Phys Rev Lett 80(12):2586–2589

Paganin DM (2006) Coherent X-ray optics, 2nd edn. Clarenden Press, Oxford, UK

Paganin DM, Labriet H, Brun E, Bérujon S (2018) Single-image geometric-flow x-ray speckle tracking. Phys Rev A 98(5):053813

Pagot E, Cloetens P, Fiedler S, Bravin A, Coan P, Baruchel J, Härtwig J, Thomlinson W (2003) A method to extract quantitative information in analyzer-based x-ray phase contrast imaging. Appl Phys Lett 82(20):3421–3423

Pellegrini C (2012) The history of X-ray free-electron lasers. Eur Phys J H 37(5):659–708

Pellegrini C (2016) X-ray free-electron lasers: from dreams to reality. Phys Scr T 169:014004

Pfeiffer F (2018) X-ray ptychography. Nat Photon 12(1):9–17

Pfeiffer F, Bech M, Bunk O, Kraft P, Eikenberry EF, Brönnimann C, Grünzweig C, David C (2008) Hard-X-ray dark-field imaging using a grating interferometer. Nat Mater 7:134–137

Pfeiffer F, Kottler C, Bunk O, David C (2007) Hard X-ray phase tomography with low-brilliance sources. Phys Rev Lett 98(10):108105

Pfeiffer F, Weitkamp T, Bunk O, David C (2006) Phase retrieval and differential phase-contrast imaging with low-brilliance X-ray sources. Nat Phys 2:258–261

Pollock HC (1983) The discovery of synchrotron radiation. Am J Phys 51(3):278–280

Raven C, Snigirev A, Snigireva I, Spanne P, Souvorov A, Kohn V (1996) Phase-contrast microtomography with coherent high-energy synchrotron x rays. Appl Phys Lett 69(13):1826–1828

Robisch A-L, Kröger K, Rack A, Salditt T (2015) Near-field ptychography using lateral and longitudinal shifts. New J Phys 17(7):073033

Robisch A-L, Salditt T (2013) Phase retrieval for object and probe using a series of defocus near-field images. Opt Express 21(20):23345–23357

Rodenburg J (2008) Ptychography and related diffractive imaging methods. Adv Imaging Electron Phys 150:87–184

Rodenburg JM, Faulkner HML (2004) A phase retrieval algorithm for shifting illumination. Appl Phys Lett 85(20):4795–4797

Rodenburg JM, Hurst AC, Cullis AG, Dobson BR, Pfeiffer F, Bunk O, David C, Jefimovs K, Johnson I (2007) Hard-X-ray lensless imaging of extended objects. Phys Rev Lett 98(3):034801

Röntgen WC (1895) Über eine neue Art von Strahlen (Vorläufige Mittheilung)," Sitzungsber. der Würzburger Physik.-Medic. Gesellsch. 9:132–141

Röntgen WC (1898) Über eine neue Art von Strahlen (Erste Mittheilung). Ann Phys (Berl) 300(1):1–11

Roseker W, Hruszkewycz SO, Lehmkühler F, Walther M, Schulte-Schrepping H, Lee S, Osaka T, Strüder L, Hartmann R, Sikorski M, Song S, Robert A, Fuoss PH, Sutton M, Stephenson GB, Grübel G (2018) Towards ultrafast dynamics with split-pulse X-ray photon correlation spectroscopy at free electron laser sources. Nat Commun 9(1):1704

Rytov SM, Kravtsov YA, Tatarskii VI (1989) Principles of statistical radiophysics. 4. Wave propagation through random media. Springer, Berlin, Germany

Shi X, Burdet N, Chen B, Xiong G, Streubel R, Harder R, Robinson IK (2019) X-ray ptychography on low-dimensional hard-condensed matter materials. Appl Phys Rev 6(1):011306

Shintake T (2007) Review of the worldwide SASE FEL development. In: 2007 IEEE particle accelerator conference (PAC), 89–93

Sidky EY, Kao C-M, Pan X (2006) Accurate image reconstruction from few-views and limited-angle data in divergent-beam CT. J X-ray Sci Tech 14:119–139

Snigirev A, Snigireva I, Kohn V, Kuznetsov S (1996) On the requirements to the instrumentation for the new generation of the synchrotron radiation sources. Beryllium windows. Nucl Instrum Methods Phys Res A 370(2):634–640

Snigirev A, Snigireva I, Kohn V, Kuznetsov S, Schelokov I (1995) On the possibilities of x-ray phase contrast microimaging by coherent high-energy synchrotron radiation. Rev Sci Instrum 66(12):5486–5492

Spiegel PK (1995) The first clinical X-ray made in America-100 years. AJR Am J Roentgenol 164(1):241–243

Stockmar M, Cloetens P, Zanette I, Enders B, Dierolf M, Pfeiffer F, Thibault P (2013) Near-field ptychography: phase retrieval for inline holography using a structured illumination. Sci Rep 3:1927

Stockmar M, Hubert M, Dierolf M, Enders B, Clare R, Allner S, Fehringer A, Zanette I, Villanova J, Laurencin J, Cloetens P, Pfeiffer F, Thibault P (2015a) X-ray nanotomography using near-field ptychography. Opt Express 23(10):12720–12731

Stockmar M, Zanette I, Dierolf M, Enders B, Clare R, Pfeiffer F, Cloetens P, Bonnin A, Thibault P (2015b) X-ray near-field ptychography for optically thick specimens. Phys Rev Appl 3(1):014005

Sutton M, Mochrie SGJ, Greytak T, Nagler SE, Berman LE, Held GA, Stephenson GB (1991) Observation of speckle by diffraction with coherent X-rays. Nature 352:608–610

Swiss Light Source (2019) Paul Scherrer Institute, "TOMCAT beamline". https://www.psi.ch/en/sls/tomcat/source. Accessed 10 July 2019

Talbot H (1836) LXXVI. Facts relating to optical science. No IV. Philos Mag 3 9(56):401–407

Teague MR (1983) Deterministic phase retrieval: a Green's function solution. J Opt Soc Am 73(11):1434–1441

Thibault P, Dierolf M, Bunk O, Menzel A, Pfeiffer F (2009) Probe retrieval in ptychographic coherent diffractive imaging. Ultramicroscopy 109(4):338–343

Thibault P, Dierolf M, Menzel A, Bunk O, David C, Pfeiffer F (2008) High-Resolution scanning x-ray diffraction microscopy. Science 321(5887):379–382

Thomson JJ (1897) Cathode rays. Royal institution proceedings 15:419–432

Thomson JJ (1897a) Cathode rays. Philos Mag 44:293–316

Thüring T, Stampanoni M (2014) Performance and optimization of X-ray grating interferometry. Phil Trans R Soc A 372(2010)

Wagner U, Parson A, Rau C (2017) Coherence length and vibrations of the coherence beamline i13 at the diamond light source. J Phys Conf Ser 849(1):012048

Wang H, Kashyap Y, Cai B, Sawhney K (2016a) High energy X-ray phase and dark-field imaging using a random absorption mask. Sci Rep 6:30581

Wang H, Kashyap Y, Sawhney K (2016b) From synchrotron radiation to lab source: advanced speckle-based X-ray imaging using abrasive paper. Sci Rep 6:20476

Wang H, Kashyap Y, Sawhney K (2016c) Quantitative X-ray dark-field and phase tomography using single directional speckle scanning technique. Appl Phys Lett 108(12):124102

Wang X, Zgadzaj R, Fazel N, Li Z, Yi SA, Zhang X, Henderson W, Chang Y-Y, Korzekwa R, Tsai H-E, Pai C-H, Quevedo H, Dyer G, Gaul E, Martinez M, Bernstein AC, Borger T, Spinks M,

Donovan M, Khudik V et al (2013) Quasi-monoenergetic laser-plasma acceleration of electrons to 2 GeV. Nat. Commun 4:1988

Wang Z, Zhu P, Huang W, Yuan Q, Liu X, Zhang K, Hong Y, Zhang H, Ge X, Gao K, Wu Z (2010) Analysis of polychromaticity effects in X-ray Talbot interferometer. Anal Bioanal Chem 397(6):2137–2141

Wang Z-T, Kang K-J, Huang Z-F, Chen Z-Q (2009) Quantitative grating-based x-ray dark-field computed tomography. Appl Phys Lett 95(9):094105

Weitkamp T, David C, Kottler C, Bunk O, Pfeiffer F (2006) Tomography with grating interferometers at low-brilliance sources. Proc SPIE 6318:63180S

Weitkamp T, Diaz A, David C, Pfeiffer F, Stampanoni M, Cloetens P, Ziegler E (2005) X-ray phase imaging with a grating interferometer. Opt Express 13(16):6296–6304

Weitkamp T, Haas D, Wegrzynek D, Rack A (2011) ANKAphase: software for single-distance phase retrieval from inline X-ray phase-contrast radiographs. J Synchrotron Rad 18(4):617–629

Wenz J, Schleede S, Khrennikov K, Bech M, Thibault P, Heigoldt M, Pfeiffer F (2015) Quantitative X-ray phase-contrast microtomography from a compact laser-driven betatron source. Nat Commun 6

Wiedorn MO, Oberthür D, Bean R, Schubert R, Werner N, Abbey B, Aepfelbacher M, Adriano L, Allahgholi A, Al-Qudami N, Andreasson J, Aplin S, Awel S, Ayyer K, Bajt S, Barák I, Bari S, Bielecki J, Botha S, Boukhelef D et al (2018) Megahertz serial crystallography. Nat Commun 9(1):4025

Wilkins SW, Gureyev TE, Gao D, Pogany A, Stevenson AW (1996) Phase-contrast imaging using polychromatic hard X-rays. Nature 384:335–337

Willmott P (2011) An introduction to synchrotron radiation: techniques and applications. Wiley, Chichester, UK

Wochner P, Gutt C, Autenrieth T, Demmer T, Bugaev V, Ortiz AD, Duri A, Zontone F, Grübel G, Dosch H (2009) X-ray cross correlation analysis uncovers hidden local symmetries in disordered matter. Proc Nat Acad Sci 106(28):11511–11514

Wolf J, Sperl JI, Schaff F, Schüttler M, Yaroshenko A, Zanette I, Herzen J, Pfeiffer F (2015) Lens-term- and edge-effect in X-ray grating interferometry. Biomed Opt Express 6(12):4812–4824

Yashiro W, Momose A (2015) Effects of unresolvable edges in grating-based X-ray differential phase imaging. Opt Express 23(7):9233–9251

Yashiro W, Terui Y, Kawabata K, Momose A (2010) On the origin of visibility contrast in x-ray Talbot interferometry. Opt Express 18(16):16890–16901

Yun W, Kirz J, Lewis SJY (2015) Structured targets for x-ray generation, US Patent US20160064175A1, filed 29 Aug 2014, and issued 21 Jan 2015

Zanette I (2011) Interférométrie X à réseaux pour l'imagerie et l'analyse de front d'ondes au synchrotron. Université de Grenoble, Grenoble, France Ph.D. thesis

Zanette I, Lang S, Rack A, Dominietto M, Langer M, Pfeiffer F, Weitkamp T, Müller B (2013) Holotomography versus X-ray grating interferometry: a comparative study. Appl Phys Lett 103(24):244105

Zanette I, Zdora M-C, Zhou T, Burvall A, Larsson DH, Thibault P, Hertz HM, Pfeiffer F (2015) X-ray microtomography using correlation of near-field speckles for material characterization. Proc Nat Acad Sci USA 112(41):12569–12573

Zanette I, Zhou T, Burvall A, Lundström U, Larsson DH, Zdora M-C, Thibault P, Pfeiffer F, Hertz HM (2014) Speckle-Based X-ray phase-contrast and dark-field imaging with a laboratory source. Phys Rev Lett 112(25):253903

Zdora M-C (2018) State of the art of X-ray speckle-based phase-contrast and dark-field imaging. J Imaging 4(5):60

Zdora M-C, Thibault P, Pfeiffer F, Zanette I (2015) Simulations of x-ray speckle-based dark-field and phase-contrast imaging with a polychromatic beam. J Appl Phys 118(11):113105

Zdora M-C, Thibault P, Zhou T, Koch FJ, Romell J, Sala S, Last A, Rau C, Zanette I (2017) X-ray phase-contrast imaging and metrology through unified modulated pattern analysis. Phys Rev Lett 118(20):203903

Zernike F (1942) Phase contrast, a new method for the microscopic observation of transparent objects. Physica 9(7):686–698
Zernike F (1955) How i discovered phase contrast. Science 121(3141):345–349

Chapter 3
Synchrotron Beamlines, Instrumentation and Contributions

Following some general information on synchrotron radiation sources at the end of the previous chapter, this chapter outlines the main features and specifications of the beamline I13 at Diamond Light Source and the beamline ID19 at the ESRF, where most of the experiments of this project were carried out. Furthermore, the main contributions to the beamlines resulting from this Ph.D. project are summarised.

3.1 Beamline I13 at Diamond Light Source

Beamline I13 at Diamond Light Source is a 250 m-long beamline, with its experimental endstation situated approximately 220 m from the source in an external building, see Fig. 2.8a in the previous chapter. The long distance from the source leads to a high spatial coherence of the X-rays at the experimental endstation, which is essential for many experiments conducted at the beamline. I13 comprises two branches, the Diamond-Manchester Imaging Branchline I13-2 and the Coherence Branchline I13-1, as can be seen in the schematic drawing in Fig. 3.1. Detailed information about the beamline is reported by Rau et al. (2011).

The Diamond-Manchester imaging branchline is dedicated to X-ray full-field imaging for various fields of applications such as biomedical imaging, geology and food sciences, but with a strong focus on materials science, in particular in collaboration with the University of Manchester. X-ray energies in the range of 8-30 keV can be used and the maximum beam size at the sample is $18 \times 8\ \text{mm}^2$ (width $\times$ height). Classic absorption as well as propagation-based phase-contrast imaging are employed for most experiments. During this Ph.D. project, X-ray grating interferometry and speckle-based phase-contrast imaging were introduced to the beamline. First user experiments with X-ray grating interferometry have been conducted by now. The setup is currently being optimised with dedicated equipment such as an upside-down sample stage and a permanent grating mounting for easy and quick installation

M.-C. Zdora, *X-ray Phase-Contrast Imaging Using Near-Field Speckles*,
Springer Theses, https://doi.org/10.1007/978-3-030-66329-2_3

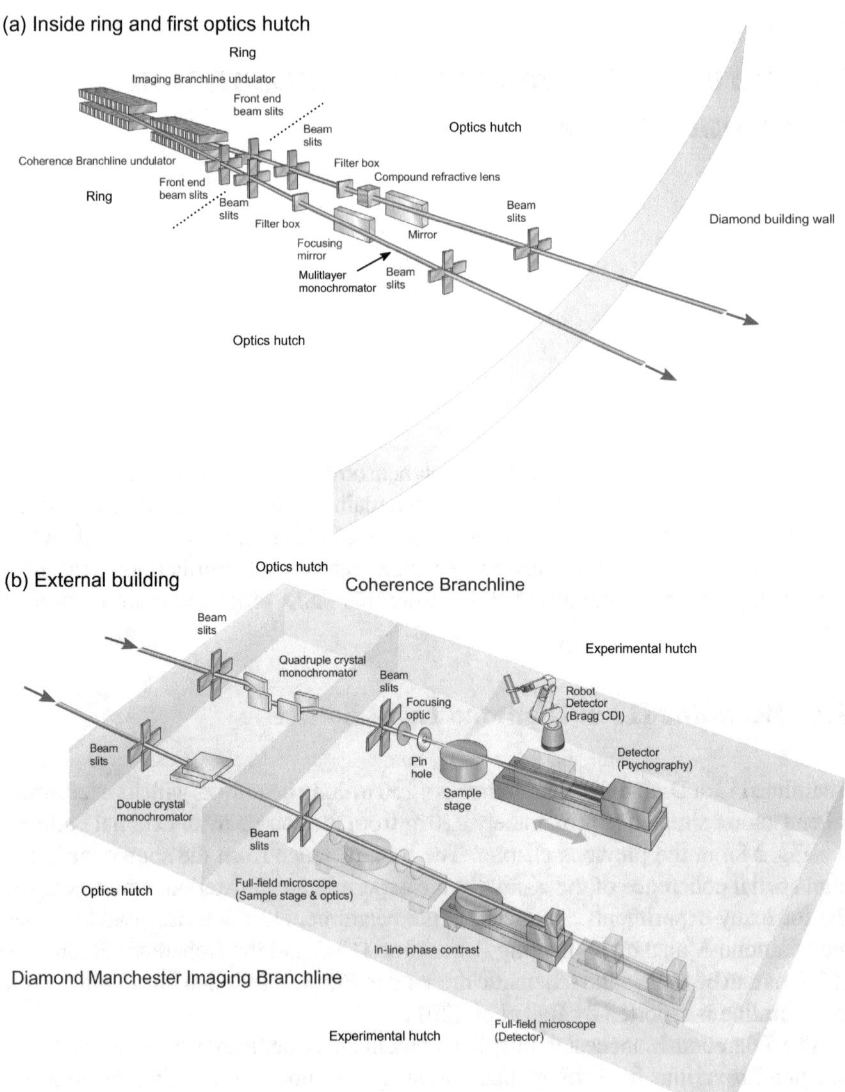

Fig. 3.1 Schematic of the beamline I13 at Diamond Light Source. The beamline consists of two branches operating in parallel: The Diamond-Manchester imaging branch and the coherence branch. Most of the work within the scope of this Ph.D. project was performed at the imaging branch. **a** Insertion devices and beam-shaping elements located inside the synchrotron ring and first optics hutch. **b** External building with second optics hutches, which house the monochromators, and experimental endstations. A multilayer monochromator was recently added to the imaging branch and is located in the first optics hutch as indicated by an arrow. Image courtesy of Diamond Light Source Ltd

of the instrument. Furthermore, several applications of X-ray speckle-based imaging have been demonstrated at I13 during this Ph.D. project. The technique is expected to attract beamline users from the fields of biomedicine, materials science and others.

The second branch is built mainly for imaging in reciprocal space, for which the diffraction pattern of the sample in the far-field is recorded. Several methods such as CDI, ptychography and Bragg ptychography are used. At the coherence branch, X-ray energies of 6–20 keV are available. As this branchline is using exclusively horizontally deflecting optics (mirror and quadrupole crystal monochromator), the spatial coherence of the X-ray beam in the vertical direction is higher than at the imaging branch. The spatial coherence in the horizontal direction at the coherence branch is increased using narrow beam-defining slits. The maximum coherent beam size at the coherence branch is approximately $200 \times 350\,\mu\text{m}^2$ (width × height) (Wagner et al. 2017).

The experiments presented in this thesis were conducted at both branches. However, most of the measurements were carried out at the imaging branch as the available energy range is more suitable and the coherence properties are sufficient for X-ray grating interferometry and speckle-based phase-contrast imaging.

3.1.1 Optical Elements

There are several optical elements installed at each branchline in optics hutches upstream of the experimental endstations. X-rays are created by two separate undulator insertion devices mounted with a canting angle of about 4 mrad for separation of the beams of the two branches.

For each branch, several sets of slits are used to shape the beam and adjust its size to match the sample dimensions. Filter boxes containing filters of different materials and thicknesses are available to reduce the flux and avoid damage of optical elements where required. They are, furthermore, used to shape the beam spectrum to a desired width and mean energy and eliminate low-energy contributions. This can be useful when a high flux is needed (ruling out the use of a monochromator), but a certain limited energy range is desired. Horizontally focussing mirrors are located downstream of the front end filter boxes for both branches. Monochromators are installed to select an X-ray beam of a defined energy when a high energy resolution is required. At the coherence branchline, a four-bounce horizontally deflecting silicon (Si) single-crystal monochromator (containing a set of Si(111) and Si(311) crystals)—located in a dedicated optics hutch outside the synchrotron building—provides monochromatic X-rays in the range of 6–20 keV. Most of the time, only one of the crystal pairs is used in a two-bounce configuration, which gives an energy resolution of approximately 10^{-4}. A vertically deflecting double-crystal monochromator is used at the imaging branchline (energy resolution: $\approx 10^{-4}$). In addition, recently a multilayer monochromator—located further upstream at about 50 m from the undulator, see arrow in Fig. 3.1—has been brought into operation at the imaging branch. It has the advantage of giving a significantly higher flux than the double-crystal monochromator albeit a larger wavelength bandwidth (energy resolution: 10^{-2}).

Downstream of the optics hutches, the experimental endstations are located, which house the setups customised for the specific experiment.

3.1.2 Detectors

Several different detector systems are available at I13 including photon-counting detectors such as the Medipix3 with Excalibur (Tartoni et al. 2012; Marchal et al. 2013) and Merlin (Plackett et al. 2013) readout systems (Quantum Detectors Ltd, UK), as well as lens-coupled charge-coupled device (CCD) and complementary metal-oxide-semiconductor (CMOS)-based camera systems. For the experiments presented here, a CCD and an sCMOS (scientific CMOS) camera coupled to an infinity-corrected optical microscope were used in combination with a scintillation screen for conversion of the incoming X-rays to visible light. A range of micropscope objectives with different magnification factors is available. The experiments shown in this thesis were conducted with a pco.4000 (CCD sensor; chip size: 4008 × 2672 pixels; pixel size: 9 μm) and a pco.edge 5.5 camera (sCMOS sensor; chip size: 2560 × 2160 pixels; pixel size: 6.5 μm), PCO AG (Kelheim, Germany), in combination with 1.25×, 2×, 4× or 10× objectives by Olympus (Tokyo, Japan) and a 2× magnifying eyepiece lens, leading to a total magnification of the system of 2.5×, 4×, 8× and 20×, respectively.

3.1.3 Developments and Setup Optimisations at I13 during this Ph.D. Project

During the course of this Ph.D. project, several X-ray phase-contrast imaging methods were employed and developed at Diamond I13:

- Propagation-based phase-contrast imaging, an established technique at the I13 imaging branch, was applied to biomedical and palaeontological specimens in collaborations with the Division of Cardiovascular Medicine, Radcliffe Department of Medicine, University of Oxford, UK, the Neurophysiology of Behaviour Laboratory at The Francis Crick Institute, UK, and the Department of Earth Sciences, University of Oxford, UK, see also Chap. 1.
- X-ray grating interferometry was implemented for the first time at I13 at both branches, see Chaps. 4 and 7. A robust setup for grating interferometry with a single grating was developed and applied to biomedical imaging. Furthermore, first users were supported in carrying out conventional grating interferometry with two gratings as well as single-grating interferometry. A dedicated setup for fast installation and easy operation of X-ray grating interferometry with one or two

gratings has been constructed at I13 as a result of contributions from this Ph.D. project, see also the following section, and is currently being commissioned.
- X-ray speckle-based imaging was performed for the first time at I13 in conventional experimental implementations (single-shot and stepping). Moreover, a new operational mode, called unified modulated pattern analysis, was developed, first demonstrated and optimised at I13, see Chap. 6. The unified modulated pattern analysis was applied to optics characterisation and biomedical imaging at I13, see Chaps. 7 and 8.

The specific roles of the Ph.D. candidate in these projects are outlined in Sect. 1.3 in Chap. 1.

3.1.4 X-ray Grating Interferometry and Speckle Imaging Equipment

After the first implementations of X-ray grating interferometry at I13 a dedicated setup has been constructed at I13, see Fig. 3.2a and b, which is currently under commissioning. It features an upside-down sample tomography stage, which allows for immersing the sample in a refractive index matching medium to reduce phase-wrapping artefacts at the edges of the specimen. A water tank is used for this purpose as successfully utilised at other beamlines (e.g. ID19, ESRF). The gratings available at I13 are shown in Fig. 3.2c and d. Other gratings temporarily provided by collaborators were also used during some of the experiments.

The setup for X-ray speckle-based imaging is straightforward and simple to assemble as only the diffuser on translational stages is required as an additional optical element. An example of a setup at I13 is shown in Fig. 6.2 in Chap. 6 (Sect. 6.2). Some of the diffusers used for X-ray speckle-based imaging in this project are shown in Fig. 3.3. For most measurements, commercially available sandpaper, consisting of silicon carbide particles on cellulose backing, was used, which is available in different grit sizes (Federation of European Producers of Abrasives 2019). For high-energy applications, e.g. geology and materials science, as shown in Chap. 9, Sect. 9.3 performed at Diamond I12, many layers of sandpaper were stacked for increasing the phase shift in the diffuser. Alternatively, a plastic box filled with sand (25 mm thickness in beam direction) was used, see Fig. 3.3b.

3.2 Beamline ID19 at the ESRF

Apart from Diamond I13, a major part of the work in this Ph.D. project was conducted at beamline ID19 at the ESRF (European Synchrotron Radiation Facility 2019). Similar to I13, ID19 is also a long beamline with its experimental endstation located

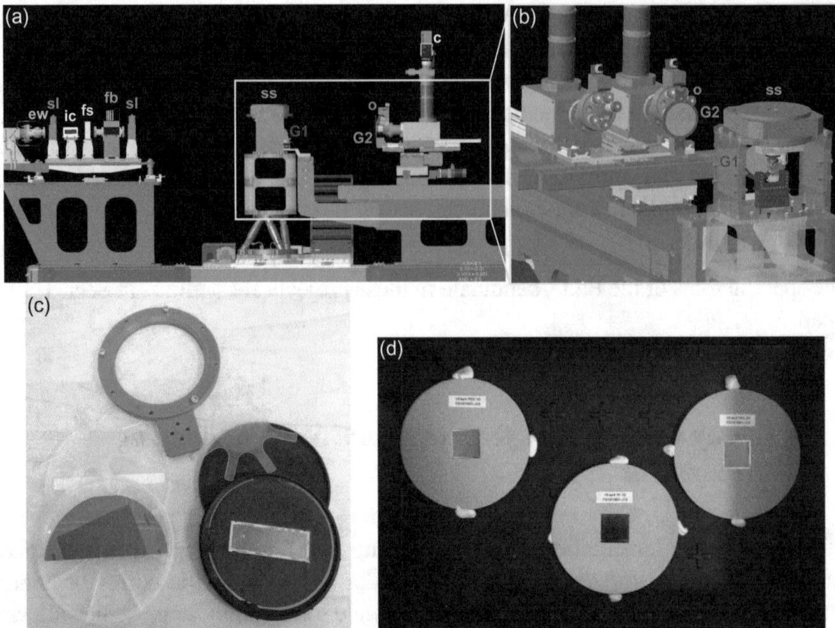

Fig. 3.2 X-ray grating interferometry equipment and newly developed setup at Diamond I13. **a**, **b** Model of the recently installed X-ray grating interferometry setup at Diamond I13. Apart from the standard beamline components upstream (ew: exit window, sl: slits, ic: ion chamber, fs: fast shutter, fb: filter box), the newly assembled grating setup components are shown (yellow box): An upside-down tomography sample stage (ss) for use with a water bath (wb), a phase grating (G1) with piezo stepping stage and an absorption grating (G2) just upstream of the objectives (o) that are scintillator-coupled via a microscope to the camera (c). Image courtesy of Ljubo Zaja, Diamond Light Source. **c** Some of the gratings used for initial experiments at I13. Left: phase grating (silicon lines on silicon wafer, fabricated at Paul Scherrer Institut, Switzerland, period: 4.785 µm, line height: 29.5 µm, π-shifting at 23 keV) borrowed from ESRF ID19 beamline and right: absorption grating (gold lines on silicon wafer, fabricated by microworks, Eggenstein-Leopoldshafen, Germany, period: 2.4 µm, line height: 80 µm, layout type: bridge design). Wafer diameter of both gratings: 100 mm. The gratings are mounted in a 3D-printed plastic grating holder (top) designed by Christian David (Paul Scherrer Institut, Switzerland) and Franz Pfeiffer (Technische Universität München, Germany). **d** Custom-made phase gratings (fabricated at Paul Scherrer Institut, Switzerland) optimised for an X-ray energy of 19 keV (1D π-shifting, 1D $\pi/2$-shifting, 2D $\pi/2$-shifting), which are about to be commissioned at I13

in an external hutch outside the synchrotron ring at a distance of 145 m from the source.

The beamline is used for imaging applications in a large range of fields such as cultural heritage, materials and engineering, medicine, life sciences, environmental sciences, physics, earth and planetary sciences and chemistry, with a focus on materials science and palaeontology. Several different techniques are implemented

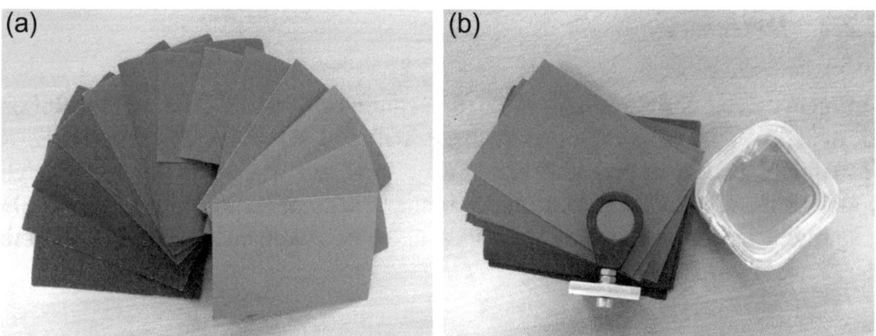

Fig. 3.3 Different diffusers used for the X-ray speckle-based imaging experiments in this thesis. **a** Sandpaper of different granularities. The grit size is chosen depending on the effective pixel size of the setup. For higher X-ray energies, typically several sheets of sandpaper are stacked. **b** Mounting of the diffusers. The sandpaper was mounted in a lens holder (left). For high-energy applications carried out at Diamond I12, a plastic container filled with sand was used as a diffuser (25 mm thickness in beam direction)

at ID19 including conventional absorption imaging, phase-contrast imaging and laminography.

There are five different insertion devices available for use at the beamline, depending on the required X-ray energy, spectrum and photon flux: four undulators and a wiggler. The achievable X-ray energy range is 10–250 keV (European Synchrotron Radiation Facility 2019). The beam size at the sample can reach a minimum of $0.1 \times 0.1\,\text{mm}^2$ and a maximum of $60 \times 15\,\text{mm}^2$ (width × height), allowing imaging of large specimens (European Synchrotron Radiation Facility 2019).

Experiments in the context of this thesis were carried out with the undulator insertion devices and at relatively low energies up to approximately 35 keV.

3.2.1 Optical Elements

The first optical hutch, located in the synchrotron building, contains diaphragms, slits and filters. The vertical double-crystal monochromator is located in the monochromator hutch in the external building just upstream of the experimental hutch. It also houses a stroboscopic shutter, i.e. an adjustable chopper, which can be used to reduce dose for white-beam operation, and a shutter. Furthermore, a multilayer monochromator is installed.

3.2.2 Detectors

Different detectors are available at ID19, including several CCD cameras (FReLoN 2k and E2V cameras developed at the ESRF) and sCMOS/CCD-based cameras (pco.dimax, pco.edge, pco.4000 cameras by PCO AG) equipped with visible light optics and scintillators. The pco.edge 5.5 camera in combination with 2.1× and 10× magnifying lens systems was used for most of the measurements taken at ID19 within the scope of this Ph.D. project.

3.2.3 Developments and Work at ID19 within this Ph.D. Project

The main work carried out at beamline ID19 by the Ph.D. candidate includes (see also contributions in Sect. 1.3, Chap. 1):

- Operation of X-ray grating interferometry using the dedicated setup developed by Zanette (2011), in particular in collaboration with the Biomaterials Science Center (BMC), University of Basel, Switzerland for beamtime support.
- The first implementation of X-ray speckle-based imaging at ID19 including:
  - The demonstration of X-ray speckle-based imaging for fast, time-resolved imaging (Olbinado et al. 2018).
  - The application of X-ray speckle-based phase tomography that goes beyond a proof-of-principle demonstration, see speckle-based virtual histology in Chap. 8.

References

European Synchrotron Radiation Facility (2019) Beamline ID19. https://www.esrf.eu/home/UsersAndScience/Experiments/StructMaterials/ID19/over.html. Accessed 10 July 2019

Federation of European Producers of Abrasives (2019) FEPA P-grit sizes coated abrasives. https://www.fepa-abrasives.com/abrasive-products/grains. Accessed 21 June 2019

Marchal J, Horswell I, Willis B, Plackett R, Gimenez EN, Spiers J, Ballard D, Booker P, Thompson JA, Gibbons P, Burge SR, Nicholls T, Lipp J, Tartoni N (2013) EXCALIBUR: a small-pixel photon counting area detector for coherent X-ray diffraction-Front-end design, fabrication and characterisation. J Phys Conf Ser 425(6):062003

Olbinado M, Grenzer J, Pradel P, Resseguier TD, Vagovic P, Zdora M-C, Guzenko V, David C, Rack A (2018) Advances in indirect detector systems for ultra high-speed hard X-ray imaging with synchrotron light. J Instrum 13(04):C04004

Plackett R, Horswell I, Gimenez EN, Marchal J, Omar D, Tartoni N (2013) Merlin: a fast versatile readout system for Medipix3. J Instrum 8(01):C01038

Rau C, Wagner U, Pešić Z, De Fanis A (2011) Coherent imaging at the Diamond beamline I13. Phys Status Solidi A 208(11):2522–2525

Tartoni N, Dennis G, Gibbons P, Gimenez E, Horswell I, Marchal J, Pedersen U, Pesic Z, Plackett R, Rau C, Somayaji R, Spiers J, Thompson J, Willis B, Angelsen C, Booker P, Burge S, Lipp J, Nicholls T, Taghavi S et al (2012) Excalibur: a three million pixels photon counting area detector for coherent diffraction imaging based on the Medipix3 ASIC. In: 2012 IEEE nuclear science symposium and medical imaging conference record (NSS/MIC), pp 530–533

Wagner U, Parson A, Rau C (2017) Coherence length and vibrations of the coherence beamline I13 at the diamond light source. J Phys Conf Ser 849(1):012048

Zanette I (2011) Interférométrie X à réseaux pour l'imagerie et l'analyse de front d'ondes au synchrotron. Ph.D. thesis. Université de Grenoble, Grenoble, France

Chapter 4
X-ray Single-Grating Interferometry

4.1 Introductory Remarks

X-ray grating interferometry (XGI) and speckle-based imaging (SBI) rely on similar principles: The modulations of a reference pattern induced by X-ray refraction, absorption and small-angle scattering in the sample are analysed computationally. The key differences between the methods are the data analysis method and the type of reference pattern, as discussed in more detail in Appendix A. However, it will be shown in Chap. 6 that the two approaches can be unified in a single analysis method that can be applied to both periodic and random reference patterns.

Another difference between the two methods is that XGI is typically performed with two (or three for the case of low-brilliance sources) gratings to make it compatible with large-pixel detectors, while SBI makes use of only a single phase modulator. For small effective pixel sizes, on the other hand, XGI measurements can be performed even with a single grating, making it conceptually similar to SBI. The single-grating implementation has only gained popularity in the recent years. It has been explored within the scope of this Ph.D. project and is presented in this chapter.

The chapter starts with a summary of the underlying principles of XGI and the concepts of XGI data acquisition and reconstruction. This is followed by the demonstration of single-grating interferometry for high-contrast, quantitative 3D biomedical imaging. In the last section of this chapter, more recent results of the Ph.D. project for single-grating interferometry at Diamond I13 beamline are presented, making use of a multilayer monochromator for faster image acquisition.

4.2 Principles of X-ray Grating Interferometry

X-ray grating interferometry relies on the use of a periodic interference pattern as a wavefront marker to measure the phase shift, absorption and small-angle scattering of the X-rays induced by a sample in the beam. As shown in the schematic setup in

M.-C. Zdora, *X-ray Phase-Contrast Imaging Using Near-Field Speckles*,
Springer Theses, https://doi.org/10.1007/978-3-030-66329-2_4

Fig. 2.4 in Chap. 2, Sect. 2.4, a phase grating G1, placed upstream or downstream of the sample, acts as a beam splitter and separates the incoming X-rays into several diffracted beams. Depending on the type of grating, the main portion of the diffracted energy is contained in either the two first diffraction orders (for a π-shifting grating) or the two first orders plus the zeroth order (for a $\pi/2$-shifting grating). The first-order diffracted beams diverge from the transmitted beam under an angle $\theta = \lambda/p_1$, where λ is the wavelength of the X-rays and p_1 the period of the phase grating (Paganin 2006). As the diffraction angles θ are on the order of microradians, the separation between the first-order diffracted beams is very small and they interfere in the detector plane producing a periodic interference pattern of typically a few micrometres period[1] if a sufficient level of transverse coherence of the beam is given. The position, period and visibility of the interference pattern can be predicted by the Talbot effect.

4.2.1 Talbot Effect

The Talbot effect is a result of Fresnel diffraction when a plane wave illuminates a periodic object, leading to self images of the object at certain distances from the object plane. These are known as Talbot distances and for an absorbing object with period p, they are given by:

$$z_T = m\frac{2p^2}{\lambda} \tag{4.1}$$

with $m = 1, 2, 3, \ldots$. Furthermore, at odd multiples of $z_T/2$, self images of the object can also be observed, but they are shifted by $p/2$.

The Talbot self-imaging effect was first reported by Henry Fox Talbot (Talbot 1836) and later linked to Fresnel diffraction by Lord Rayleigh (Rayleigh 1881). While these discoveries were reported for the visible light regime, it was demonstrated two decades ago that the Talbot effect also applies to hard X-rays (Cloetens et al. 1997), which is the basic underlying principle for XGI.

Furthermore, the Talbot effect translates from absorbing to phase-shifting objects (Guigay 1971). For the latter, the periodic phase shift in the object is converted into periodic intensity variations along the propagation direction of the photons, which repeat themselves at fractions of the Talbot distance defined in Eq. (4.1) known as fractional Talbot distances. While for the absorbing case, the Talbot images are direct self images of the object, this is not necessarily the case for phase objects and the interference patterns at the fractional Talbot distances are generally not exact images of the phase or intensity profile in the object plane (Guigay 1971). They do, however, have a direct relationship with the phase profile of the object. The intensity distribution at a certain distance from the phase object can be determined mathematically. Guigay (1971) showed that for an object with transmission function

[1]For a π-shifting grating, interference between the -1, $+1$ and 0 order beams occurs.

that is periodic with period p, the intensity at a quarter of the first Talbot distance, $z_1/4$, is given by:

$$I(x; z = z_1/4) = 1 + \sin[\phi(x) - \phi(x + p/2)]. \tag{4.2}$$

For XGI, the Talbot effect for phase objects is exploited to produce a self image of a beam-splitter phase grating in the detector plane, which is subsequently used as a reference pattern to track the X-ray wavefront. Phase gratings for XGI are usually designed to be binary and to have a duty cycle $\gamma = 0.5$. The latter is defined as the ratio between the width of the grating lines and the period of the grating and $\gamma = 0.5$ has been shown to lead to the best sensitivity of the setup (Modregger et al. 2011). For this case, the transmission function $\mathcal{T}$ of a purely phase-shifting grating with period p_1 is $\mathcal{T} = 1$ between the grating structures $(0 \leq (x \bmod p_1) < p_1/2)$ and $\mathcal{T} = \exp(i\phi)$ in the area of the grating lines $(p_1/2 \leq (x \bmod p_1) < p_1)$, i.e. the phase shift of the X-rays after passing through the grating lines is ϕ. Through Eq. (4.2), the intensity pattern generated at the fractional Talbot distance $z_{1/4} = \frac{1}{4}\frac{2p_1^2}{\lambda}$ is then given by (Guigay 1971):

$$\begin{aligned} I(x; z = z_1/4) &= 1 - \sin(\phi) \quad \text{for} \quad 0 \leq (x \bmod p_1) < p_1/2 \\ I(x; z = z_1/4) &= 1 + \sin(\phi) \quad \text{for} \quad p_1/2 \leq (x \bmod p_1) < p_1. \end{aligned} \tag{4.3}$$

The contrast of the interference pattern can be quantified in terms of the visibility v defined as:

$$v = \frac{I_{\max} - I_{\min}}{I_{\max} + I_{\min}}, \tag{4.4}$$

where $I_{\max}$ and $I_{\min}$ are the maximum and minimum intensities of the pattern.

From Eqs. (4.3) and (4.4) it can directly be understood that at the fractional Talbot distance $z_{1/4} \equiv z_1/4$ the visibility is at its maximum ($v = 1$) for a grating introducing a phase shift of $\phi = \pi/2$ and at its minimum ($v = 0$—equivalent to no contrast) for a phase shift of $\phi = \pi$. Analogous calculations can be performed for other fractional Talbot distances. It can be derived that for a $\pi/2$-shifting grating the interference pattern has maximum contrast $v = 1$ and the same period p_1 as the grating itself at odd multiples of the fractional Talbot distance $z_{1/4}$, i.e. $z_{3/4}, z_{5/4}, \ldots$. However, the patterns are laterally shifted by $p_1/2$. A π-shifting grating, on the other hand, produces an image of maximum visibility $v = 1$ and with a period $p_1/2$ at odd multiples of $z_{1/16}$. Therefore, it is convenient to define the nth fractional Talbot distance f_n as follows (Weitkamp et al. 2006):

$$f_n = n\frac{p_1^2}{2\eta^2\lambda}, \tag{4.5}$$

where $n = 0, 1, 2, \ldots$, $\eta = 1$ for a $\pi/2$-shifting grating and $\eta = 2$ for a π-shifting grating. Maximum visibility is reached at odd fractional Talbot orders $(n = 1, 3, 5, \ldots)$.

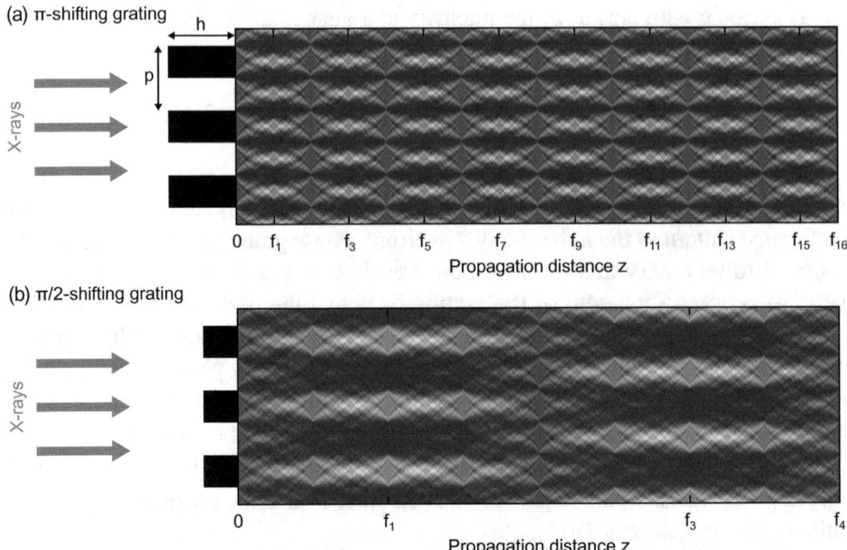

Fig. 4.1 Visualisation of the Talbot effect. Simulated Talbot carpets produced by **a** a π-shifting and **b** a $\pi/2$-shifting phase grating with period p from just behind the grating up until the first Talbot distance z_T. Maximum intensity of the interference pattern is observed at odd fractional Talbot distances. For the π-shifting case, the period of the pattern is $p/2$. For the $\pi/2$-shifting case, the period of the interference pattern is p and a lateral shift occurs at $0.5z_T$

The evolution of the interference pattern with increasing propagation distance from the phase grating to the detector plane can be visualised as a Talbot carpet.

Figure 4.1a shows the Talbot carpet of a π-shifting phase grating (duty cycle: 0.5) with period p for distances from 0 to the first Talbot distance. It can be seen that maximum intensity of the pattern is reached for odd fractional Talbot distances and the period of the interference pattern is $p/2$. For the case of a $\pi/2$-shifting phase grating with the same period, see Fig. 4.1b, the period of the pattern is p and there is a lateral shift of the interference pattern occurring at half the Talbot distance. The odd fractional Talbot distances of maximum intensity are found at much larger distances than for the π-shifting grating. Hence, π-shifting gratings are often preferred as they provide a higher flexibility in the setup dimensions. They are, however, more challenging to fabricate as grating lines of larger height are required to achieve a phase shift of π while the period remains on the order of micrometres. Gratings with a high aspect ratio (ratio of grating line height and width) are significantly more difficult to fabricate.

In the explanations above, a monochromatic plane wave with perfect spatial coherence was assumed to be incident on the grating. In case of a spherical wavefront, the magnification of the interference pattern should be taken into account. It leads to a magnified period of the pattern downstream of the grating and longer fractional Talbot distances compared to the case of a plane-wave illumination (Zanette 2011).

Furthermore, a reduced spatial coherence of the X-rays due to an extended source size leads to a significant decrease in visibility and blurring of the interference pattern (Weitkamp et al. 2006). The intensity observed in the detection plane can then be expressed as a convolution of the intensity obtained with a point source and the projected intensity distribution of the source in the observation plane. The latter can be described as a Gaussian distribution with a width that increases with the distance from the phase grating. As a consequence, the interference pattern in the detection plane resembles a sinusoidal function.

A decreased temporal coherence, i.e. a polychromatic X-ray beam, results in a loss in visibility of the interference pattern, which is, however, usually not the limiting factor for the quality of the measurements, even at laboratory tube sources with a broad spectrum (Engelhardt et al. 2008), as discussed in Sect. 2.4.5 of Chap. 2.

4.2.2 Phase-Stepping Scan and Signal Extraction

In the previous section, the creation of a line interference pattern at the fractional Talbot distances as a result of X-ray refraction by the beam-splitter phase grating G1 was discussed. In this section, it will be described how this interference pattern is used to perform X-ray phase-contrast imaging.

When a sample is inserted into the beam path, the incident X-rays undergo absorption as well as refraction in the specimen as described in Sect. 2.2. As a result of refraction, the wavefront is laterally shifted leading to a displacement of the interference pattern in the detector plane, which is directly related to the refraction angle of the X-rays.

As a small effective detector pixel size in the range of a micrometre or less would be required to directly resolve the fine interference pattern of only a few micrometres period, an absorption or analyser grating G2 is usually introduced downstream and placed just in front of the detector, as shown in Fig. 2.4. The absorption grating G2 has the same period p_2 as the interference pattern created by G1 in the detector plane and contains absorbing rather than phase-shifting lines. This translates the fringe position into an intensity variation in the detector pixels, which allows for using detector pixel sizes that are significantly larger than the grating period while still achieving a high phase sensitivity and good spatial resolution. Although nowadays detector systems with small effective pixel sizes are readily available, it can be beneficial to use a larger pixel size to increase the field of view for imaging of larger samples. Furthermore, applications away from the synchrotron are often conducted with medical detectors that have pixel sizes in the range of tens to hundreds of micrometres.

The absorption grating G2 is typically made from highly absorbing gold lines on a silicon wafer (David et al. 2007). The line pattern is defined with a photo lithography process, followed by gold plating. The thickness of the lines must be sufficient to lead to minimal transmission of X-rays for the design energy of the interferometer, in order to ensure a high visibility. The production of absorption gratings is challenging as typically lines with high aspect ratios are required to achieve a high X-ray absorption as well as a small period. The phase grating G1, on the other hand, should absorb as little of the X-ray beam as possible, but lead to a phase shift of π or $\pi/2$ at the design energy. G1 is typically produced by etc.hing the structures into a silicon substrate (David et al. 2007; Baborowski et al. 2014) or through X-ray lithography and electroplating (LIGA) (Noda et al. 2007; Koch 2017).

Data acquisition is usually conducted by performing a phase-stepping scan, for which one of the gratings is translated perpendicularly to the grating lines in several evenly-spaced small steps over at least one period of the interference pattern (Weitkamp et al. 2005), as illustrated in Fig. 4.2.[2] The same scan is performed with and without the sample in the beam and a sample and a reference phase-stepping curve are obtained for each pixel, showing the intensity variations during the scan. By comparing reference and sample curves pixel-wise, three complementary image signals can be extracted: An attenuation image, similar to the image that would have been obtained without the grating, which visualises the absorption of X-rays in the sample,[3] a differential phase signal quantifying the refraction, which is directly related to the first derivative of the phase shift in the specimen, and a dark-field signal giving information about small-angle scattering from unresolved features.

Under conditions of full transverse coherence, the phase-stepping curve can be seen as a convolution of the box function describing the interference pattern in the detector plane on the one hand and the box function representing the absorbing analyser grating on the other hand, resulting in a triangular curve (Bech 2009). However, the partial coherence of the source leads in practice to blurring that can be described mathematically by the convolution of the triangular function with a Gaussian that accounts for the demagnified source, as mentioned above. Therefore, the phase-stepping curves can usually be approximated by sinusoidal functions. This allows for an efficient reconstruction of the image signals via Fourier analysis, which has the advantage of being significantly faster than other methods operating in real space. However, reconstruction in real space, e.g. by using curve fitting procedures, is necessary in cases where the phase-stepping curve is well approximated by a sinusoidal or the steps are not evenly spaced.

In the sinusoidal approximation, the reference and sample phase-stepping curves in a pixel (m, n) of the image and at the grating position x_g can be expressed as:

$$I^{\text{ref,sam}}(m, n; x_g) = a_0^{\text{ref,sam}}(m, n) + a_1^{\text{ref,sam}}(m, n) \cos\left(\frac{2\pi}{p_2}x_g + \phi_1^{\text{ref,sam}}(m, n)\right), \tag{4.6}$$

[2]The phase-stepping scan is here shown for scanning of G2, but can in principle also be performed with G1. It should, however, be considered that this is not necessarily completely equivalent due to grating imperfections, the alignment of stepping stages etc.

[3]As the detector is placed some distance away from the sample, the attenuation image typically also contains contributions from Fresnel diffraction showing as edge-enhancement fringes.

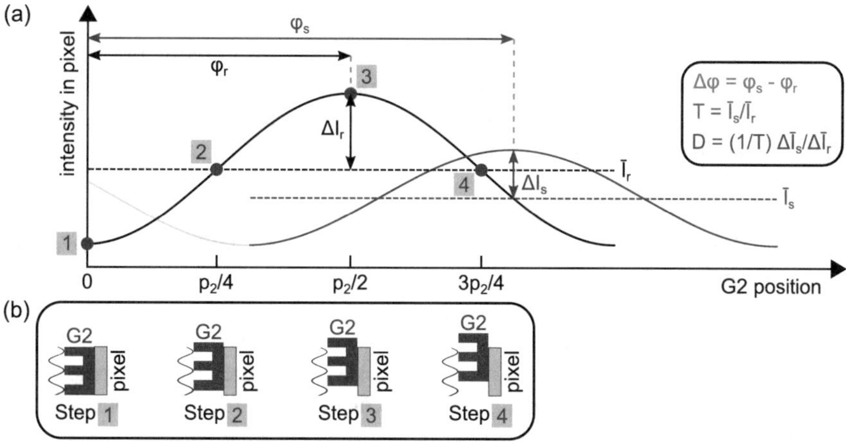

Fig. 4.2 Principle of the phase-stepping implementation of X-ray grating interferometry. Phase stepping is performed here in four steps over one period of the interference pattern. **a** The intensity recorded in each pixel oscillates with the lateral position of the stepped grating, depending on the relative position of the interference fringe pattern and the absorption grating G2, see panel (**b**). When a sample is placed in the beam (magenta curve), the phase-stepping curve is displaced by $\Delta\phi$ due to refraction, reduced in mean intensity by T due to absorption and reduced in visibility by D due to small-angle scattering in the specimen

with the zeroth and first order amplitude coefficients $a_0^{\text{ref,sam}}$ and $a_1^{\text{ref,sam}}$, the phase coefficients $\phi_1^{\text{ref,sam}}$ and the period p_2 of the analyser grating G2. This is equivalent to a Fourier series with only the zeroth and first order components (Pfeiffer et al. 2008).

The transmission signal T and attenuation A in a pixel[4] are given by the change in mean intensity of the phase-stepping curve and can be obtained with the model in Eq. (4.6) as the ratio of zeroth-order Fourier coefficients a_0^{sam} and a_0^{ref} of the sample and reference scan:

$$A = 1 - T = 1 - \frac{a_0^{\text{sam}}}{a_0^{\text{ref}}}. \tag{4.7}$$

The differential phase signal $\partial\Phi/\partial x$ in the horizontal direction x, i.e. perpendicular to the grating lines, is directly related to the refraction angle α_x in the horizontal via $\partial\Phi/\partial x = (2\pi/\lambda)\,\alpha_x$. It can be determined in small-angle approximation from the phases ϕ_{sam} and ϕ_{ref} of the sample and reference phase-stepping curves. The differential phase shift of the X-rays in the specimen is determined by the phase difference of the two phase-stepping curves:

$$\frac{\partial\Phi}{\partial x} = \frac{2\pi}{\lambda}\alpha_x = \frac{p_2}{\lambda z}\left(\phi_{\text{sam}} - \phi_{\text{ref}}\right), \tag{4.8}$$

[4]The pixel denotation (m, n) is dropped from here on for simplicity.

where p_2 is the period of G2 and z the inter-grating distance between G1 and G2. Due to the 1D character of the line interference pattern, a typical grating interferometry setup is only sensitive to the phase shift and small-angle scattering in one direction (here x). However, to achieve sensitivity in both directions, two line gratings can be mounted in crossed configuration or a 2D chequerboard grating can be used (Zanette et al. 2010; Morimoto et al. 2015a).

The differential phase signal is very sensitive to small density differences in the sample and can hence often reveal features that are not visible in a conventional absorption image. However, in some cases particular sample properties can lead to artefacts in the differential phase images. These include high attenuation, strong scattering and strong refraction or sharp edges in the sample. For the first, the low photon count after the sample leads to a high noise level in the phase stepping curve, making retrieval of the shift of the phase-stepping curves difficult or impossible. For the second, the visibility drops to close to zero and hence the phase of the stepping curve cannot be determined. For the last, the large displacement of the sample phase-stepping curve can exceed one period of the interference pattern while only shifts in the interval $[-\pi, \pi)$ can be measured due to the periodicity of the pattern. As a result, sharp phase jumps can be observed in these cases, which is known as phase-wrapping artefact. To avoid phase wrapping at the air-sample interface, the specimen is often placed in a bath with a refractive-index-matching liquid, such as water. Furthermore, various approaches have been developed to correct for this effect, see e.g. Ghiglia and Pritt (1998), Jerjen et al. (2011), Haas et al. (2011), Epple et al. (2013), Hahn et al. (2015), Rodgers et al. (2018).

The visibility v, as defined in Sect. 4.2.1, is equivalent to the amplitude of the phase-stepping curve normalised by the mean value. The visibilities in the reference and sample scans can hence be expressed as the ratio of the first- and zeroth-order Fourier coefficients: $v_{\text{sam}} = a_1^{\text{sam}}/a_0^{\text{sam}}$ and $v_{\text{ref}} = a_1^{\text{ref}}/a_0^{\text{ref}}$, respectively. The dark-field signal D describes the loss in visibility of the intensity pattern, which is mainly due to small-angle scattering in the sample, and is consequently given by:

$$D = \frac{v_{\text{sam}}}{v_{\text{ref}}} = \frac{a_1^{\text{sam}}}{a_0^{\text{sam}}}\frac{a_0^{\text{ref}}}{a_1^{\text{ref}}} = \frac{1}{T}\frac{a_1^{\text{sam}}}{a_1^{\text{ref}}}. \tag{4.9}$$

4.2.3 Angular Sensitivity and Spatial Resolution

Two important measures of the quality of the differential phase images are the angular sensitivity and the spatial resolution. The angular sensitivity describes the sensitivity of the differential phase signal measurement and is commonly defined as the standard deviation of the refraction angle signal in a region of interest (ROI)—typically 150 × 150 pixels—in a homogeneous background region of the image outside sample.

It is inversely proportional to the period of the grating and scales with the X-ray propagation distance (Donath et al. 2009). The angular sensitivity is, furthermore, dependent on the positioning of the gratings and the sample, as discussed in detail by Donath et al. (2009). Moreover, the visibility of the interference pattern and the detector noise and dynamic range play an important role (Revol et al. 2010; Modregger et al. 2011). The best achievable angular sensitivities that have been reported for XGI are in the range of tens of nanoradians (Pfeiffer et al. 2007; Schulz et al. 2010; Modregger et al. 2011). For phase tomography scans (see Chap. 2, Sect. 2.5.4), the angular sensitivity of the projections has an influence on the density resolution in the 3D tomogram, i.e. the smallest resolvable density difference. Density resolutions of better than 1 mg/cm^3 havthe smallest resolvable e been reported in the literature for XGI (Pfeiffer et al. 2007, 2009; Schulz et al. 2010; Zanette et al. 2013b).

The spatial resolution depends on a number of factors, including the detector resolution, the period of the analyser grating G2 and the separation of the two first order diffracted beams created by the phase grating (Weitkamp et al. 2005; Zanette 2011). This separation is:

$$s_{+1-1} = 2\frac{\lambda}{p_1}z, \tag{4.10}$$

where λ is the X-ray wavelength, z the inter-grating distance and p_1 the period of the phase grating G1. If the resolution of the detector system is better than this separation, this effect leads to blurring in the reconstructed images. It should be noted that this consideration is valid for a π-shifting phase grating that leads to separation of the beam in predominantly the two first diffraction orders, as mentioned above. For a $\pi/2$- or $3\pi/2$-shifting grating, apart from the two first orders, also the zeroth order contains a significant portion of the diffracted energy, which reduces the effective lateral beam shear (Bérujon et al. 2012; Bérujon 2013).

Typically, the spatial resolution of XGI data is on the order of tens of micrometres. The highest spatial resolution values reported in the literature are at around 10 µm or slightly less (Schulz et al. 2010; Herzen et al. 2011; Zanette et al. 2013a; Hipp et al. 2017) for conventional implementations, but can reach a few microns for single-grating setups (Hipp et al. 2017).

Generally, there is a trade-off between a high spatial resolution and a high angular sensitivity. In practical implementations, the propagation distance z and grating period p_1 should be chosen carefully, in order to avoid blurring due to the separation of the diffracted beams while also maintaining a high sensitivity.

4.2.4 *Grating Interferometry at Laboratory Sources*

As X-ray grating-based phase-contrast imaging relies on interference effects, a degree of spatial coherence of the X-ray beam is required for the classic implementation called Talbot interferometer. This previously restricted the method mainly to synchrotron sources, impeding its widespread use. However, a decade ago it was shown that the implementation at conventional laboratory X-ray sources with large spot sizes is possible by introducing a third grating known as source grating G0 (Pfeiffer et al. 2006). The grating G0 is placed directly behind the X-ray source and splits the beam into several coherent line sourcelets, which are mutually incoherent, but interfere constructively in the observation plane. For this, the period p_0 of the source grating needs to be:

$$p_0 = p_1 \frac{l}{z}, \tag{4.11}$$

where p_1 is the period of G1, l is the G0-G1 distance and z the inter-grating distance between G1-G2. The setup using the three gratings G0, G1, and G2 is called Talbot-Lau interferometer (Clauser and Li 1994; Pfeiffer et al. 2006). The data acquisition and analysis processes are the same as for the two-grating interferometer.

4.2.5 Single-Shot Grating Interferometry

Apart from the most common data acquisition scheme based on phase stepping, which requires recording multiple images at several equidistant grating steps, XGI can also be performed in single-shot mode. This is particularly suitable for cases, where imaging speed and/or dose-efficiency are the main concern. Several approaches have been proposed for single-shot grating-based imaging, such as moiré imaging (David et al. 2002; Momose et al. 2003), spatial harmonic analysis (Wen et al. 2010) and the use of a direct-conversion detector without G2 (Kagias et al. 2016).

Amongst these, moiré-fringe analysis is the most widespread method. A periodic moiré-fringe pattern can be generated by misaligning the two gratings G1 and G2 slightly against each other by rotating one of them around the optical axis. The modulation of the pattern when introducing a sample into the beam path can then be analysed from a single reference and sample image set. As the period of the moiré carrier fringes is much larger than the period of the Talbot line interference pattern, this method is limited by a reduced spatial resolution compared to the phase-stepping approach, but benefits from much shorter scan times and increased robustness of the setup.

4.2.6 Grating Interferometry without Absorption Grating G2

Although grating interferometry is commonly implemented using a phase grating G1 and an absorption grating G2, similar measurements and data analysis can be performed with only a single grating, G1, if the period of the interference pattern can be directly resolved with the detector system. Such a single-grating setup has the advantage of providing a better signal-to-noise ratio and higher dose efficiency as the high X-ray absorption by G2, typically around 50%, is avoided. Moreover, the setup with a single grating is simpler and more robust than the conventional implementation, does not require alignment of the two gratings and is less susceptible to instabilities.

Since a requirement for removing G2 from the setup is that the interference pattern can be directly resolved, the period of G1 needs to be sufficiently large compared to the pixel size. Hence, typically phase gratings with larger periods are chosen for single-grating imaging than for a two-grating setup, for which the periods are not directly coupled to the pixel size. As the fractional Talbot distance, see Eq. (4.5), increases with the square of the period p_1 of the phase grating, smaller fractional Talbot distances will be used in practice for a single-grating setup. Both the propagation distance and the period of the phase grating should be chosen carefully considering the trade-off in sensitivity, spatial resolution and field of view, as mentioned in Sect. 4.2.3. In particular, the separation of the diffracted beams can become significant for small grating periods and high-resolution detectors.

From direct comparisons of single- and double-grating interferometry measurements it was concluded that the single-grating approach can give a higher spatial resolution, while the double-grating setup can provide a higher phase sensitivity resulting in a better density resolution when performed in tomography (Hipp et al. 2017; Schulz et al. 2017b; Thalmann et al. 2017; Balles et al. 2018). A first demonstration of single-grating phase tomography in phase-stepping mode yielded a density resolution of 9 mg/cm^3 at a spatial resolution of 8 μm (Takeda et al. 2007).

In the last years, the single-grating approach has seen increasing interest and a number of setups have been reported in the literature using either a single phase or absorption grating to create an interference pattern combined with a phase-stepping scan (Hipp et al. 2016; Balles et al. 2016; Hipp et al. 2017; Thalmann et al. 2017; Schulz et al. 2017b; Balles et al. 2018; Yang et al. 2019). Experimental implementations of single-shot single-grating measurements for high-speed imaging have been proposed using 1D (Wen et al. 2010; Wang et al. 2009) and 2D phase and absorption gratings (Itoh et al. 2011; Morgan et al. 2012; Rizzi et al. 2013).

Apart from imaging applications, it has been shown that a single-grating setup can, furthermore, be successfully used for near-error-free wavefront measurements and metrology (Grizolli et al. 2017; Inoue et al. 2018).

All of the above results were obtained with X-ray beams of sufficient spatial coherence to observe the self-image of the G1 phase grating without G0. This is the case for synchrotron sources, but also achievable at high-end laboratory sources such as the liquid-metal-jet source (see Chap. 2, Sect. 2.6.1), for which a single-grating setup has been demonstrated as well (Balles et al. 2016, 2018). Moreover, analogous to the classic Talbot-Lau interferometer with three gratings, a G2-less setup can also be realised at conventional tube source with low spatial coherence by adding a source grating G0. For the case of tube sources, the magnification by the cone beam geometry contributes to a small effective pixel size in the G1 plane, making it possible to directly resolve the interference pattern. For this purpose, the G0-G1 distance needs to be reduced sufficiently so that the pattern appears highly magnified in the detector plane. This has first been demonstrated by Momose et al. almost a decade ago with the idea of making grating interferometry more accessible by avoiding the difficulties in the production of G2 (Momose et al. 2011). However, due to limitations of the G0-G1 distance, this first proof-of-principle demonstration had a source-detector distance of almost 7 m, which is too long for practical use. Equivalent setups with smaller dimensions were proposed later by Morimoto et al. who replaced the source grating G0 by a source with multi-line metal targets embedded in a diamond substrate, which allowed for a much more compact arrangement (Morimoto et al. 2014). A similar setup was also realised with a 2D metal target and 2D phase grating for 2D differential phase sensitivity (Morimoto et al. 2015a, b). Other approaches for G2-less grating interferometry make use of novel detector technologies such as hybrid-pixel detectors that give sub-pixel resolution, e.g. the MÖNCH detector (Cartier et al. 2016) or the Medipix detector (Krejci et al. 2010, 2011).

A single-grating setup with phase-stepping acquisition was developed and implemented at Diamond I13 beamline in 2015 as part of this Ph.D. project (more details on the contributions in Sect. 1.3, Chap. 1). First experimental results were obtained for high-quality phase-contrast imaging of biomedical specimens. These were published in a peer-reviewed article (Zdora et al. 2017) and are shown in the following section of this chapter. Furthermore, during this Ph.D. project single-grating interferometry was implemented at Diamond I13 using a multilayer monochromator to achieve fast high-resolution grating-based imaging. First results are presented in Sect. 4.4. The data obtained with a phase-stepping implementation and analysis obtained at this setup was also compared with a single-shot analysis based on the spatial harmonic concept (Wen et al. 2010). The results are reported in a conference proceedings publication (Marathe et al. 2017), for which the author of this thesis provided the raw data and the results from the phase-stepping analysis.

4.3 X-ray Phase Microtomography with a Single Grating for High-Throughput Investigations of Biological Tissue

As part of this Ph.D. project, XGI was implemented for the first time at the beamline I13 at Diamond Light Source. A single-grating setup, as introduced in the previous section was chosen for this purpose, as it is more robust than the common implementation with two gratings and allows for shorter scan times and reduced dose to the sample. XGI tomographies of biological samples embedded in paraffin wax were acquired during this Ph.D. project, such as a human brain specimen (Khimchenko et al. 2016) and a mouse embryo (Zdora et al. 2017). The latter is presented in detail in this section, which is a modified excerpt of the original paper published as (Zdora et al. 2017): M.-C. Zdora, J. Vila-Comamala, G. Schulz, A. Khimchenko, A. Hipp, A. C. Cook, D. Dilg, C. David, C. Grünzweig, C. Rau, P. Thibault, and I. Zanette, "X-ray phase microtomography with a single grating for high-throughput investigations of biological tissue," Biomed. Opt. Express **8**(2), 1257–1270 (2017), licensed under CC BY 4.0.

4.3.1 Background and Motivation

The three-dimensional visualisation of biological tissues is the basis for a large number of biomedical studies, such as investigations of diseases and developmental defects. For this, it is required to obtain high-contrast, high-resolution images of the specimens in a reasonably short time and without the need for elaborate sample preparation.

Several techniques have been developed for the investigation of biological samples. Histology (Baldock et al. 2016; Graham et al. 2015)—for long the gold standard for tissue investigation—can achieve excellent 2D resolution and contrast, but is a destructive technique, needs extensive sample preparation including staining agents and suffers from tissue deformations making registration of slices and hence 3D reconstruction of the volume cumbersome. Drawbacks of other available methods include the need for staining procedures (absorption X-ray micro computed tomography in combination with contrast agents (Metscher 2009; Wong et al. 2012)), the limitation in spatial resolution (micro magnetic resonance imaging (Schneider and Bhattacharya 2004; Zamyadi et al. 2010)), the requirement of transparency and small size of the sample (optical projection tomography (Sharpe 2003; Anderson et al. 2013)) and a relatively low contrast and limited penetration depth (optical coherence tomography (Tschernig et al. 2013)). High-resolution episcopic microscopy (HREM) has recently seen increased interest, in particular for imaging of embryos in developmental studies (Weninger et al. 2006; Dunlevy et al. 2010; Mohun and Weninger 2011; Geyer et al. 2012). The technique is based on automated physical sectioning of the resin-embedded sample in layers of a few micrometres thickness (Weninger et al. 2006). After each slicing step, the surface of the sample block is imaged with a visible light microscope using fluorescent illumination, which leads to clear visualisation of the tissue against the resin that contains fluorescent dyes. A volume data set might then be obtained by combining together the slice series.

Most of the drawbacks of the commonly employed methods can be overcome by tomography based on X-ray phase-contrast imaging (XPCI), which delivers high-contrast, high-resolution three-dimensional information about the density composition of the sample in a non-destructive way and without the need for contrast agents (Fitzgerald 2000; Momose et al. 1996). As outlined in Chap. 2, Sect. 2.2, the interaction of X-rays with the sample can be described by the complex refractive index $n = 1 - \delta + i\beta$, in which δ and β account for the phase shift and absorption in the specimen, respectively (Attwood 1999). For biological soft tissues with small density differences, the refractive index decrement δ can be up to three orders of magnitude larger than the imaginary part β of the refractive index and XPCI can deliver images with much higher contrast compared to conventional absorption imaging in these cases.

Among the different XPCI methods (see Chap. 2, Sect. 2.4), X-ray grating interferometry (David et al. 2002; Momose et al. 2003; Weitkamp et al. 2005) achieves excellent contrast and quantitative information about the density distribution of biological soft-tissue samples when employed in tomographic mode (Momose et al.

2006; Pfeiffer et al. 2007; Zhu et al. 2010; Hoshino et al. 2012). The potential of XGI for histopathological, diagnostic, or structural and morphological studies of soft tissues with high contrast and high resolution has been evaluated against a range of other commonly used methods such as conventional absorption-based microtomography (Schulz et al. 2010) and mammography (Scherer et al. 2015), magnetic resonance imaging (Schulz et al. 2012), histology (Zanette et al. 2013b) and cryotome-based planar epi-illumination imaging (Tapfer et al. 2013), while a direct comparison with the state-of-the-art HREM technique had not been reported before.

As explained above in Sect. 4.2.6, the conventional XGI setup comprising a beam-splitter phase grating G1 and an absorption grating G2, can be simplified by omitting G2 in cases where the spatial resolution of the detector system is sufficient to directly resolve the interference pattern created by G1. This allows for more dose- and time-efficient imaging with a simple arrangement and avoids the costly and challenging fabrication of G2. Although there have been a number of previous implementations of this approach, the single-grating setup has received increased attention only very recently and systematic studies on the properties of single-grating interferometry have been performed in the last years (Hipp et al. 2017; Thalmann et al. 2017; Schulz et al. 2017b; Balles et al. 2018).

At the same time, a setup for XGI with a single grating was developed and implemented at the I13-2 Diamond-Manchester Branchline at Diamond Light Source, UK for the purpose of biomedical imaging applications. In the following, this setup is presented and the reconstructed phase volumes are compared with results obtained by the state-of-the art HREM technique, demonstrating the high potential of single-grating XGI for quantitative biological imaging.

4.3.2 *Materials and Methods*

4.3.2.1 Experimental Setup at Diamond I13

Measurements were conducted with an XGI setup with a single grating (G1) installed at the I13-2 Diamond-Manchester Imaging Beamline at Diamond Light Source (Rau et al. 2011). The X-ray radiation is produced by a 2 m-long undulator insertion device (23 mm period) located approximately 220 m upstream of the experimental hutch. A monochromatic X-ray beam of energy 19 keV was extracted with a silicon (111) double-crystal monochromator for the measurements.

The results presented in the following were obtained with the setup shown in Fig. 4.3. The general concept and properties of a single-grating setup were described previously in Sect. 4.2.6 and considerations on the spatial resolution and angular sensitivity of XGI can be found in Sect. 4.2.3.

The sample was mounted on a high-precision rotation stage and the phase grating was placed approximately 5 cm further downstream. The grating was fabricated at the Institute for Microstructure Technology (IMT) of the Karlsruhe Institute of Technology (KIT) using LIGA-processing (Reznikova et al. 2008). It was made of

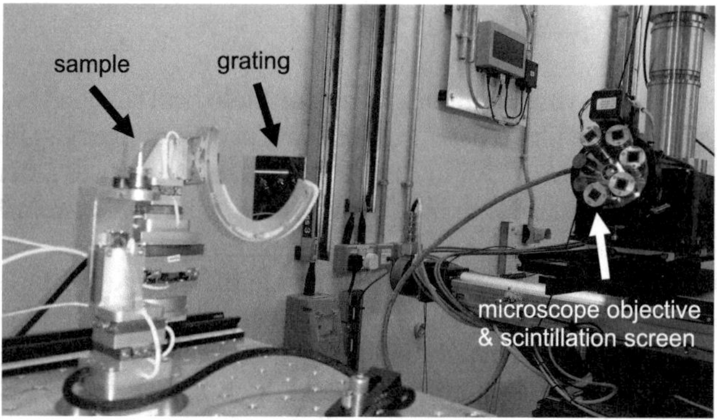

Fig. 4.3 Experimental setup for X-ray grating-based imaging using a single phase grating. The grating (period: 10 µm) was mounted at a distance of 5 cm downstream of the sample. A high-resolution X-ray detector (effective pixel size: 1.13 µm) is placed 72 cm further downstream to record the interference pattern created by the grating. The phase shift of the X-ray wavefront induced by the sample results in a distortion of the interference pattern. Figure reprinted with permission from Zdora et al. (2017), licensed under CC BY 4.0

nickel and had a nominal period of $p = 10\,\mu\text{m}$. The grating lines were 10.4 µm high, which yields a phase shift of $3\pi/2$ at an energy of approximately 20 keV.[5]

The detector system was positioned at a distance of $z = 72\,\text{cm}$ downstream of the grating, where the interference pattern produced by the grating showed maximum contrast and had the same period as the grating itself. The detector system was assembled in-house and consisted of a scintillation screen coupled to an infinity-corrected optical microscope and a pco.4000 CCD camera (chip size: 4008×2672 pixels, pixel size: 9 µm). The scintillation screen was a double-side polished $CdWO_4$ crystal of 150 µm thickness supplied by Hilger Crystals Ltd. A 4× objective (Olympus UPLSAPO 4X) with a numerical aperture of 0.16 was used in combination with a tube lens and a relay optics system to achieve optimal imaging performance over the large chip of the pco.4000 CCD camera. This led to an 8-fold total optical magnification and an effective pixel size of 1.13 µm. The resolution of the detector system at 10% MTF (modulation transfer function), determined from a star test pattern measurement, was approximately 433 cycles/mm, equivalent to a smallest resolvable feature size of about 2.3 µm (full period resolution).

[5]This differs slightly from the X-ray beam energy of the measurement, which was chosen to optimise preceding experiments. However, as an X-ray grating interferometer is largely achromatic (Engelhardt et al. 2008; Thüring and Stampanoni 2014; Wang et al. 2010), its performance was not impacted significantly by the slight difference between beam energy and grating design energy.

4.3.2.2 Sample Preparation and Choice

The potential of XGI with a single grating for biological imaging was evaluated on a wild-type mouse embryo (strain C57BL/6N) at estimated embryonic day (ED) 13.5 with the head removed.[6] Animal maintenance, husbandry and procedures were carried out in accordance with British Home Office regulations (Animals Scientific Procedures Act 1986).

The embryo was fixed overnight in 4% paraformaldehyde (PFA) in phosphate buffered saline (PBS) and then dehydrated by increasing concentration of ethanol (30–100%). It was subsequently cleared using histoclear (National Diagnostics) for one hour, and embedded in paraffin overnight after several wax washes at 60 °C. The paraffin was trimmed down to a small cylinder around the sample. For the measurements, the paraffin cylinder was mounted on a translation and tomographic rotation stage as shown in Fig. 4.3.

Mice are often used as models to study embryonic development and disorders because they are genetically and physiologically similar to humans, their genes can be easily manipulated and they reproduce fast (Merlo et al. 2001; Bier and Mcginnis 2016). Recently, mice have been used extensively as model animal systems for genetic knock-out studies, where a certain gene of interest is turned off, in order to learn more about its function by comparison with wild-type individuals. One aim of these studies is to identify lethal gene knock-outs to understand the significance and function of genes critical for embryonic survival and normal development (Mohun et al. 2013; Wilson et al. 2016; Deciphering the Mechanisms of Developmental Disorders 2016). The information from mouse model investigations can then help to give insight into the causes of human congenital abnormalities to develop methods for early diagnosis and therapy (Nguyen and Xu 2008; Deciphering the Mechanisms of Developmental Disorders 2016).

For these types of studies, high-throughput imaging of a large number of specimens is essential and it is important to visualise the inner structure of the embryos with high contrast and high resolution to detect abnormalities. In the following, it is demonstrated that XGI with a single grating is a suitable candidate for this purpose and shows advantages over existing methods.

4.3.2.3 Data Acquisition and Analysis

Data acquisition and analysis for the single-grating setup can be carried out following the same phase-stepping and Fourier analysis approach as in conventional two-grating interferometry (Weitkamp et al. 2005), which is described in Sect. 4.2.2.

Equation (4.8) describes the relationship between the differential phase shift $\partial\Phi/\partial x$ perpendicular to the grating lines, the refraction angle α in the horizontal direction and the phases ϕ_s and ϕ_r of the sample and reference intensity curves of

[6] The head was removed as part of another study on specimens from the same batch.

the phase-stepping scan. For the case of a single grating, the G2 period p_2 is simply replaced by the period p of the interference pattern in the image plane and we get:

$$\frac{\partial \Phi}{\partial x} = \frac{2\pi}{\lambda}\alpha = \frac{p}{\lambda z}(\phi_s - \phi_r), \tag{4.12}$$

with the grating-detector distance z and the X-ray wavelength λ.

Phase stepping of the grating was performed in 5 steps over one grating period, and the acquisition time was 2 s per frame. A total of 1201 sample projections were acquired for evenly spaced viewing angles of the sample over 180 degrees rotation and reference scans were taken every 200 angles. The total exposure time of the tomography scan was approximately 3.5 h including overhead time due to motor movement and data transfer.

From each phase-stepping scan, a differential phase and a transmission signal were retrieved by Fourier analysis. The phase volume was reconstructed from the differential phase projections using the FBP algorithm with a Hilbert filter, which incorporates the integration step (Pfeiffer et al. 2007), see Chap. 2, Sect. 2.5.4. The phase shift Φ is related to the refractive index decrement δ via Eq. (2.28) and hence the phase volume delivers the 3D distribution of δ in the sample. For the investigation of biomedical specimens, the electron density ρ_{el} is the typical quantity of interest. It is in good approximation directly proportional to δ for materials containing low-Z elements at the given photon energy (Guinier 1994; Als-Nielsen and McMorrow 2011), see Eq. (2.31). Moreover, ρ_{el} is approximately proportional to the mass density of the object, so that the phase volume gives information about the 3D density distribution in the specimen, see Eq. (2.32).

From the same data set, also an absorption tomogram, similar to the one that would have been obtained without grating in the beam, can be retrieved from the reconstructed transmission signal. The transmission T can be expressed in terms of the linear attenuation coefficient μ, which is directly related to the imaginary part β of the refractive index via $\mu = 4\pi\beta/\lambda$, see Eq. (2.27). The tomographic reconstruction of the absorption volume showing the 3D distribution of μ was performed with the standard FBP algorithm including logarithmisation, see Sect. 2.5.3 in Chap. 2. The transmission/absorption projections and volume show a signal similar to the one that would have been measured inline without the grating interferometer. It should be noted that under conditions not optimised for absorption imaging, this does not correspond to the pure absorption signal, but is also influenced by propagation effects. In the following, this signal is therefore referred to as propagation signal.

Three-dimensional volume rendering of the mouse embryo was performed from the phase tomogram using Avizo 9.1.1 software.

4.3.3 *Results*

4.3.3.1 Interference Pattern

The interference pattern produced by the phase grating at the first fractional Talbot distance (Cloetens et al. 1997) is shown in Fig. 4.4a.

The visibility—determined via Fourier analysis as the ratio of the first and zeroth order Fourier coefficients of the phase stepping curve, see Eq. (4.4)—is approximately 29% in the central part of the image. It is limited here by the lateral coherence of the X-ray beam reduced by the finite source size, roughness and instabilities of the optical components in the beam and defects of the scintillation screen.

A ROI of the line pattern taken from the centre of the field of view can be seen in Fig. 4.4b and a horizontal profile averaged along the vertical direction is plotted across this area in Fig. 4.4c.

4.3.3.2 Differential Phase and Propagation Projections

From the phase-stepping scans with and without sample in the beam, the refraction angle of the X-rays in the specimen and the propagation signal were determined as described in the previous section for each of the 1201 viewing angles of the tomography scan. Examples of the reconstructed projections are shown in Fig. 4.5.

In the differential phase signal in Fig. 4.5a, not only the outlines of the embryo and its internal structures such as the peritoneal cavity (indicated by white arrows) are clearly visible, but also structures within the organs can be observed. Two air bubbles in the paraffin wax lead to strong refraction. On the left side of the field of

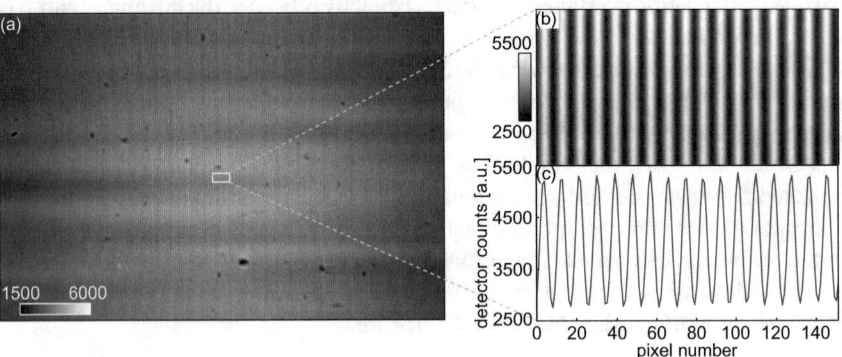

Fig. 4.4 Reference line interference pattern. **a** Interference pattern observed at the first fractional Talbot distance. The visibility in the centre of the field of view is approximately 29%. **b** ROI (152×76 pixels) in the centre of the image. **c** Horizontal profile through (**b**) averaged along the vertical direction. The grey values represent arbitrary intensity units. Figure reprinted with permission from Zdora et al. (2017), licensed under CC BY 4.0

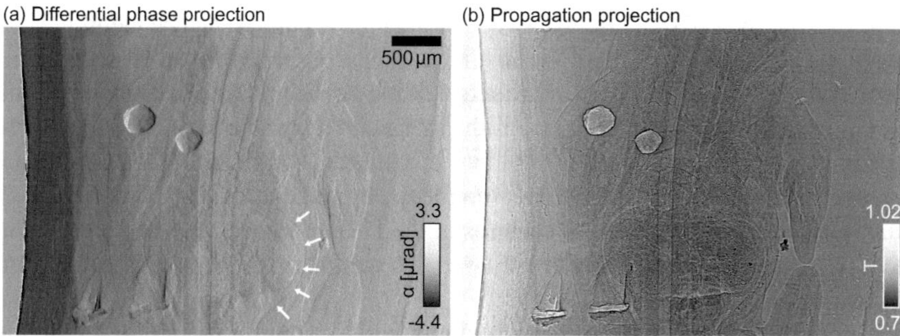

Fig. 4.5 Multimodal 2D projections. One of the 1201 projections showing **a** the differential phase signal measured as the refraction angle of the incoming X-rays induced by the sample and **b** the propagation signal, consisting of absorption and edge-enhancement effects. Figure reprinted with permission from Zdora et al. (2017), licensed under CC BY 4.0

view, the edge of the paraffin embedding can be seen causing some phase wrapping (see Sect. 4.2.2). The projection in Fig. 4.5b is similar to the one that would have been obtained without the grating in the beam and contains a combination of absorption and edge-enhancement effects arising upon propagation.

As mentioned previously, the angular sensitivity is a typical quantity of interest to asses the quality of the phase-contrast images. It is a measure of the resolution of the angular deviation measurement and is commonly estimated from the differential phase projections by taking the standard deviation of a region in the background area without sample, see Sect. 4.2.3. The angular sensitivity of the measurements presented here was calculated from a 150×150 pixels area in the surrounding paraffin and is approximately 110 nrad.

4.3.3.3 Phase Volume

Slices through the phase tomogram are presented in Fig. 4.6b, d and f. They are compared to the equivalent images of a second mouse embryo sample (with head) at the same gestational age (ED13.5) obtained with the state-of-the-art HREM method, see Fig. 4.6c, e and g (*HREM data provided by Deciphering the Mechanisms of Developmental Disorders (*http://dmdd.org.uk/*), a programme funded by the Wellcome Trust with support from the Francis Crick Institute, is licensed under a Creative Commons Attribution Non-Commercial Share Alike licence*).

The orientations of the slicing planes—sagittal, transverse and frontal—are illustrated on a photograph of a mouse embryo sample in Fig. 4.6a. The phase tomogram slices shown in Fig. 4.6b, d, f provide detailed insight into the inner features of the sample: not only the central part of the embryo's body can be clearly distinguished from the surrounding paraffin using the phase information, but also the different organs are visualised with high contrast as highlighted by the labelled arrows in Fig. 4.6b, f. In particular, the spine (A), the heart ventricles (B) and atria (C), the lungs (E), the liver (F), and intestines (G) of the mouse embryo can be identified and their inner structure is unveiled. Within the heart, it is possible to even distinguish the atrioventricular cushions (D) from the surrounding atrial and ventricular myocardium (B, C). In the frontal slice in Fig. 4.6f, the stomach (H) can be seen. The transverse slice in Fig. 4.6d reveals the fine structure of the heart ventricles. When comparing the results from XGI and HREM measurements, it can be noted that for the sagittal, transverse and frontal planes all features in the HREM images can be identified in the XGI slices. Moreover, they are shown with comparable image quality in terms of contrast and resolution.

A non-uniformity of the paraffin background is visible in the phase tomogram slices in Fig. 4.6. This can be attributed to the effects of phase wrapping and region-of-interest tomography. Strong refraction at the air-paraffin interface causes phase wrapping at the edge of the wax block. This could be avoided, for example, by placing the sample in a water bath during the measurement. Furthermore, parts of the paraffin block moved out of the field of view for some projections of the tomography scan. This leads to a cupping artefact visible in the transverse slices of the reconstructed phase volume. In future studies, care will be taken to avoid region-of-interest tomography by ensuring the whole paraffin block fits into the field of view. This could be done by reducing the size of the paraffin block or choosing a lower magnification. However, while the first was difficult because of the large size of the specimen, the latter was not possible for the presented setup as a detector system with an objective of lower magnification would have been unable to resolve the grating interference pattern. The artefact observed in the upper left part of Fig. 4.6b and upper right part of Fig. 4.6f is caused by phase wrapping at the edges of two air bubbles in the paraffin wax (clearly visible in the projections in Fig. 4.5).

From the phase volume, one can extract the 3D shape of the mouse embryo from the surrounding paraffin by volume rendering. Figure 4.7a shows the rendered embryo and the frontal, sagittal and transverse slicing planes through the centre of the sample (grey areas show paraffin).

The limbs and outer shape of the specimen are nicely visualised. Furthermore, it can be seen that the removal of the head and embedding in paraffin led to slight damage of the embryo skin and cracking of outer areas, which is also visible in Fig. 4.6d, f. In Fig. 4.7b–d, cuts of the embryo and the corresponding central slices of the phase volume are shown. Animations sliding through all frontal, sagittal and transverse slices of the embryo volume can be found in the Supplementary Material of the article published in Biomedical Optics Express (Zdora et al. 2017).

The volume of the propagation-based signal, reconstructed from the same data set as the phase volume, is included in Fig. 4.8 for completeness. Note that it is not an absorption volume as it would be obtained from a scan optimised for the measurement of the imaginary part β of the refractive index, see Sect. 2.5.3 in Chap. 2. The volume in Fig. 4.8 is mainly dominated by edge-enhancement fringes occurring in this imaging regime, as already observed in the projections. Phase retrieval was not performed here, but a single-distance algorithm such as the one proposed by

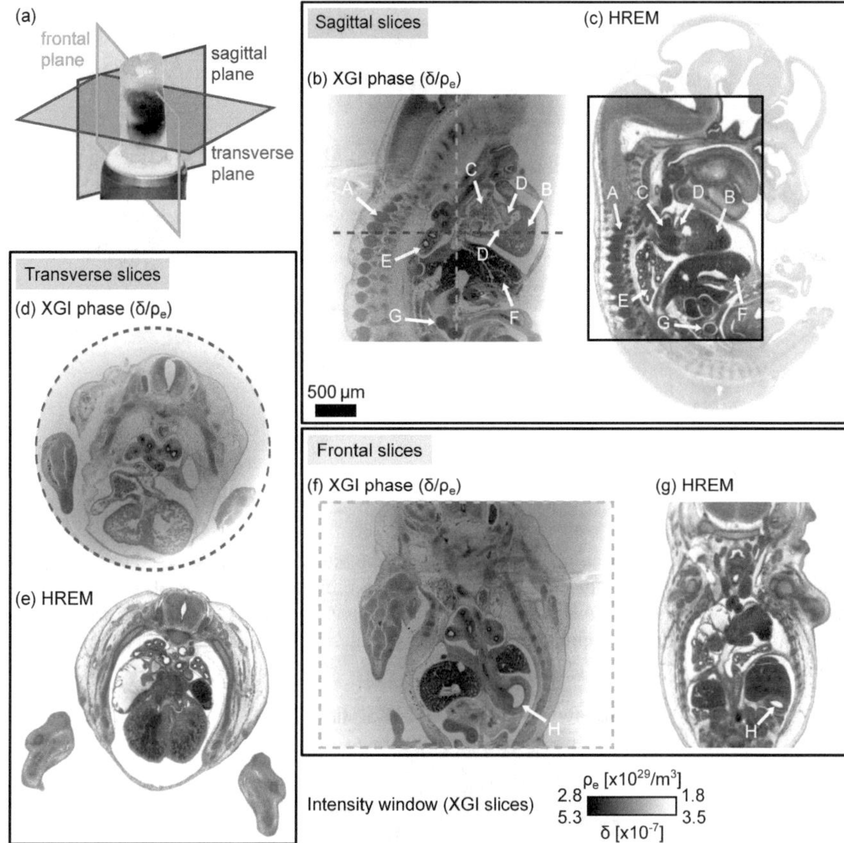

Fig. 4.6 Virtual slices through the 3D phase volume. Sagittal, transverse and frontal slices through the phase tomogram of a mouse embryo (head removed) embedded in paraffin wax in panels **b**, **d** and **f** compared to the results obtained with HREM of a different specimen at the same gestational stage (ED13.5) in panels **c**, **e** and **g**. Organs within the embryo are indicated by white arrows and labelled: A: spine, B: heart ventricles, C: heart atria, D: atrioventricular cushions, E: lungs, F: liver, G: intestines, H: stomach. Panel **a** shows the definition of the sectioning planes on a photograph of an embryo specimen. *HREM data provided by Deciphering the Mechanisms of Developmental Disorders (*http://dmdd.org.uk/*), a programme funded by the Wellcome Trust with support from the Francis Crick Institute, is licensed under a Creative Commons Attribution Non-Commercial Share Alike licence*. Figure reprinted with permission from Zdora et al. (2017), licensed under CC BY 4.0

Paganin et al. (2002) could be applied to the data in future studies. However, this approach does not deliver quantitative density information and, as the signal is based on the Laplacian of the phase shift, the sensitivity is expected to be lower than for the differential phase information.

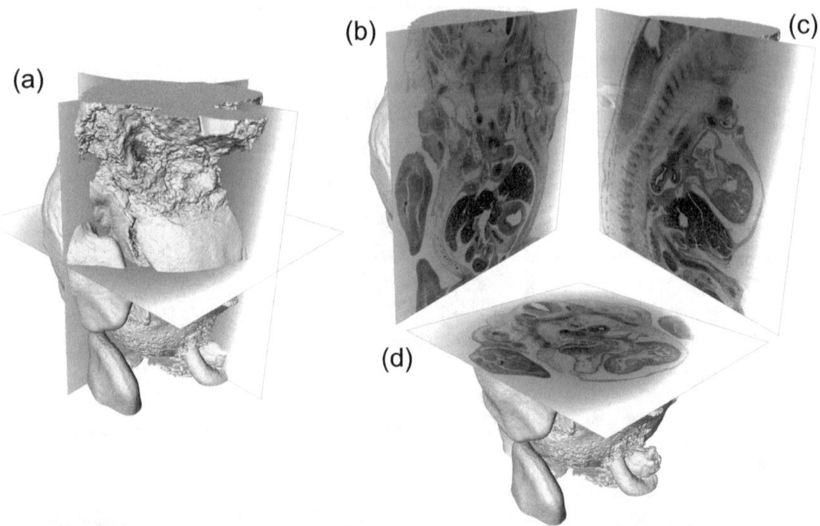

Fig. 4.7 3D volume visualisation. **a** 3D rendering of the mouse embryo extracted from the phase volume (orthogonal slicing planes through the centre are indicated). **b** Frontal, **c** sagittal and **d** transverse cuts through the embryo showing the corresponding slices through the whole phase volume (including paraffin). Figure reprinted with permission from Zdora et al. (2017), licensed under CC BY 4.0

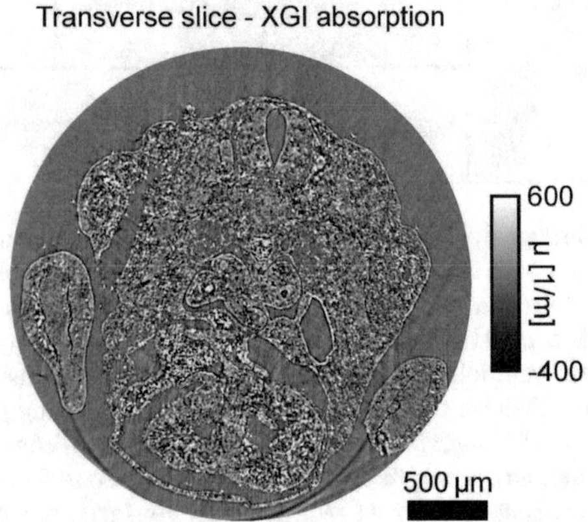

Fig. 4.8 Propagation volume. Transverse slice through the propagation volume which shows a combination of absorption and edge-enhancement effects. It was retrieved from the same data set as the phase volume in Fig. 4.6. The dominating edge enhancement effects at the edges of features also result in negative values of μ. Figure reprinted with permission from Zdora et al. (2017), licensed under CC BY 4.0

4.3.4 Discussion

4.3.4.1 Qualitative and Quantitative Analysis of the XGI Phase Volume

As observed in the previous section, the phase signal shows the internal structure of the specimen in great detail due to its high sensitivity to small density differences. A qualitative assessment of the reconstructed phase volume can be conducted by visual inspection of the slices. The small electron density differences between the various types of tissue are clearly visualised in the phase signal in Fig. 4.6b, d, f showing the organs of the mouse embryo with high contrast.

An advantage of XGI over other phase-sensitive imaging methods is that it provides quantitative information on the refraction properties of the sample under study. The quantity reconstructed in the phase tomogram is the distribution of the decrement δ of the complex refractive index or the electron density ρ_{el}. To illustrate the separation of the small density differences in the sample, a histogram is plotted in Fig. 4.9b, showing the distribution of δ and ρ_{el} in a ROI of the sagittal slice of the phase volume containing mainly organs (see Fig. 4.9a). Regions of 10000 pixels centred in the different organs in Fig. 4.9a were chosen (coloured boxes) to determine an estimate for the δ-values of the organs. Values were calculated for liver (red), lung (green), heart ventricles (blue) and the cavities inside the mouse embryo (orange) and are listed in Table 4.1. The peaks for the different organs of the embryo can be separated in the histogram in Fig. 4.9b, see arrows, which confirms the good contrast of the phase volume.

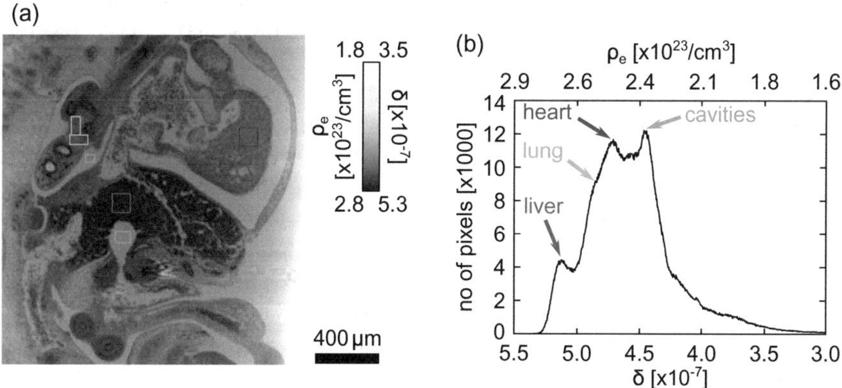

Fig. 4.9 Quantitative measurement and analysis of the electron density distribution in the mouse embryo. **a** ROI from the sagittal slice through the phase volume in Fig. 4.6b containing the organs of the mouse embryo. Small areas within the organs are chosen to determine the corresponding δ-values (coloured boxes). **b** Histogram of the slice in (**a**) showing a separation of the peaks for the different organs. Figure reprinted with permission from Zdora et al. (2017), licensed under CC BY 4.0

Table 4.1 Refractive index decrements δ and electron densities ρ_{el} of the different organs of the mouse embryo measured in the small regions of interest (coloured boxes) in Fig. 4.9a. Table reprinted with permission from Zdora et al. (2017), licensed under CC BY 4.0

Organ	Liver	Lung	Heart	Cavities
$\delta[\times 10^{-7}]$	5.17 ± 0.03	4.84 ± 0.07	4.71 ± 0.04	4.44 ± 0.03
$\rho_{el}[\times 10^{23}/\text{cm}^3]$	2.71 ± 0.02	2.53 ± 0.04	2.47 ± 0.02	2.32 ± 0.02

For further quantitative analysis, the δ-values for liver and paraffin wax were determined from the whole phase volume by calculating the mean value in three regions of interest of $100 \times 100 \times 100$ pixels located within the material. This analysis could not be performed for all of the organs as some of them cover only a small region of the volume. For liver, a mean value of $\delta_{\text{liver}} = (5.10 \pm 0.05) \times 10^{-7}$ is determined, which agrees within the error margins with the result retrieved from the 2D ROI in the sagittal slice listed in Table 4.1. For paraffin, a value of $\delta_{\text{paraffin}} = (4.15 \pm 0.18) \times 10^{-7}$ is obtained in an area close to the sample. For a rough comparison of the value for δ of liver tissue, the properties of a liver phantom used in radiotherapy (Liver Equivalent Electron Density Plug of the Electron Density Phantom Model 062M supplied by Computerized Imaging Reference Systems, Inc., see www.cirsinc.com) can be used. The nominal electron density of the liver equivalent is $\rho_{el\ \text{liver}} = 3.516 \times 10^{23}$ electrons/cm^3, which can be converted to $\delta_{\text{liver, 19keV}} = 6.71 \times 10^{-7}$ using Eq. (2.31). A theoretical value of $\delta_{\text{paraffin, 19keV}} = 6.26 \times 10^{-7}$ for paraffin is obtained from Henke et al. (1993) using the chemical formula $C_{25}H_{52}$ and density $\rho = 0.95\,\text{g/cm}^3$. The experimentally determined values are on the same order of magnitude as the theoretically calculated ones, which confirms the validity of our measurements. However, discrepancies from the literature values can be observed. This can be mainly attributed to changes in the properties of the sample induced by the effects of preparation and embedding. As the specimen was infused by paraffin, the values obtained here represent a mixture of tissue and paraffin wax. To separate the influence of the setup from the sample properties, a calibration sample could be used in future experiments.

Moreover, phase-wrapping artefacts and the effects of region-of-interest tomography—as discussed in Sect. 4.3.3.3—affect the quantitative measurement of the refraction index decrement and lead to a deviation from the literature values. In future studies, this will be prevented by scanning the sample in a water bath and adjusting the sample dimensions and optical magnification to avoid region-of-interest tomography.

The spatial resolution of the reconstructed projection images is affected by the optical elements in the X-ray path, the resolution of the detector system, and the separation of the diffracted beams—created by the phase grating—in the observation

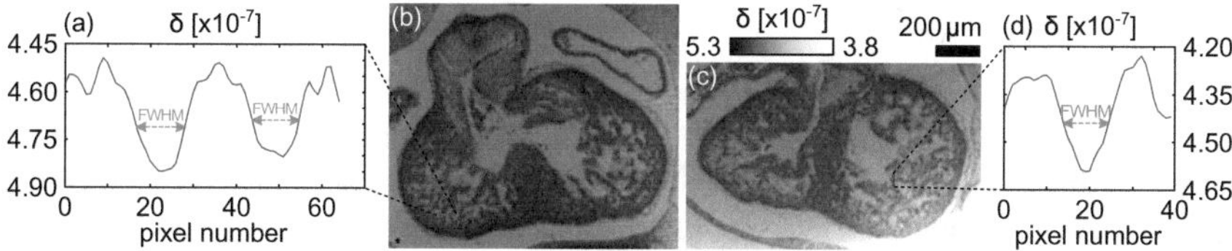

Fig. 4.10 Spatial resolution estimation. Region of the phase volume showing the ventricles of the mouse embryo heart in **b** transverse and **c** frontal view. Line profiles across the smallest discernible structures in **a** the transverse slice and **d** the frontal slice. The average FWHM of the peaks (FWHM $\approx 12\,\mu\text{m}$) is taken as a measure for the spatial resolution. Figure reprinted with permission from Zdora et al. (2017), licensed under CC BY 4.0

plane, as discussed in Sect. 4.2.3. In this case, the separation was $2\lambda z/p \approx 9.4\,\mu\text{m}$ with the wavelength $\lambda = 0.065\,\text{nm}$, the grating-detector distance $z = 0.72\,\text{m}$, and the grating period $p = 10\,\mu\text{m}$.

To estimate the upper limit of the spatial resolution in the reconstructed phase volume after the FBP step, the full width at half maximum (FWHM) of the smallest discernible structures in the phase volume slices is used, as proposed by Zanette et al. (2010). Structures in the transverse and frontal slices of the heart were chosen for this analysis and the selected features are indicated by red lines in Fig. 4.10b and c, respectively. From the corresponding line profiles in Fig. 4.10a, d, a FWHM of $11.8\,\mu\text{m}$ was measured for the frontal slice and FWHM values of $12.7\,\mu\text{m}$ and $12.3\,\mu\text{m}$ for the first and second feature (first and second peak) in the transverse slice, respectively.

4.3.4.2 Comparison of XGI and HREM Results

As illustrated in Fig. 4.6, the quality of the phase tomogram retrieved in this study is comparable to the results obtained with the HREM method that is commonly used for this type of sample. While HREM can achieve excellent results, XGI has the major advantages of being a non-destructive technique and not requiring any contrast agents while delivering a similar image quality at a comparable scan time. As the specimen is not treated with staining solutions and is not physically sectioned, tissue distortions are minimised and the sample preparation time is significantly reduced. In contrast to HREM, XGI delivers quantitative volumetric information about the sample, important e.g. for studies of diseases at different stages and developmental defects, for which quantitative density and volume measurements of the organs and other tissue types are of interest.

4.3.5 Conclusions

The results demonstrate the high potential of X-ray grating interferometry with a single grating in phase-stepping mode for X-ray phase-contrast imaging of biological samples and they have been evaluated against data obtained from high-resolution episcopic microscopy measurements.

The single-grating XGI setup installed at I13-2 beamline at Diamond Light Source showed high performance and stability. The phase volume of a mouse embryo embedded in paraffin wax obtained with this instrument reveals fine details of the organs at high contrast and resolution, comparable with state-of-the-art HREM images. Data acquisition can be performed in the short period of a few hours, maximising sample throughput and statistics of the results. As XGI with a single grating is a non-destructive method without the need for contrast agents, which can be implemented with a simple, robust and flexible setup, it provides a promising alternative to HREM for the 3D visualisation of the inner structure of biological specimens, especially for quantitative density and volume investigations.

For future studies the current setup will be optimised further to increase the visibility of the interference pattern, avoid artefacts induced by region-of-interest tomography, phase wrapping and optics instabilities and to quantify the effect of sample preparation. Moreover, it will be possible to speed up the data collection process by using a multilayer monochromator or polychromatic beam allowing for shorter exposure times through higher photon flux and by optimising the data acquisition scheme, e.g. by taking the rotation axis as fast tomographic axis or employing an interlaced stepping method (Zanette et al. 2012).

The high-quality images obtained with X-ray single-grating phase-contrast imaging will allow conducting high-throughput morphological, pathological and diagnostic investigations of biomedical specimens as an alternative or complementary method to HREM.

4.4 Single-Grating Phase Tomography of Biological Tissue Using a Multilayer Monochromator

Following the first implementation of X-ray single-grating interferometry at Diamond I13 presented in the previous section, the setup was further optimised to speed up data acquisition. This is particularly important to achieve high throughput for studies requiring a large number of samples for statistical analysis. For this purpose, a single-grating interferometer was installed at Diamond I13 using an X-ray beam from a multilayer monochromator (MLM). The setup and first results obtained during this Ph.D. project are presented in the following.

4.4.1 Background and Motivation

Often XGI at synchrotrons is performed using a DCM, which has the advantage of providing an X-ray beam with high degree of longitudinal coherence (energy resolution: $\approx 10^{-4}$). However, the use of a DCM comes at the cost of a significant decrease in photon flux as only a very small part of the X-ray beam is selected. On the other hand, using the fully polychromatic beam will have a negative effect on the visibility of the interferometer and will impose a high dose on the sample due to the presence of low energies in the beam spectrum. A solution can be the use of an MLM, which selects a quasi-monochromatic beam with a broader wavelength bandwidth (energy resolution: $\approx 10^{-2}$) from the incoming X-rays. It provides sufficient monochromaticity for a good visibility of the grating interference pattern and at the same time delivers a higher X-ray flux, which allows for significantly reducing the exposure time (in the present case by about one order of magnitude).

In this section, the first XGI setup using an MLM at Diamond I13 is presented and preliminary results for biomedical imaging applications are shown. It is demonstrated that the installed single-grating setup allows for retrieving high-contrast images of biological specimens at short exposure times, making it attractive for high-throughput biological and biomedical research, e.g. developmental and gene knock-out studies.

The results in this section are not published elsewhere, but the reconstructed differential phase projections were included in a comparative study of the phase-stepping and spatial-harmonic single-shot image processing techniques for grating interferometry (Marathe et al. 2017).

4.4.2 Materials and Methods

4.4.2.1 Experimental Setup

The single-grating setup installed at Diamond I13-2 beamline is shown in Fig. 4.11. Quasi-monochromatic X-rays of an energy of 15 keV were selected from the full undulator spectrum by using an MLM. The X-ray beam impinged onto the sample, the isolated heart of a mouse embryo (estimated embryonic day ED13.5) embedded in paraffin,[7] which was mounted on a tomography stage.

[7]For more details about the mouse embryo see Sect. 4.3.2.2. The sample presented in this section is the isolated heart of another mouse embryo specimen prepared the same way and at the same gestational stage as the one in Sect. 4.3.2.2.

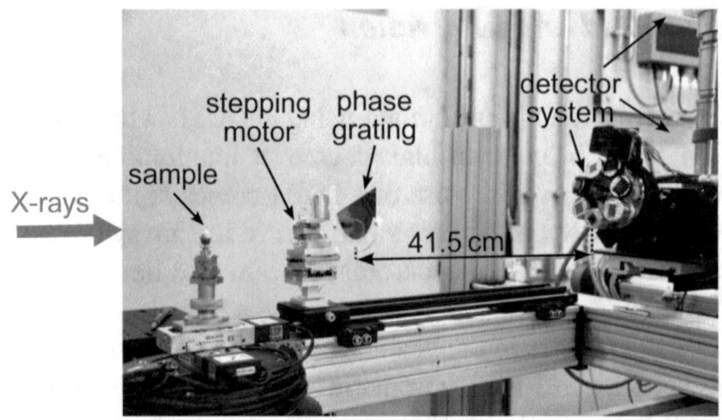

Fig. 4.11 Experimental setup for X-ray single-grating interferometry at Diamond I13-2 beamline using a multilayer monochromator

The phase grating G1 was placed about 10 cm downstream of the sample.[8] It was mounted on a high-precision piezo scanning motor to enable precise and accurate phase stepping. The phase grating (silicon lines on silicon wafer, fabricated at Paul Scherrer Institut, Switzerland) had a nominal period of $p_1 = 4.785\,\mu\text{m}$ and line height of $h = 29.5\,\mu\text{m}$, see Sect. 3.1.4 in Chap. 3. It was originally designed as a π-shifting grating for a conventional two-grating setup optimised for an energy of 23 keV. However, for the case presented here, it was used as a $3\pi/2$-shifting grating at 15 keV to produce a fine interference pattern of period $p = p_1$ at the fractional Talbot distances, which could be directly resolved by the detector.

The detector system consisted of a pco.edge 5.5 camera (sCMOS sensor of size: 2560×2160 pixels, pixel size: 6.5 µm), coupled to a tube lens and a relay optics system, a $4\times$ magnifying objective (Olympus UPLSAPO 4X) with numerical aperture of 0.16 and a scintillation screen. This resulted in a total 8-fold magnification and an effective pixel size of 0.81 µm. The detector was located at a distance of 41.5 cm downstream of G1, which is close to the third fractional Talbot distance.

4.4.2.2 Data Acquisition and Analysis

First, the period of the interference pattern was calibrated by stepping the grating over a range of several periods and plotting the phase-stepping intensity curve for the average signal in a 10×10 pixels area in the centre of the reference pattern. This process was repeated for several scanning ranges and the period of the pattern was determined to be 4.75 µm.

For imaging the mouse embryo heart, a tomography scan with 2001 projections over 180° of sample rotation (angular increment: 0.09°) was acquired. For each projection, the grating G1 was scanned horizontally—perpendicular to the grating

[8] The sample could not be positioned closer to the grating due to limitations in the setup. However, this does not have an impact on the sensitivity of the system in the case of a parallel beam geometry. It has been shown that for a diverging (point) source the sensitivity decreases with increasing sample-G1 distance (Donath et al. 2009). However, for a parallel beam (as given here in good approximation), this is only true for the case of placing the sample downstream of G1. If the sample is located upstream of G1, the sensitivity is independent of the sample position.

lines—in five steps over one period of the interference pattern with a step size of 0.95 μm. A set of five reference phase-stepping scans without the sample in the beam was taken every 100 sample projections. The average of the five sets was used as a new reference for the following 100 sample images to account for instabilities in the reference background. Five dark images without X-ray exposure were recorded before and after the tomography scan and the average dark image was subtracted from each of the sample and reference frames before further processing. The exposure time for a single frame was 100 ms and the total scan time was approximately 30 minutes including overhead time due to motor movement. This is short compared to the exposure times in the range of several seconds that were required for a similar setup using a DCM (see previous section). However, it is still longer than similar measurements performed at I13 beamline without the use of a grating, e.g. using single-distance propagation-based phase-contrast imaging, which can be performed in only a few minutes (Rau et al. 2017, 2019). On the other hand, the latter cannot provide multimodal and quantitative information and is less sensitive to small density differences, see also discussion in Sect. 2.4.6, Chap. 10.

For each projection, the transmission and the differential phase signals were extracted from the phase-stepping scans with and without the sample, as described in Sects. 4.2.2 and 4.3.2.3. Tomographic transmission and phase volumes were reconstructed from the projections using the methods explained in Sects. 2.5 and 4.3.2.3.

It should be noted once again that—as discussed in Sect. 4.3.2.3—the transmission projections and transmission volume do not show the pure absorption of X-rays in the sample, but rather a combination of absorption and propagation effects. In the previous section this signal was hence referred to as "propagation signal". Here, the denotation "transmission signal" is kept despite propagation effects also being present in the images.

4.4.3 Results

4.4.3.1 Interference Pattern

An example of the reference interference pattern produced by the phase grating at the third fractional Talbot distance is shown in Fig. 4.12a for the first grating position of the phase-stepping scan. The bright and dark low-frequency horizontal stripes are caused by the surface roughness of the MLM.

Furthermore, instabilities of the MLM lead to some remaining grating lines in the visibility map, as can be observed in Fig. 4.12b. Distortions of the optical microscope objective in the detector system and possibly grating imperfections may also have contributed to this effect. Instabilities and low photon counts, furthermore, reduce the visibility of the interferometer, see Fig. 4.12b. The mean visibility over the entire field of view is $(8.6 \pm 2.2)\%$. This is low compared to typical values from double-grating interferometers, see e.g. Zanette (2011). It is, moreover, lower than for the measurements presented for the single-grating setup in the previous section, see

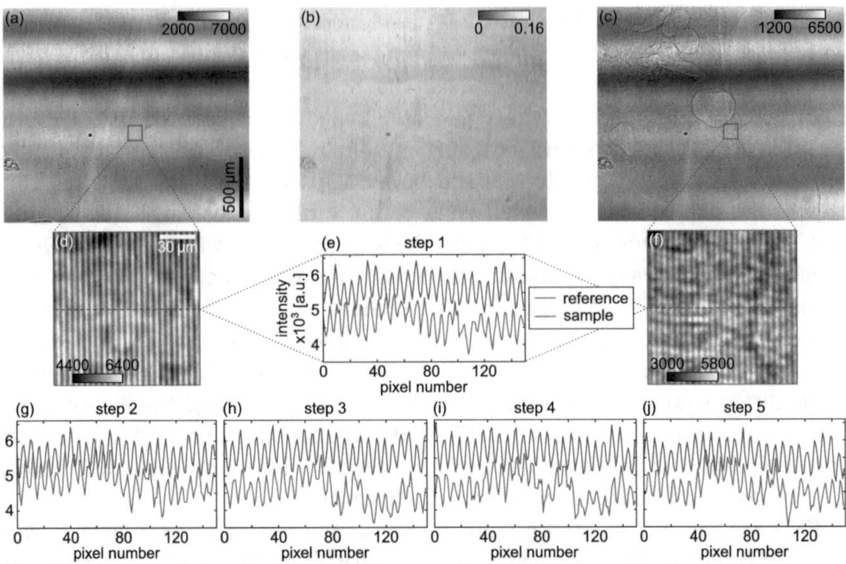

Fig. 4.12 Interference patterns and visibility map. Grating interference pattern **a** without and **c** with the sample in the beam. **b** Visibility map of the field of view. The visibility is significantly affected by MLM instabilities. **d, f** Regions of interest (150 × 150 pixels) of the reference and sample interference patterns in panels (**a**) and (**c**), respectively. **e**, **g–j** Line profiles through the centre of the ROIs for each of the five grating positions of the phase-stepping scan. The presence of the sample leads to a reduction in intensity due to absorption and a shift of the pattern due to refraction

Fig. 4.4. Apart from instabilities of the MLM, this is due to the small period of the interference pattern that is less than 6 pixels across and the large point spread function (PSF) of the detector system (mainly due to the relatively thick scintillation screen), which limits the resolution of the instrument in this configuration. However, the visibility values achieved here are comparable to the ones obtained with a similar setup at DESY, PETRA III synchrotron, for which an average visibility of 8% has been reported (Hipp 2018), which was operated at an X-ray energy of 20 keV with an effective pixel size of 2.4 µm, a phase grating with period $p_1 = 10\,\mu\text{m}$ ($3\pi/2$-shifting with line height $h = 10.4\,\mu\text{m}$) and a grating-detector distance of 40 cm. A ROI of 150 × 150 pixels in the reference interference pattern is shown in Fig. 4.12d. It reveals the fine line interference pattern that is directly resolved by the detector system. A line profile through the centre of Fig. 4.12d is plotted as a blue curve in Fig. 4.12e. Figure 4.12c shows the interference pattern at the first grating step of the phase-stepping scan with the sample in the beam. The corresponding ROI and a line profile through its centre can be found in Fig. 4.12f, e (red line), respectively. Absorption in the sample leads to a decrease in intensity and refraction to a displacement of the line pattern compared to Fig. 4.12a, d. Corresponding line plots through ROIs of the interference patterns for all of the five grating steps are plotted in Fig. 4.12e, g–j. The gradual shift of the line pattern upon stepping can be observed, as expected.

Intensity fluctuations occurring between the five steps, in particular for the second step (Fig. 4.12g), are caused by drift and vibrations of the MLM. This resulted in artefacts in the reconstructed signals as discussed in the following subsection.

4.4.3.2 Differential Phase and Transmission Projections

The differential phase and transmission signals of one of the 2001 projections are shown in Fig. 4.13.

The displacement of the sample phase-stepping curves with respect to the reference curve, $\Delta\phi = \phi_{\mathrm{sam}} - \phi_{\mathrm{ref}}$, is directly related to the differential phase shift of the X-rays via Eq. (4.8) in Sect. 4.2.2. The signal $\Delta\phi$ is shown in Fig. 4.13a. As discussed in Sect. 4.3.2.3, the transmission signal in Fig. 4.13b contains absorption and propagation effects.

The outlines of the heart inside the paraffin block as well as some inner structure can be detected in the differential phase as well as the transmission signals. Air bubbles in the paraffin wax lead to strong refraction, leading to phase wrapping in Fig. 4.13a and strong edge-enhancement fringes in Fig. 4.13b.

The reconstructed signals are corrupted by a number of artefacts, such as remaining grating lines and low-frequency horizontal stripes. These are a result of the strong intensity fluctuations occurring between the steps of the phase-stepping scan, which are caused mainly by MLM instabilities, but also distortions by the microscope objective and possibly grating imperfections.

Usually, the angular sensitivity is determined from a small ROI in the background region of the refraction angle signal, where the phase shift is homogeneous (air, water, immersion liquid etc.), see Sect. 4.2.3. For the present case of a local (region-of-interest) tomography with no air background, this is not possible. Therefore, the

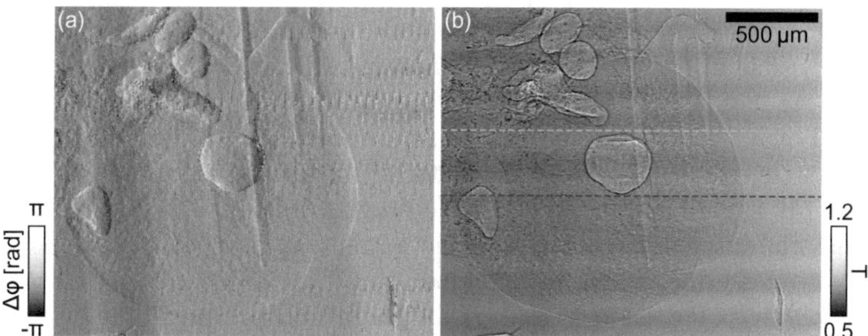

Fig. 4.13 Reconstructed signals of one of the 2001 projections of the tomography scan. **a** Differential phase signal and **b** transmission signal. The latter contains not only absorption, but also propagation effects showing as edge enhancement. The dashed lines indicate the location of the short-axis slices through the volumes that are shown in Fig. 4.14. Remaining grating lines in the differential phase signal are due to MLM instabilities, distortions caused by the microscope objective and grating imperfections

angular sensitivity is here estimated from a region within the paraffin surrounding the specimen. This is not an accurate measurement as the paraffin block is cylindrical and not completely homogeneous, but it can be seen as an upper limit and gives an idea of the order of magnitude of the achievable angular sensitivity. The standard deviation of the signal was determined in ROIs of 50 × 50 pixels and the angular sensitivity estimated this way lies between 0.4–0.6 µrad.

The spatial resolution was here determined directly from the phase volume as shown in the following section.

4.4.3.3 Phase and Transmission Volumes

Figure 4.14 shows various short-axis and long-axis slices through the phase and transmission volumes that were reconstructed from the differential phase and transmission projections. The locations of the short-axis slices are indicated in the projections in Fig. 4.13b. The phase slices are shown in Fig. 4.14a, c and the transmission slices in Fig. 4.14b, d.

The transmission volume shows only weak contrast as the absorption signal is not very sensitive to the small density differences that are typically encountered in soft biological tissue. However, as noted before, the scan was not optimised for absorption imaging and is strongly affected by edge-enhancement fringes.

The short-axis slices through the phase volume in Fig. 4.14a, c reveal the detailed structure of the mouse embryo heart. The right and left ventricles (RV, LV), the interventricular septum (IVS) and the right atrium (RA) can clearly be distinguished from the background. The apparent separation of the left ventricle in two parts in Fig. 4.14c is due to the oblique cutting angle through the heart. Moreover, details like the aorta (AO), the pulmonary artery (PA) and even the pulmonary valve (PV) can be identified. Long-axis slices through the phase volume are shown in Fig. 4.14k–n and their locations are indicated by the coloured dashed lines in Fig. 4.14b. Apart from the features already identified in the short-axis slices, additionally the mitral valve (MV) is visible in Fig. 4.14n. A series of short-axis cuts at different height steps, between the slice in Fig. 4.14a and the slice in Fig. 4.14c, is presented in Fig. 4.14e–j for a ROI of 905 × 905 pixels around the aorta and pulmonary artery. Going further from basal (top) to apical (bottom) direction, the aortic valve becomes clearly visible in Fig. 4.14f, g, until it opens up into the left ventricle in Fig. 4.14i, j.

Some artefacts of different origin affect the quality of the reconstructed slices. Air bubbles trapped in the paraffin as well as the edge of the wax block itself (indicated by the black arrow in Fig. 4.14a, b) lead to strong refraction resulting in phase-wrapping artefacts showing as pronounced streaking in the phase volume slices. Furthermore, the paraffin block did not remain in the field of view for all projections of the scan and the effects of local tomography cause additional low-frequency artefacts like cupping. As discussed in Sect. 4.3.4.1, phase wrapping at the paraffin edges could be avoided by immersing the sample in a refractive-index-matching liquid, e.g. a water bath. This could not be realised for the scan presented here as no suitable sample stage was available for upside-down mounting. Such a stage has now been

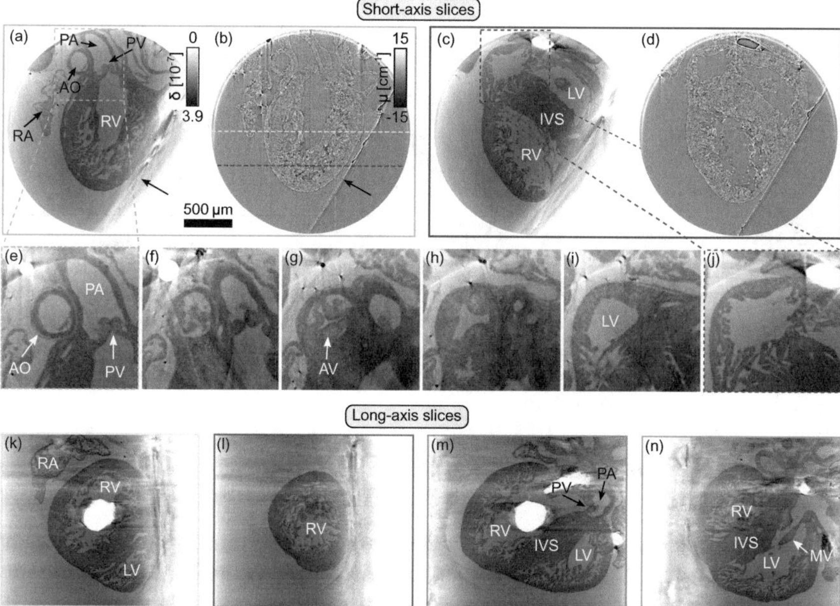

Fig. 4.14 Virtual slices through the phase and transmission volumes. **a, c** Short-axis slices through the phase volume and **b, d** corresponding slices through the transmission volume. As expected, the contrast of the transmission volume is low, but it should be noted that the scan was not optimised for absorption imaging. **e–j** Regions of interest around the aorta (AO) and pulmonary valve (PV) in the short-axis phase slices moving from basal to mid planes. The aortic valve (AV) can clearly be distinguished, which further down opens into the left ventricle (LV). **k–n** Long-axis slices through the phase volume as indicated by the coloured lines in panel (**b**). The phase slices show fine details of the tissue structure as labelled: RV: right ventricle, LV: left ventricle, IVS: interventricular septum, RA: right atrium, AO: aorta, PA: pulmonary artery, PV: pulmonary valve, MV: mitral valve. Streaking artefact and non-uniform background are caused by phase wrapping at the edges of the paraffin block and the effects of local tomography

purchased and commissioned at I13, see Chap. 3, Sect. 3.1.4. Analogous to the data set in Sect. 4.3.4.1, the spatial resolution of the phase slices was estimated by analysing a line profile through one of the smallest visible features. While the true achievable spatial resolution will in general be better, this provides an upper limit of the resolution. The short-axis slice in Fig. 4.14a was used for this analysis, which is shown again in Fig. 4.15a. A ROI visualising the fine structure of the heart muscle is presented in Fig. 4.15b and a line across one of the smallest features was selected. Its profile is plotted in Fig. 4.15c. The FWHM of the inverse peak is about 10.9 pixels, equivalent to a feature size of approximately 8.8 μm.

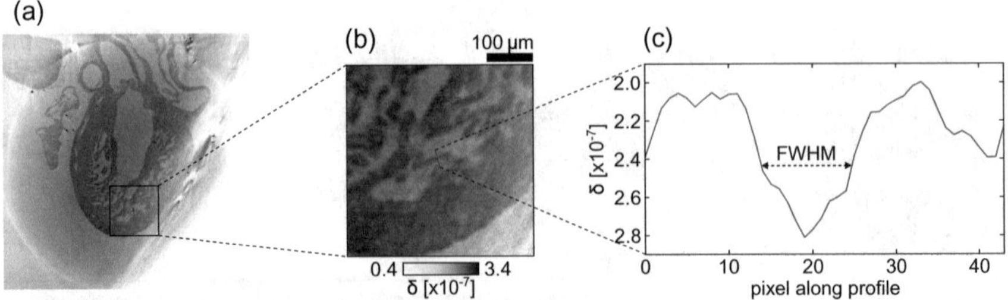

Fig. 4.15 Spatial resolution estimation. **a** Short-axis slice through the phase volume chosen for estimating the spatial resolution of the measurement. **b** ROI of panel (**a**) showing the fine structure of the heart muscle. **c** Line profile across one of the smallest features, as indicated by the red line in panel (**b**). The FWHM of the peak (FWHM ≈8.8 μm) is taken as an upper limit of the resolution

4.4.4 Discussion and Conclusion

The first implementation of an X-ray single-grating interferometry setup at Diamond I13 beamline using an MLM and its application for biomedical imaging were presented. The use of an MLM made it possible to use exposure times that were one order of magnitude shorter than for a similar setup using the DCM thanks to a significantly higher photon flux.

Despite some artefacts arising mainly due to monochromator instabilities, the phase volume showed high contrast and allowed to identify detailed features of a mouse embryo heart. The good image quality in combination with a simple setup and short exposure time make this implementation a promising candidate for studies on biological specimens, especially when a high sample throughput is of importance. This will in particular be interesting for anatomical and developmental studies on animal models, as also discussed in Sect. 4.3.5.

For future measurements, the quality of the images can be further improved by overcoming the various sources of artefacts encountered for the scan presented here. Final commissioning of the MLM at Diamond I13-2 beamline is expected to increase its stability, which will reduce artefacts caused by fluctuations in the flat field. Furthermore, phase-wrapping artefacts at the edges of the paraffin block can be avoided by placing the sample in a water bath during the measurement. This will be possible with the new dedicated setup containing an upside-down hanging sample tomography stage, which is currently under commissioning at Diamond I13-2, as discussed in Sect. 3.1 of Chap. 3. Phase wrapping at the air bubbles inside the wax can be reduced by taking care during specimen embedding. Additionally, the wax cylinder can be trimmed down closer to the sample to fit the entire block inside the field of view and avoid local tomography. After optimisation of the setup, it will be a promising tool for high-contrast and fast imaging of specimens with small density differences, providing 3D information about the sample's anatomy and quantitative density distribution.

4.5 Concluding Remarks

XGI is the first method that was further developed and applied during this Ph.D. project. Instead of the conventional implementation with two gratings, a single-grating setup was installed at Diamond I13-2 beamline, which is simpler and more robust. Furthermore, thanks to the removal of the analyser grating, the exposure time can be reduced significantly compared to a conventional setup. On the other hand, it should be considered that this approach can only be used for relatively small samples on the order of a few millimetres across. This is due to the limitations in the field of view imposed by the small pixel size, which is necessary to directly resolve the fine grating interference pattern.

Single-grating interferometry has seen increased interest in the recent years due to its simplicity, robustness, flexibility and dose- and cost-efficiency, but not many applications have been reported to this date. It was shown in this chapter that single-grating interferometry has great potential for biomedical imaging and can provide high-contrast 3D density maps of specimens using a simple setup.

The implementation of single-grating interferometry at Diamond I13 was performed in two different configurations: with a DCM for optimised visibility of the interference pattern increasing the sensitivity of the instrument and with an MLM for optimised acquisition speed allowing for high sample throughput. While both setups delivered high-contrast images of biomedical specimens, the first had an approximately four times better angular sensitivity, but the latter could be operated with an exposure time that was an order of magnitude shorter. It should, however, be noted that the components used in the two setups, such as gratings, camera, microscope objectives and sample, were not the same and they can therefore not be compared in a quantitative manner. In particular, the sensitivity of an XGI setup is determined by a range of factors including the effective pixel size, the type of grating and the propagation distance, amongst others.

The XGI implementations at Diamond I13 are currently being optimised further and a customised setup has recently been constructed, which amongst other components includes an upside-down tomography stage compatible with the use of a water bath for refractive-index-matching, see Sect. 3.1 in Chap. 3. First user experiments with XGI at I13 have already been carried out with the preliminary setups (including improvised upside-down sample mounting) for single- and double-grating interferometry, see Khimchenko et al. (2016); Schulz et al. (2016, 2017a).

XGI is a powerful method for phase-sensitive and multimodal imaging. In this chapter, applications of the phase-contrast signal for biomedical research were presented, which is the most prominent field of XGI (Stutman et al. 2011; Stampanoni et al. 2011; Bravin et al. 2012; Sztrókay et al. 2013; Pfeiffer et al. 2013; Tapfer et al. 2012; Momose et al. 2014). However, there are many other fields of applications such as materials science (Sarapata et al. 2015; Yang et al. 2016), food science (Einarsdóttir et al. 2014; Miklos et al. 2015) and archaeology (Ludwig et al. 2018).

Furthermore, there has been extensive research on the use of the dark-field signal, e.g. for biomedical and clinical applications such as studying the structure of teeth (Jensen et al. 2010; Jud et al. 2016) and bones (Potdevin et al. 2012; Jud et al. 2017; Baum et al. 2015) and for the detection of pulmonary emphysema (Schleede et al. 2012; Yaroshenko et al. 2013; Meinel et al. 2014; Hellbach et al. 2014; Willer et al. 2018). Applications in materials science include the identification of cracks (Lauridsen et al. 2015), the visualisation of fibre orientation (Malecki et al. 2014; Revol et al. 2013; Prade et al. 2017), the quantification of water uptake in cement (Yang et al. 2014; Prade et al. 2016) and the detection of changes in the micro-structure of a material on sub-pixel length scales (Revol et al. 2011; Jerjen et al. 2013). Such studies might in the future also be performed at the XGI setups described in this chapter.

As outlined above, the potential of XGI for quantitative phase-contrast and dark-field imaging has been demonstrated in a large number of studies in the last decade. However, a main limitation of the technique in its most common implementation with 1D line gratings is given by its sensitivity in only one direction. It can lead to artefacts and loss of information, in particular for samples with strong anisotropy and especially for the dark-field signal. While this can be tackled by using two line gratings mounted in crossed configuration or a 2D chequerboard grating (Zanette et al. 2010; Morimoto et al. 2015a), the latter are challenging and costly to fabricate and the first approach requires precise alignment. It will be shown in the next chapters of this thesis that an alternative capable of 2D sensing is provided by the latest X-ray multimodal imaging method, namely X-ray speckle-based imaging, which makes use of a 2D random interference pattern as a wavefront marker.

References

Als-Nielsen J, McMorrow D (2011) Elements of modern x-ray physics, 2nd edn. Wiley, Chichester, UK

Anderson GA, Wong MD, Yang J, Henkelman RM (2013) 3D imaging, registration, and analysis of the early mouse embryonic vasculature. Dev Dynam 242(5):527–538

Attwood D (1999) Soft x-rays and extreme ultraviolet radiation: principles and applications. Cambridge University Press, Cambridge, UK

Baborowski J, Revol V, Kottler C, Kaufmann R, Niedermann P, Cardot F, Neels DA, Despont M (2014) High aspect ratio, large area silicon-based gratings for x-ray phase contrast imaging. In: 2014 IEEE 27th international conference on micro electro mechanical systems (MEMS), pp 490–493

Baldock R, Bard J, Davidson DR, Morriss-Kay G (eds) (2016) Kaufman's atlas of mouse development supplement. Academic Press, Boston, MA, USA

Balles A, Fella C, Dittmann J, Wiest W, Zabler S, Hanke R (2016) X-ray grating interferometry for 9.25 keV design energy at a liquid-metal-jet source. AIP Conf Proc 1696:020043

Balles A, Muller D, Dittmann J, Fella C, Hanke R, Zabler S (2018) Computed tomography from a single grating x-ray interferometer at a laboratory liquid-metal-jet source. Microsc Microanal 24(S2):152–153

Baum T, Eggl E, Malecki A, Schaff F, Potdevin G, Gordijenko O, Garcia EG, Burgkart R, Rummeny EJ, Noël PB, Bauer JS, Pfeiffer F (2015) X-ray dark-field vector radiography–a novel technique for osteoporosis imaging. J Comput Assist Tomogr 39(2):286–289

Bech M (2009) X-ray imaging with a grating interferometer. University of Copenhagen, Copenhagen, Denmark Ph.D. thesis

Bérujon S (2013) At-Wavelength metrology of hard x-ray synchrotron beams and optics: method developments and applications. Université de Grenoble, Grenoble, France Ph.D. thesis

Bérujon S, Wang H, Ziegler E, Sawhney K (2012) Shearing interferometer spatial resolution for at-wavelength hard x-ray metrology. AIP Conf Proc 1466(1):217–222

Bier E, Mcginnis W (2016) Epstein's inborn errors of development: the molecular basis of clinical disorders of morphogenesis, chapter I.3 model organisms in the study of development and disease, model organisms in the study of development and disease. Oxford University Press, Oxford, UK

Bravin A, Coan P, Suortti P (2012) X-ray phase-contrast imaging: from pre-clinical applications towards clinics. Phys Med Biol 58(1):R1–R35

Cartier S, Kagias M, Bergamaschi A, Wang Z, Dinapoli R, Mozzanica A, Ramilli M, Schmitt B, Brückner M, Fröjdh E, Greiffenberg D, Mayilyan D, Mezza D, Redford S, Ruder C, Schädler L, Shi X, Thattil D, Tinti G, Zhang J et al (2016) Micrometer-resolution imaging using MÖNCH: towards G_2-less grating interferometry. J Synchrotron Rad 23(6):1462–1473

Clauser JF, Li S (1994) Talbot-vonLau atom interferometry with cold slow potassium. Phys Rev A 49(4):R2213–R2216

Cloetens P, Guigay JP, De Martino C, Baruchel J, Schlenker M (1997) Fractional Talbot imaging of phase gratings with hard x-rays. Opt Lett 22(14):1059–1061

David C, Bruder J, Rohbeck T, Grünzweig C, Kottler C, Diaz A, Bunk O, Pfeiffer F (2007) Fabrication of diffraction gratings for hard x-ray phase contrast imaging. Microelectron Eng 84(5):1172–1177. Proceedings of the 32nd international conference on micro- and nano-engineering

David C, Nöhammer B, Solak HH, Ziegler E (2002) Differential x-ray phase contrast imaging using a shearing interferometer. Appl Phys Lett 81(17):3287–3289

Deciphering the Mechanisms of Developmental Disorders (2016). http://dmdd.org.uk/

Donath T, Chabior M, Pfeiffer F, Bunk O, Reznikova E, Mohr J, Hempel E, Popescu S, Hoheisel M, Schuster M, Baumann J, David C (2009) Inverse geometry for grating-based x-ray phase-contrast imaging. J Appl Phys 106(5):054703

Dunlevy L, Bennett M, Slender A, Lana-Elola E, Tybulewicz VL, Fisher EM, Mohun T (2010) Down's syndrome-like cardiac developmental defects in embryos of the transchromosomic Tc1 mouse. Cardiovasc Res 88(2):287–295

Einarsdóttir H, Nielsen MS, Miklos R, Lametsch R, Feidenhans'l R, Larsen R, Ersbøll BK (2014) Analysis of micro-structure in raw and heat treated meat emulsions from multimodal x-ray microtomography. Innov Food Sci Emerg Technol 24:88–96

Engelhardt M, Kottler C, Bunk O, David C, Schroer C, Baumann J, Schuster M, Pfeiffer F (2008) The fractional Talbot effect in differential x-ray phase-contrast imaging for extended and polychromatic x-ray sources. J Microsc 232(1):145–157

Epple FM, Potdevin G, Thibault P, Ehn S, Herzen J, Hipp A, Beckmann F, Pfeiffer F (2013) Unwrapping differential x-ray phase-contrast images through phase estimation from multiple energy data. Opt Express 21(24):29101–29108

Fitzgerald R (2000) Phase-Sensitive x-ray imaging. Phys Today 53(7):23–26

Geyer SH, Maurer B, Pötz L, Singh J, Weninger WJ (2012) High-Resolution episcopic microscopy data-based measurements of the arteries of mouse embryos: evaluation of significance and reproducibility under routine conditions. Cells Tissues Organs 195(6):524–534

Ghiglia D, Pritt M (1998) Two-dimensional phase unwrapping: theory, algorithms, and software. Wiley-Interscience publication, Wiley, New York, NY, USA

Graham E, Moss J, Burton N, Armit C, Richardson L, Baldock R (2015) The atlas of mouse development eHistology resource. Development 142(11):1909–1911

Grizolli W, Shi X, Kolodziej T, Shvyd'ko Y, Assoufid L (2017) Single-grating Talbot imaging for wavefront sensing and x-ray metrology. Proc SPIE 10385:1038502

Guigay J (1971) On fresnel diffraction by one-dimensional periodic objects, with application to structure determination of phase objects. Opt Acta 18(9):677–682

Guinier A (1994) X-ray diffraction in crystals, imperfect crystals, and amorphous bodies. Dover books on physics series, Dover, New York, NY, USA

Haas W, Bech M, Bartl P, Bayer F, Ritter A, Weber T, Pelzer G, Willner M, Achterhold K, Durst J, Michel T, Prümmer M, Pfeiffer F, Anton G, Hornegger J (2011) Phase-unwrapping of differential phase-contrast data using attenuation information. Proc SPIE 7962:79624R

Hahn D, Thibault P, Fehringer A, Bech M, Koehler T, Pfeiffer F, Noël PB (2015) Statistical iterative reconstruction algorithm for x-ray phase-contrast CT. Sci Rep 5(1):10452

Hellbach K, Yaroshenko A, Meinel FG, Bech M, Eickelberg O, Bamberg F, Nikolaou K, Reiser MF, Yildirim AÖ, Pfeiffer F (2014) Quantitative in vivo x-ray dark-field radiography for early pulmonary emphysema diagnosis. Eur Respir J 44:(Suppl 58)

Henke BL, Gullikson EM, Davis JC (1993) X-ray interactions: photoabsorption, scattering, transmission, and reflection at E = 50–30,000 eV, Z = 1–92. Atomic Data and Nuclear Data Tables 54(2):181–342

Herzen J, Donath T, Beckmann F, Ogurreck M, David C, Mohr J, Pfeiffer F, Schreyer A (2011) X-ray grating interferometer for materials-science imaging at a low-coherent wiggler source. Rev Sci Instrum 82(11):113711

Hipp A, Herzen J, Hammel JU, Lytaev P, Schreyer A, Beckmann F (2016) Single-grating interferometer for high-resolution phase-contrast imaging at synchrotron radiation sources. Proc SPIE 9967:996718

Hipp A, Moosmann J, Herzen J, Hammel JU, Schreyer A, Beckmann F (2017) High-resolution grating interferometer for phase-contrast imaging at PETRA III. Proc SPIE 10391:1039108

Hipp AC (2018) High-resolution grating-based phase-contrast imaging for synchrotron radiation sources. Universität Hamburg, Hamburg, Germany Ph.D. thesis

Hoshino M, Uesugi K, Yagi N (2012) Phase-contrast x-ray microtomography of mouse fetus. Biol Open 1(3):269–274

Inoue T, Matsuyama S, Kawai S, Yumoto H, Inubushi Y, Osaka T, Inoue I, Koyama T, Tono K, Ohashi H, Yabashi M, Ishikawa T, Yamauchi K (2018) Systematic-error-free wavefront measurement using an x-ray single-grating interferometer. Rev Sci Instrum 89(4):043106

Itoh H, Nagai K, Sato G, Yamaguchi K, Nakamura T, Kondoh T, Ouchi C, Teshima T, Setomoto Y, Den T (2011) Two-dimensional grating-based x-ray phase-contrast imaging using Fourier transform phase retrieval. Opt Express 19(4):3339–3346

Jensen TH, Bech M, Zanette I, Weitkamp T, David C, Deyhle H, Rutishauser S, Reznikova E, Mohr J, Feidenhans'l R, Pfeiffer F (2010) Directional x-ray dark-field imaging of strongly ordered systems. Phys Rev B 82(21):214103

Jerjen I, Revol V, Brunner AJ, Schuetz P, Kottler C, Kaufmann R, Luethi T, Nicoletti G, Urban C, Sennhauser U (2013) Detection of stress whitening in plastics with the help of x-ray dark field imaging. Polym Test 32(6):1094–1098

Jerjen I, Revol V, Schuetz P, Kottler C, Kaufmann R, Luethi T, Jefimovs K, Urban C, Sennhauser U (2011) Reduction of phase artifacts in differential phase contrast computed tomography. Opt Express 19(14):13604–13611

Jud C, Braig E, Dierolf M, Eggl E, Günther B, Achterhold K, Gleich B, Rummeny E, Noël P, Pfeiffer F, Muenzel D (2017) Trabecular bone anisotropy imaging with a compact laser-undulator synchrotron x-ray source. Sci Rep 7(1):14477

Jud C, Schaff F, Zanette I, Wolf J, Fehringer A, Pfeiffer F (2016) Dentinal tubules revealed with x-ray tensor tomography. Dent Mater 32(9):1189–1195

Kagias M, Cartier S, Wang Z, Bergamaschi A, Dinapoli R, Mozzanica A, Schmitt B, Stampanoni M (2016) Single shot x-ray phase contrast imaging using a direct conversion microstrip detector with single photon sensitivity. Appl Phys Lett 108(23):234102

Khimchenko A, Schulz G, Deyhle H, Thalmann P, Zanette I, Zdora M-C, Bikis C, Hipp A, Hieber SE, Schweighauser G, Hench J, Müller B (2016) X-ray micro-tomography for investigations of brain tissues on cellular level. Proc SPIE 9967:996703

Koch FJ (2017) X-ray optics made by x-ray lithography: process optimization and quality control. Karlsruher Institut für Technologie (KIT), Karlsruhe, Germany Ph.D. thesis

Krejci F, Jakubek J, Kroupa M (2010) Hard x-ray phase contrast imaging using single absorption grating and hybrid semiconductor pixel detector. Rev Sci Instrum 81(11):113702

Krejci F, Jakubek J, Kroupa M (2011) Single grating method for low dose 1-D and 2-D phase contrast x-ray imaging. J Instrum 6(01):C01073

Lauridsen T, Willner M, Bech M, Pfeiffer F, Feidenhans'l R (2015) Detection of sub-pixel fractures in x-ray dark-field tomography. Appl Phys A 121(3):1243–1250

Ludwig V, Seifert M, Niepold T, Pelzer G, Rieger J, Ziegler J, Michel T, Anton G (2018) Non-Destructive testing of archaeological findings by grating-based x-ray phase-contrast and dark-field imaging. J Imaging 4(4):58

Malecki A, Potdevin G, Biernath T, Eggl E, Willer K, Lasser T, Maisenbacher J, Gibmeier J, Wanner A, Pfeiffer F (2014) X-ray tensor tomography. EPL 105(3):38002

Marathe S, Zdora M-C, Zanette I, Cipiccia S, Rau C (2017) Comparison of data processing techniques for single-grating x-ray Talbot interferometer data. Proc SPIE 10391:103910S

Meinel F, Yaroshenko A, Hellbach K, Bech M, Nikolaou K, Eickelberg O, Reiser MF, Pfeiffer F, Önder Yildirim A (2014) Improved diagnosis of pulmonary emphysema using in vivo dark-field radiography. Pneumologie 68(6):A67

Merlo G, Altruda F, Poli V (2001) Mice as experimental organisms. Wiley, Hoboken, NJ, USA

Metscher BD (2009) MicroCT for developmental biology: a versatile tool for high-contrast 3D imaging at histological resolutions. Dev Dyn 238(3):632–640

Miklos R, Nielsen MS, Einarsdóttir H, Feidenhans'l R, Lametsch R (2015) Novel x-ray phase-contrast tomography method for quantitative studies of heat induced structural changes in meat. Meat Sci 100:217–221

Modregger P, Pinzer BR, Thüring T, Rutishauser S, David C, Stampanoni M (2011) Sensitivity of x-ray grating interferometry. Opt Express 19(19):18324–18338

Mohun T, Adams DJ, Baldock R, Bhattacharya S, Copp AJ, Hemberger M, Houart C, Hurles ME, Robertson E, Smith JC, Weaver T, Weninger W (2013) Deciphering the Mechanisms of Developmental Disorders (DMDD): a new programme for phenotyping embryonic lethal mice. Dis Model Mech 6(3):562–566

Mohun TJ, Weninger WJ (2011) Imaging heart development using high-resolution episcopic microscopy. Curr Opin Genet Dev 21:573–578

Momose A, Kawamoto S, Koyama I, Hamaishi Y, Takai K, Suzuki Y (2003) Demonstration of x-ray talbot interferometry. Jpn J Appl Phys 42:L866–L868

Momose A, Kuwabara H, Yashiro W (2011) X-ray phase imaging using Lau effect. Appl Phys Express 4(6):066603

Momose A, Takeda T, Itai Y, Hirano K (1996) Phase-contrast x-ray computed tomography for observing biological soft tissues. Nat Med 2:473–475

Momose A, Yashiro W, Kido K, Kiyohara J, Makifuchi C, Ito T, Nagatsuka S, Honda C, Noda D, Hattori T, Endo T, Nagashima M, Tanaka J (2014) X-ray phase imaging: from synchrotron to hospital. Philos. Trans R Soc A 372(2010):20130023

Momose A, Yashiro W, Takeda Y, Suzuki Y, Hattori T (2006) Phase tomography by x-ray talbot interferometry for biological imaging. Jpn J Appl Phys 45(6R):5254

Morgan KS, Paganin DM, Siu KKW (2012) X-ray phase imaging with a paper analyzer. Appl Phys Lett 100(12):124102

Morimoto N, Fujino S, Ohshima K, Harada J, Hosoi T, Watanabe H, Shimura T (2014) X-ray phase contrast imaging by compact Talbot-Lau interferometer with a single transmission grating. Opt Lett 39(15):4297–4300

Morimoto N, Fujino S, Ito Y, Yamazaki A, Sano I, Hosoi T, Watanabe H, Shimura T (2015a) Design and demonstration of phase gratings for 2D single grating interferometer. Opt Express 23(23):29399–29412

Morimoto N, Fujino S, Yamazaki A, Ito Y, Hosoi T, Watanabe H, Shimura T (2015b) Two dimensional x-ray phase imaging using single grating interferometer with embedded x-ray targets. Opt Express 23(13):16582–16588

Nguyen D, Xu T (2008) The expanding role of mouse genetics for understanding human biology and disease. Dis Model Mech 1(1):56–66

Noda D, Tanaka M, Shimada K, Hattori T (2007) Fabrication of diffraction grating with high aspect ratio using x-ray lithography technique for x-ray phase imaging. Jpn J Appl Phys 46(2):849–851

Paganin D, Mayo SC, Gureyev TE, Miller PR, Wilkins SW (2002) Simultaneous phase and amplitude extraction from a single defocused image of a homogeneous object. J. Microsc. 206(1):33–40

Paganin DM (2006) Coherent x-ray optics, 2nd edn. Clarenden Press, Oxford, UK

Pfeiffer F, Bech M, Bunk O, Kraft P, Eikenberry EF, Brönnimann C, Grünzweig C, David C (2008) Hard-x-ray dark-field imaging using a grating interferometer. Nat Mater 7:134–137

Pfeiffer F, Bunk O, David C, Bech M, Duc GL, Bravin A, Cloetens P (2007) High-resolution brain tumor visualization using three-dimensional x-ray phase contrast tomography. Phys Med Biol 52(23):6923–6930

Pfeiffer F, David C, Bunk O, Poitry-Yamate C, Grütter R, Müller B, Weitkamp T (2009) High-sensitivity phase-contrast tomography of rat brain in phosphate buffered saline. J Phys Conf Ser 186(1):012046

Pfeiffer F, Herzen J, Willner M, Chabior M, Auweter S, Reiser M, Bamberg F (2013) Grating-based x-ray phase contrast for biomedical imaging applications. Z Med Phys 23(3):176–185

Pfeiffer F, Weitkamp T, Bunk O, David C (2006) Phase retrieval and differential phase-contrast imaging with low-brilliance x-ray sources. Nat Phys 2:258–261

Potdevin G, Malecki A, Biernath T, Bech M, Jensen TH, Feidenhans'l R, Zanette I, Weitkamp T, Kenntner J, Mohr J, Roschger P, Kerschnitzki M, Wagermaier W, Klaushofer K, Fratzl P, Pfeiffer F (2012) X-ray vector radiography for bone micro-architecture diagnostics. Phys Med Biol 57(11):3451–3461

Prade F, Fischer K, Heinz D, Meyer P, Mohr J, Pfeiffer F (2016) Time resolved x-ray dark-field tomography revealing water transport in a fresh cement sample. Sci Rep 6:29108

Prade F, Schaff F, Senck S, Meyer P, Mohr J, Kastner J, Pfeiffer F (2017) Nondestructive characterization of fiber orientation in short fiber reinforced polymer composites with x-ray vector radiography. NDT E Int 86:65–72

Rau C, Bodey A, Storm M, Cipiccia S, Marathe S, Zdora M-C, Zanette I, Wagner U, Batey D, Shi X (2017) Micro- and nano-tomography at the DIAMOND beamline I13L imaging and coherence. Proc SPIE 10391:103910T

Rau C, Storm M, Marathe S, Bodey AJ, Zdora M-C, Cipiccia S, Batey D, Shi X, Schroeder SML, Das G, Loveridge M, Ziesche R, Connolly B (2019) Fast multi-scale imaging using the Beamline I13L at the Diamond Light Source. Proc SPIE 11113:111130P

Rau C, Wagner U, Pešić Z, De Fanis A (2011) Coherent imaging at the Diamond beamline I13. Phys Status Solidi A 208(11):2522–2525

Rayleigh L (1881) XXV. On copying diffraction-gratings, and on some phenomena connected therewith. Philos Mag 5 11(67):196–205

Revol V, Jerjen I, Kottler C, Schütz P, Kaufmann R, Lüthi T, Sennhauser U, Straumann U, Urban C (2011) Sub-pixel porosity revealed by x-ray scatter dark field imaging. J Appl Phys 110(4):044912

Revol V, Kottler C, Kaufmann R, Straumann U, Urban C (2010) Noise analysis of grating-based x-ray differential phase contrast imaging. Rev Sci Instrum 81(7):073709

Revol V, Plank B, Kaufmann R, Kastner J, Kottler C, Neels A (2013) Laminate fibre structure characterisation of carbon fibre-reinforced polymers by x-ray scatter dark field imaging with a grating interferometer. NDT E Int 58:64–71

Reznikova E, Mohr J, Boerner M, Nazmov V, Jakobs PJ (2008) Soft x-ray lithography of high aspect ratio SU8 submicron structures. Microsyst Technol 14:1683–1688

Rizzi J, Mercère P, Idir M, Silva PD, Vincent G, Primot J (2013) X-ray phase contrast imaging and noise evaluation using a single phase grating interferometer. Opt Express 21(14):17340–17351

Rodgers G, Schulz G, Deyhle H, Marathe S, Bikis C, Weitkamp T, Müller B (2018) A quantitative correction for phase wrapping artifacts in hard x-ray grating interferometry. Appl Phys Lett 113(9):093702

Sarapata A, Ruiz-Yaniz M, Zanette I, Rack A, Pfeiffer F, Herzen J (2015) Multi-contrast 3D x-ray imaging of porous and composite materials. Appl Phys Lett 106(15):154102

Scherer K, Willer K, Gromann L, Birnbacher L, Braig E, Grandl S, Sztrókay-Gaul A, Herzen J, Mayr D, Hellerhoff K, Pfeiffer F (2015) Toward clinically compatible phase-contrast mammography. PLoS One 10(6)

Schleede S, Meinel FG, Bech M, Herzen J, Achterhold K, Potdevin G, Malecki A, Adam-Neumair S, Thieme SF, Bamberg F, Nikolaou K, Bohla A, Yildirim AÖ, Loewen R, Gifford M, Ruth R, Eickelberg O, Reiser M, Pfeiffer F (2012) Emphysema diagnosis using x-ray dark-field imaging at a laser-driven compact synchrotron light source. Proc Natl Acad Sci 109(44):17880–17885

Schneider JE, Bhattacharya S (2004) Making the mouse embryo transparent: identifying developmental malformations using magnetic resonance imaging. Birth Defects Res C Embryo Today 72(3):241–249

Schulz G, Götz C, Deyhle H, Müller-Gerbl M, Zanette I, Zdora M-C, Khimchenko A, Thalmann P, Rack A, Müller B (2016) Hierarchical imaging of the human knee. Proc SPIE 9967:99670R

Schulz G, Götz C, Müller-Gerbl M, Zanette I, Zdora M-C, Khimchenko A, Deyhle H, Thalmann P, Müller B (2017a) Multimodal imaging of the human knee down to the cellular level. J Phys Conf Ser 849(1):012026

Schulz G, Thalmann P, Khimchenko A, Müller B (2017b) Grating-based tomography applications in biomedical engineering. Proc SPIE 10391:1039117

Schulz G, Waschkies C, Pfeiffer F, Zanette I, Weitkamp T, David C, Müller B (2012) Multimodal imaging of human cerebellum–merging x-ray phase microtomography, magnetic resonance microscopy and histology. Sci Rep 2:826

Schulz G, Weitkamp T, Zanette I, Pfeiffer F, Beckmann F, David C, Rutishauser S, Reznikova E, Müller B (2010) High-resolution tomographic imaging of a human cerebellum: comparison of absorption and grating-based phase contrast. J R Soc Interface 7(53):1665–1676

Sharpe J (2003) Optical projection tomography as a new tool for studying embryo anatomy. J Anat 202(2):175–181

Stampanoni M, Wang Z, Thüring T, David C, Roessl E, Trippel M, Kubik-Huch RA, Singer G, Hohl MK, Hauser N (2011) The first analysis and clinical evaluation of native breast tissue using differential phase-contrast mammography. Investig Radiol 46(12):801–806

Stutman D, Beck TJ, Carrino JA, Bingham CO (2011) Talbot phase-contrast x-ray imaging for the small joints of the hand. Phys Med Biol 56(17):5697–5720

Sztrókay A, Herzen J, Auweter SD, Liebhardt S, Mayr D, Willner M, Hahn D, Zanette I, Weitkamp T, Hellerhoff K, Pfeiffer F, Reiser MF, Bamberg F (2013) Assessment of grating-based x-ray phase-contrast CT for differentiation of invasive ductal carcinoma and ductal carcinoma in situ in an experimental ex vivo set-up. Eur Radiol 23(2):381–387

Takeda Y, Yashiro W, Suzuki Y, Aoki S, Hattori T, Momose A (2007) X-ray phase imaging with single phase grating. Jpn J Appl Phys 46(3):L89–L91

Talbot H (1836) LXXVI. Facts relating to optical science. No IV. Philos Mag 3 9(56):401–407

Tapfer A, Bech M, Velroyen A, Meiser J, Mohr J, Walter M, Schulz J, Pauwels B, Bruyndonckx P, Liu X, Sasov A, Pfeiffer F (2012) Experimental results from a preclinical x-ray phase-contrast CT scanner. Proc Natl Acad Sci 109(39):15691–15696

Tapfer A, Bech M, Zanette I, Symvoulidis P, Stangl S, Multhoff G, Molls M, Ntziachristos V, Pfeiffer F (2013) Three-dimensional imaging of whole mouse models: comparing nondestructive x-ray phase-contrast micro-CT with cryotome-based planar epi-illumination imaging. J Microsc 253(1):24–30

Thalmann P, Bikis C, Hipp A, Müller B, Hieber SE, Schulz G (2017) Single and double grating-based x-ray microtomography using synchrotron radiation. Appl Phys Lett 110(6):061103

Thüring T, Stampanoni M (2014) Performance and optimization of x-ray grating interferometry. Phil Trans R Soc A 372:2010

Tschernig T, Thrane L, Jørgensen T, Thommes J, Pabst R, Yelbuz TM (2013) An elegant technique for ex vivo imaging in experimental research–optical coherence tomography (OCT). Ann Anat 195(1):25–27

Wang Z, Huang Z, Zhang L, Kang K, Zhu P (2009) Fast x-ray phase-contrast imaging using high resolution detector. IEEE Trans Nucl Sci 56(3):1383–1388

Wang Z, Zhu P, Huang W, Yuan Q, Liu X, Zhang K, Hong Y, Zhang H, Ge X, Gao K, Wu Z (2010) Analysis of polychromaticity effects in x-ray Talbot interferometer. Anal Bioanal Chem 397(6):2137–2141

Weitkamp T, David C, Kottler C, Bunk O, Pfeiffer F (2006) Tomography with grating interferometers at low-brilliance sources. Proc SPIE 6318:63180S

Weitkamp T, Diaz A, David C, Pfeiffer F, Stampanoni M, Cloetens P, Ziegler E (2005) X-ray phase imaging with a grating interferometer. Opt Express 13(16):6296–6304

Wen HH, Bennett EE, Kopace R, Stein AF, Pai V (2010) Single-shot x-ray differential phase-contrast and diffraction imaging using two-dimensional transmission gratings. Opt Lett 35(12):1932–1934

Weninger WJ, Geyer SH, Mohun TJ, Rasskin-Gutman D, Matsui T, Ribeiro I, da Costa LF, Izpisúa-Belmonte JC, Müller GB (2006) High-resolution episcopic microscopy: a rapid technique for high detailed 3D analysis of gene activity in the context of tissue architecture and morphology. Anat Embryol (Berl) 211(3):213–221

Willer K, Fingerle AA, Gromann LB, De Marco F, Herzen J, Achterhold K, Gleich B, Muenzel D, Scherer K, Renz M, Renger B, Kopp F, Kriner F, Fischer F, Braun C, Auweter S, Hellbach K, Reiser MF, Schroeter T, Mohr J et al (2018) X-ray dark-field imaging of the human lung-a feasibility study on a deceased body. PLoS One 13(9):1–12

Wilson R, McGuire C, Mohun T (2016) The DMDD project, Deciphering the mechanisms of developmental disorders: phenotype analysis of embryos from mutant mouse lines. Nucleic Acids Res 44(D1):D855–D861

Wong MD, Dorr AE, Walls JR, Lerch JP, Henkelman RM (2012) A novel 3D mouse embryo atlas based on micro-CT. Development 139(17):3248–3256

Yang F, Griffa M, Bonnin A, Mokso R, Di Bella C, Münch B, Kaufmann R, Lura P (2016) Visualization of water drying in porous materials by x-ray phase contrast imaging. J Microsc 261(1):88–104

Yang F, Prade F, Griffa M, Jerjen I, Di Bella C, Herzen J, Sarapata A, Pfeiffer F, Lura P (2014) Dark-field x-ray imaging of unsaturated water transport in porous materials. Appl Phys Lett 105(15):154105

Yang J, Zong F, Lei Y, Huang J, Liu J, Guo J (2019) X-ray dark-field imaging with a single absorption grating. J Phys D: Appl Phys 52(19):195401

Yaroshenko A, Meinel FG, Bech M, Tapfer A, Velroyen A, Schleede S, Auweter S, Bohla A, Yildirim AÖ, Nikolaou K, Bamberg F, Eickelberg O, Reiser MF, Pfeiffer F (2013) Pulmonary emphysema diagnosis with a preclinical small-animal x-ray dark-field scatter-contrast scanner. Radiology 269(2):427–433

Zamyadi M, Baghdadi L, Lerch JP, Bhattacharya S, Schneider JE, Henkelman RM, Sled JG (2010) Mouse embryonic phenotyping by morphometric analysis of MR images. Physiol Genomics 42A(2):89–95

Zanette I, Weitkamp T, Donath T, Rutishauser S, David C (2010) Two-Dimensional x-ray grating interferometer. Phys Rev Lett 105(24):248102

Zanette I (2011) Interférométrie X à réseaux pour l'imagerie et l'analyse de front d'ondes au synchrotron. Université de Grenoble, Grenoble, France Ph.D. thesis

Zanette I, Bech M, Rack A, Le Duc G, Tafforeau P, David C, Mohr J, Pfeiffer F, Weitkamp T (2012) Trimodal low-dose x-ray tomography. Proc Natl Acad Sci USA 109(26):10199–10204

Zanette I, Lang S, Rack A, Dominietto M, Langer M, Pfeiffer F, Weitkamp T, Müller B (2013a) Holotomography versus x-ray grating interferometry: a comparative study. Appl Phys Lett 103(24):244105

Zanette I, Weitkamp T, Le Duc G, Pfeiffer F (2013b) X-ray grating-based phase tomography for 3D histology. RSC Adv 3(43):19816–19819

Zdora M-C, Vila-Comamala J, Schulz G, Khimchenko A, Hipp A, Cook AC, Dilg D, David C, Grünzweig C, Rau C, Thibault P, Zanette I (2017) X-ray phase microtomography with a single grating for high-throughput investigations of biological tissue. Biomed Opt Express 8(2):1257–1270

Zhu P, Zhang K, Wang Z, Liu Y, Liu X, Wu Z, McDonald SA, Marone F, Stampanoni M (2010) Low-dose, simple, and fast grating-based X-ray phase-contrast imaging. Proc Natl Acad Sci USA 107(31):13576–13581

Chapter 5
Principles and State of the Art of X-ray Speckle-Based Imaging

5.1 Introductory Remarks

X-ray speckle-based imaging is the second major project of this Ph.D. project and its main focus. Conceptually, X-ray grating interferometry and speckle-based imaging are based on the same principle: The sample-induced distortions of a reference pattern are analysed to retrieve information about the properties of the specimen. However, the creation of the reference pattern and the way the information is extracted from the acquired data differ for the two methods. Which of the two techniques is more suitable, depends on the goal of the specific experiment and the given experiment constraints, as discussed in Sect. 2.4.6 of the Conclusions. The differences, advantages and limitations of X-ray grating interferometry and speckle-based imaging are summarised in Appendix A.

This chapter gives an overview of the working principles, state of the art and recent developments of X-ray speckle-based imaging to establish a background and understanding for the following chapters. As part of this Ph.D. project, the author wrote an extensive literature review on X-ray speckle-based imaging. This chapter contains adapted sections of the original review paper published in the special issue "Phase-contrast and dark-field imaging" of the Journal of Imaging (Zdora 2018): M.-C. Zdora, "State of the Art of X-ray Speckle-Based Phase-Contrast and Dark-Field Imaging," J. Imaging **4**(5), 60 (2018), licensed under CC BY 4.0.

5.2 Background and Introduction

X-ray speckle-based imaging (Bérujon et al. 2012a, b; Morgan et al. 2012) is the most recent addition to the group of phase-sensitive imaging methods that have been developed over the last decades. X-ray speckle-based imaging (Bérujon et al. 2012a),

M.-C. Zdora, *X-ray Phase-Contrast Imaging Using Near-Field Speckles*,
Springer Theses, https://doi.org/10.1007/978-3-030-66329-2_5

as well as grating interferometry (Pfeiffer et al. 2008) and analyser-based imaging (Pagot et al. 2003), allow one to reconstruct, in addition to the phase-contrast signal, also the so-called dark-field image, which is a measure of small-angle scattering from features in the sample that cannot be resolved directly (Nesterets 2008; Yashiro et al. 2010). The dark-field signal can deliver valuable complementary information about the specimen and has recently been used increasingly for medical applications (Baum et al. 2015; Ando et al. 2016; Schleede et al. 2012; Meinel et al. 2014; Yaroshenko et al. 2015, 2016; Hellbach et al. 2016; Noël et al. 2017; Gromann et al. 2017; Willer et al. 2018) and materials science (Revol et al. 2011, 2013; Lauridsen et al. 2015; Yang et al. 2014; Prade et al. 2015, 2016; Schaff et al. 2017).

Since it was first proposed a few years ago, X-ray speckle-based phase-contrast and dark-field imaging has drawn significant attention due to its simple, robust and flexible experimental arrangement, cost-effectiveness and relatively low spatial and temporal coherence requirements. These properties also led to the swift translation of the technique to polychromatic laboratory sources (Zanette et al. 2014; Zhou et al. 2015; Wang et al. 2016b) and its extension from 2D projection imaging to 3D tomography implementation (Zanette et al. 2015; Wang et al. 2015a). Furthermore, it has been shown that, in addition to its great potential for phase-contrast and dark-field imaging for the investigation and visualisation of specimens, X-ray near-field speckle can be employed in the field of metrology for highly precise and accurate X-ray optics characterisation, beam phase sensing and beam coherence measurements (Bérujon et al. 2012a, 2014, 2015, 2017; Kashyap et al. 2015; Wang et al. 2017b).

Despite being a relatively recent approach, the rapid development and increasing interest in the X-ray near-field speckle method promise a widespread implementation and expanding range of applications in the near future.

This review provides an overview of the principles and state of the art of the X-ray speckle-based imaging and metrology technique. Starting from the basic concept of X-ray near-field speckle, the different experimental implementations with their advantages, limitations and challenges are discussed, followed by a more detailed description of the proposed dark-field reconstruction approaches. Subsequently, further progress such as the translation to laboratories and the extension to tomographic imaging is illustrated and applications of the technique are shown. In the end, a summary and outlook for anticipated future developments are given and the most recent progresses and applications of the technique are outlined.

5.3 Basic Principles of X-ray Speckle-Based Multimodal Imaging

5.3.1 X-ray Speckle as a Wavefront Marker

A speckle pattern is created when (partially) coherent light is incident on an object consisting of randomly distributed scatterers. The phenomenon of speckle has been explored extensively for laser light (Dainty 1984a; Goodman 2007). Laser speckle is on the one hand often an undesired effect e.g. for laser-based displays (Chellappan et al. 2010; Kuratomi et al. 2010; Pan and Shih 2014) and in coherent optical imaging (Liba et al. 2017; Schmitt et al. 1999). On the other hand, it has found many applications such as speckle imaging in astronomy (Bates 1982; Horch 1995; Dainty 1984b), electronic speckle pattern interferometry for stress, strain and vibration measurements of rough surfaces (Jones and Wykes 1989; Høgmoen and Pedersen 1977; Ennos 1975; Sharp 1989; Yang et al. 2014) and dynamic speckle for the investigation of biological processes (Aizu and Asakura 1991, 1996; Fujii et al. 1985; Zdora et al. 1994; Rabal and Braga 2008; Boas and Dunn 2010). Even the use of laser speckle for eye testing has been demonstrated (Mohon and Rodemann 1973).

The phenomenon of speckle exists in the far- as well as the near-field regime. However, it is important to note that the properties of these two types of speckle patterns are fundamentally different. While far-field speckles are linked to properties of the illuminating beam, its dimensions and wavelength, it has been shown that in the near field the properties of the speckles are closely related to the scattering features themselves and the speckle size is independent of the propagation distance and the energy of the beam (Giglio et al. 2000, 2001, 2004; Brogioli et al. 2002; Cerbino 2007; Gatti et al. 2008; Magatti et al. 2009).

Just over ten years ago, it was demonstrated that the concept of near-field speckle can be directly transferred from the optical to the X-ray regime and the same criteria and properties apply (Cerbino et al. 2008).

Here, it should be noted that the first applications of X-ray near-field speckle were reported already a few years earlier, although not explicitly classifying the observed effects as near-field speckle. Suzuki et al. (2002) and Kitchen et al. (2004) report on speckle produced by lung tissue when imaging small animals using the propagation-based phase-contrast technique. They show that the arising speckle can enhance the visible appearance of the lungs in the acquired images and they explain the occurrence of the speckle by multiple refraction of the X-ray beam in the alveoli of the lung and subsequent free-space propagation. It was later recognised as near-field speckle and explored further for the measurement of lung air volume (Leong et al. 2013). Kim and Lee (2006) demonstrate the characterisation of blood flow by means of cross-correlation of X-ray near-field speckle created by scattering off the blood cells. Furthermore, X-ray near-field speckles have been employed for the determination of size distributions in opaque materials of granular or porous structure (Carnibella et al. 2012, 2013; Leong et al. 2019).

For X-rays, the near-field regime is much more accessible than for visible light due to their short wavelength. Therefore, it is easily possible to record X-ray near-field speckle created by illuminating small randomly scattering structures with an X-ray beam. Thanks to the special properties of near-field speckle, the speckle size can be controlled by the size of the scattering particles and distortions of the speckle pattern upon propagation are only determined by the shape of the wavefront (Cerbino et al. 2008). The above properties make X-ray near-field speckle suitable for use as a wavefront marker for sensing the beam phase. This was soon realised and the first demonstrations of X-ray speckle-based imaging followed (Morgan et al. 2012; Bérujon et al. 2012b).

The principle of X-ray speckle imaging is simple: an object in the X-ray beam will lead to a distortion of the X-ray wavefront, which can be observed as a modulation of the speckle pattern. The object can be a sample to be investigated, but also optical elements in the experimental setup resulting in desired or undesired changes to the beam. By tracking the modulations of the speckle pattern, the differential phase shift of the X-ray wavefront can be obtained, which allows one to determine the refractive properties of the object. Additionally, the transmission and small-angle scattering information of the specimen can be retrieved. Typically, commercially distributed sandpaper, consisting of small silicon carbide grains, or biological filter membranes with µm-sized pores are used as diffusers to produce a speckle pattern. They are available in different grain sizes (Federation of European Producers of Abrasives 2019) and with different pore sizes, respectively, which allows for controlling the speckle size and visibility (Aloisio et al. 2015). Other materials containing small scattering features such as finely ground sand, glass or similar and even simple cardboard could also be used as diffusers.

The speckle size is an important property of the speckle pattern that has an influence on the quality of the reconstructed multimodal images obtained with the speckle-based technique. As the speckles have an irregular, random shape and a distribution of sizes, only an average speckle size can be estimated. This is done, for example, by determining the spatial frequency at which the power spectrum shows a maximum in Fourier space (Cerbino et al. 2008; Zhou et al. 2015; Wang et al. 2016b, c) or via 2D autocorrelation of the speckle pattern in real space (Goodman 1984; Alexander et al. 1994; Hamed 2013; Zanette et al. 2015; Zdora et al. 2015, 2017b; Aloisio et al. 2015). Generally, well-defined, small speckles that can be resolved easily and cover a few pixels in the detector plane are desired. The speckle size sets a limit for the achievable spatial resolution for the single-shot speckle-tracking reconstruction approach (see Sect. 5.4.1), whereas it is not as crucial for implementations based on diffuser stepping (see Sects. 5.4.2 and 5.4.3).

Another important characteristic of the speckle pattern is the speckle visibility or contrast, which is of significant importance for the reconstruction result. A high visibility of the speckle pattern is essential for a successful operation of the reconstruction algorithm. The speckle visibility v is commonly defined in one of the following ways:

$$v = \frac{\sigma_I}{\bar{I}}, \tag{5.1}$$

where σ_I and $\bar{I}$ are the standard deviation and the mean intensity value, respectively, of the speckle pattern in a small region of interest (typically around 150×150 pixels), as e.g. used by Zanette et al. (2015), Zdora et al. (2015), Bérujon and Ziegler (2016), Wang et al. (2016a, b), or:

$$v = \frac{I_{\text{max}} - I_{\text{min}}}{I_{\text{max}} + I_{\text{min}}}, \tag{5.2}$$

where I_{max} and I_{min} are the maximum and minimum intensity values of the speckle pattern in a region of interest, as e.g. used by Zanette et al. (2014), Wang et al. (2015c), Zdora et al. (2017a), or

$$v = \frac{I_{\text{max}} - I_{\text{min}}}{2\bar{I}}, \tag{5.3}$$

as e.g. used by Zhou et al. (2015), Bérujon and Ziegler (2016).

It should be noted that the above ways to quantify the visibility of the pattern will give different values and should not be directly compared to each other. Generally Eqs. (5.2) and (5.3) will result in a higher v. This is due to the fact that in these definitions, which use the maximum and minimum intensity values, extreme outliers strongly influence and artificially increase the measured visibility. Furthermore, the impact of outliers makes the quantification of visibility using Eqs. (5.2) and (5.3) somewhat unstable as the measured visibility values will change significantly for different realisations of the speckle map from the same setup and even for different regions of a single speckle image. Equation (5.1), on the other hand, will give a more reliable and stable result as outliers have less effect on the visibility calculation.

5.3.2 *Differential Phase, Transmission and Dark-Field Signals*

As mentioned above, near-field speckle can be used to obtain information about the phase shift of X-rays in an object. In addition to the phase-contrast signal, the method also allows for the reconstruction of the sample's X-ray transmission and small-angle scattering properties, which can carry valuable complementary information. The principles of image formation for the different signals are outlined in this section.

To keep the explanations and formulas simple for the sake of clarity, in the following we only consider a parallel beam as it is given at synchrotron X-ray sources to a good approximation. However, the concepts discussed here can easily be applied to diverging sources, as mostly encountered in the laboratory (Zanette et al. 2014; Wang et al. 2016b) or for microscopy applications with a magnifying geometry implemented e.g. with a Fresnel zone plate (Bérujon et al. 2013a). The concepts and

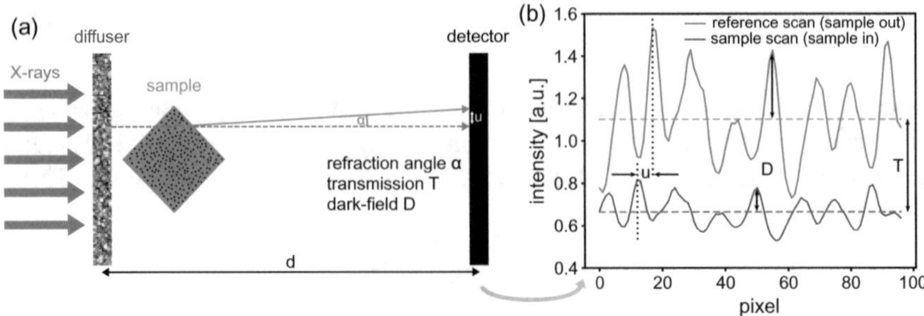

Fig. 5.1 Principle of X-ray speckle-based imaging. **a** Schematic of an X-ray speckle imaging experiment. X-rays illuminate a diffuser, creating a random near-field speckle pattern in the detector plane. When an absorbing, phase-shifting and scattering sample is placed in the beam, the reference pattern is modulated in intensity, position and visibility. **b** Line plot through a few pixels of a reference (blue) and corresponding sample (red) speckle pattern visualising the drop in intensity (dashed horizontal lines) due to absorption $A = (1 - T)$, the displacement u due to refraction of the X-rays by the angle α and the reduction in amplitude (after transmission correction) due to small-angle scattering D. Figure reprinted with permission from Zdora (2018), licensed under CC BY 4.0

reconstruction approaches presented here still hold in these cases, but one needs to take into account the magnification of the speckle pattern and sample. A magnifying geometry allows one to significantly increase the spatial resolution by decreasing the effective pixel size in the sample plane, while maintaining a high angular sensitivity that can be influenced by the distances between the (secondary) source, sample and diffuser.

The basic setup for a speckle imaging experiment is shown in Fig. 5.1a.

An X-ray beam impinges on a diffuser, e.g. a piece of sandpaper, producing a random speckle pattern, the reference interference pattern, in the detector plane. When a sample is inserted into the beam, the speckle pattern is modulated by the presence of the sample and this sample interference pattern is recorded by the detector. The modulation appears in three ways as illustrated in Fig. 5.1b: The speckles are displaced in the horizontal x and vertical y directions by a vector $\mathbf{u} = (u_x, u_y)$ due to refraction in the specimen; the mean intensity changes due to absorption; and the visibility of the pattern, i.e. the amplitude after taking into account the absorption, is reduced due to small-angle scattering from unresolved features. From these effects, the refraction angle $\alpha = (\alpha_x, \alpha_y)$ related to the differential phase shift, the transmission T (or absorption $A = 1 - T$) and the dark-field signal D, respectively, can be retrieved in a quantitative manner. The reconstruction is performed in real space and pixel-by-pixel using different analysis methods, depending on the experimental implementation, see Sect. 5.4.

The displacement of the speckle pattern when a phase-shifting sample is inserted into the beam is visualised in Fig. 5.2 for a phantom sample consisting of a silicon sphere on a wooden toothpick. When looking at a region of interest in the top part of the sphere, it can be observed that the speckles are shifted by several pixels, as can

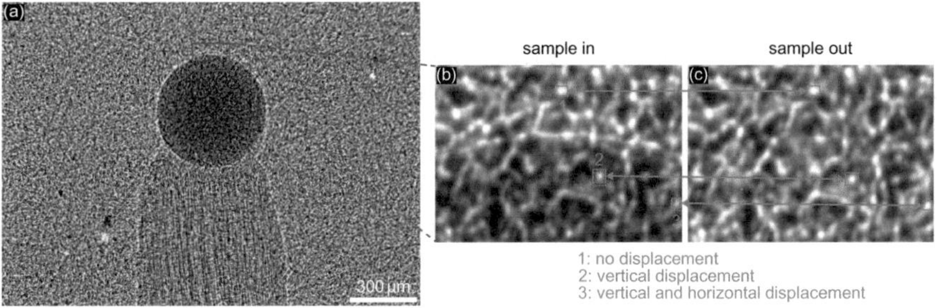

Fig. 5.2 Near-field speckles as a wavefront marker. **a** Speckle pattern created by sheets of sandpaper with a sample (silicon sphere of 480 µm diameter glued onto a wooden toothpick) in the beam. Region of interest in **b** the sample and **c** the reference interference pattern. The dashed orange boxes in (**b**) indicate the corresponding position of the marked speckles in the reference image. A displacement of the speckles in (**b**) with respect to (**c**) can be observed in the sphere (boxes labelled 2, 3), but not in the air background region (box labelled 1). Figure reprinted with permission from Zdora (2018), licensed under CC BY 4.0

be seen from comparing Fig. 5.2b and c, while outside the sphere no displacement is detected.

It should be noted that the order of the diffuser and the sample is not crucial for a setup with a parallel beam in combination with the commonly employed experimental implementations used for imaging applications, which are based on comparing reference and sample interference images. However, for the case of diverging beams, the magnification of speckle and sample should be carefully considered. Furthermore, for acquisition schemes that operate without the use of a reference speckle pattern, as often applied e.g. for the characterisation of X-ray mirrors (see Sect. 5.8.1), the two configurations give different information (see Sect. 5.4.2.3).

The transmission signal is reconstructed from the ratio of the local intensities in the sample and reference interference patterns and is similar to the image that would have been obtained without any optical elements and just the sample in the beam. It shows the absorption of the X-rays in the sample, but typically also contains contributions from edge enhancement effects that arise upon propagation in the near-field regime, as already discussed for the case of X-ray grating interferometry. This image modality is referred to as absorption, transmission or attenuation signal in this and the following chapters.

The dark-field is given as the local ratio of sample and reference visibilities, i.e. it measures the local reduction in amplitude of the speckle pattern after correcting for the transmission. The dark-field signal can be obtained with different analysis approaches, as discussed in Sect. 5.5.

The phase information about the sample is delivered as the differential phase signal, measuring the first derivative of the phase shift induced by the specimen. The direct output from the reconstruction is the displacement **u** of the speckle pattern, which can be converted into the refraction angle α of the X-rays. With the speckle method, the displacement can be measured in the horizontal and vertical directions separately from a single data set and the refraction angle $\alpha = (\alpha_x, \alpha_y)$ is obtained. As already mentioned for X-ray grating interferometry in the previous chapter (see Eq. (4.8)), the refraction angle is directly related to the differential phase shift $(\partial\Phi/\partial x, \partial\Phi/\partial y)$:

$$\begin{aligned} \frac{\partial\Phi}{\partial x} &= \frac{2\pi}{\lambda}\alpha_x \\ \frac{\partial\Phi}{\partial y} &= \frac{2\pi}{\lambda}\alpha_y, \end{aligned} \tag{5.4}$$

where λ is the X-ray wavelength. The information from the differential phase signals in the two orthogonal directions can then be combined via phase integration to obtain the phase shift Φ of the wavefront. This can be done with various methods e.g. Fourier-based approaches (Kottler et al. 2007; Morgan et al. 2012; Frankot and Chellappa 1988; Bérujon et al. 2013a), two-dimensional numerical integration using least-squares minimisation (Harker and O'Leary 2008; Bérujon et al. 2015) or matrix inversion (Bérujon and Ziegler 2016).

5.3.3 *Practical Experimental Considerations*

For a practical implementation of the speckle-based technique, there are a few main points to consider. They can be found throughout this thesis in the current and following chapters, but are summarised in this section.

The method relies on the use of the near-field speckle pattern as a wavefront marker. For this purpose, speckles should be fully resolved with the detector system to achieve a well-defined speckle pattern of good contrast, yet not too large, in particular for the single-shot implementation of speckle imaging (see Sect. 5.4.1), as discussed in Sect. 5.3.1. Hence, one should aim for a speckle size in the range of a few times the effective pixel size. As mentioned in Sect. 5.3.1, a high speckle visibility improves the quality of the reconstructed images.

The size and visibility of the speckles can be controlled by the type of diffuser used to create the interference pattern. As discussed above, in the near field the speckle dimensions are directly related to the size of the scattering features. Commonly used diffuser materials are commercial abrasive paper and biological filter membranes. The size of the silicon carbide grains of the sandpaper or the membrane pores can be chosen depending on the desired speckle size. Generally, any object containing small scattering particles can be used and the exploration of further suitable diffuser materials is subject of ongoing investigations, see Chap. 9, Sect. 9.2.

Regarding the X-ray source, the requirements imposed by the speckle-based technique are moderate. As demonstrated in Sect. 5.6 (and later in this thesis in Chap. 9, Sect. 9.4), X-ray speckle-based imaging can be performed at polychromatic laboratory sources (Zanette et al. 2014) and the demands on the temporal coherence are low (Morgan et al. 2012; Zdora et al. 2015). For the creation of speckle, which is based on interference effects, a certain degree of spatial coherence of the X-ray source is required and the coherence length of the X-ray beam at the diffuser should be larger than the size of the scatterers for an optimum speckle pattern. Reduced spatial coherence leads to a blurring of the interference pattern, which will deteriorate the reconstructed images. It has been demonstrated that micro-focus laboratory sources provide sufficient spatial coherence for successful implementation of X-ray speckle-based imaging (Zanette et al. 2014). Alternatively, it has been proposed to use a random pattern created by absorption rather than interference effects to track the beam wavefront under conditions of low spatial coherence (Wang et al. 2016a, b), see Sect. 5.6.

Another point to consider is setup instabilities during image acquisition. Instabilities in the diffuser or beam position, caused e.g. by mechanical or thermal instabilities of the diffuser mounting or of optical elements in the beam, lead to a displacement of the speckle pattern in the detector plane. This displacement can be corrected for, if the sample is smaller than the field of view, by realigning the sample and reference images to match. For the speckle-scanning methods (see Sect. 5.4.2), this can, however, give rise to artefacts due to the change in effective step size.

Speckle-based phase-contrast imaging is quite robust against intensity fluctuations of the X-ray beam. As long as the speckle visibility is sufficiently high, intensity changes between reference and sample scans do not have a strong impact on the measured refraction signal. They can be observed in the transmission and dark-field images, but corrections can also be performed for these modalities.

5.3.4 Related Techniques

The principle of observing the modulations of an X-ray reference pattern to get information about an object is not new and other established X-ray imaging methods rely on the same phenomenon. For example, X-ray grating interferometry uses a 1D (David et al. 2002; Momose et al. 2003; Weitkamp et al. 2005) or 2D (Jiang et al. 2008; Zanette et al. 2010) phase grating to produce a periodic reference interference pattern in the detector plane. As explained in detail in Chap. 4, typically the fine pattern cannot be resolved directly and a second so-called analyser grating, placed in front of the detector, is used in combination with a phase-stepping or moiré fringe acquisition approach to translate the pattern into measurable intensity variations in the detector pixels. It has been shown that X-ray grating interferometry in phase-stepping implementation can in fact be described as a special case of the speckle-scanning mode in Sect. 5.4.2 (Bérujon et al. 2012a). A direct experimental comparison of speckle-based imaging and grating interferometry can be found by (Kashyap et al. 2016b). The use of a 1D (Bennett et al. 2010) or 2D (Wen et al. 2010) transmission grid pattern for the analysis of the sample-induced changes to the reference pattern instead of a phase grating has also been reported for single-shot imaging with the spatial-harmonic technique. More recently, this approach was employed with a wire mesh and optimised for improving spatial resolution (He et al. 2019). This method has also been translated to a low-coherence conventional X-ray source combined with focussing polycapillary optics (Sun et al. 2019).

The reconstruction processes of the described grating-based methods all exploit the periodic nature of the interference pattern and are based on Fourier transformation. An alternative reconstruction approach is the analysis by cross-correlation in real space. This was demonstrated with a 1D periodic phase grating (Morgan et al. 2011b), a 2D attenuation grid (Morgan et al. 2011a) and a 2D phase grid (Morgan et al. 2013). In contrast to the above mentioned grid method by Wen et al. (Wen et al. 2010), this approach enables quantitative single-shot imaging of objects with features similar or smaller than the grid pitch as well as objects larger than the field of view. Furthermore, in principle, a periodic structure is not necessary for this approach and the grid pattern can be replaced by a random interference pattern. In this sense, the single-grid method can be seen as a precursor to X-ray single-shot speckle tracking, discussed in Sect. 5.4.1, which uses the same analysis concept.

Compared to the approaches using gratings and grids, speckle-based imaging does not suffer from phase-wrapping effects that can occur for periodic reference patterns. However, artefacts can still arise at the edges of the sample or sample features due to the strong distortions of the speckle pattern in this region, particularly caused by the presence of edge enhancement fringes. This can be reduced by a shorter propagation distance, which, however, also affects the sensitivity of the measurement, see Sect. 5.4.4. Furthermore, methods have been investigated to mitigate these artefacts, e.g. by considering the effect of the second derivative of the wavefront (Bérujon et al. 2015) or attempting to eliminate the edge effect from the image before reconstruction (Wang et al. 2017).

As with the single-grid real-space method, large samples with periodic structures of any pitch can be imaged with the speckle-based technique using an easily implemented experimental setup that does not require precise alignment as would be necessary for the two gratings in X-ray grating interferometry. Moreover, the setup for speckle imaging is flexible and the propagation distance can be chosen in the near field without any restrictions that are imposed by the fractional Talbot distances (Cloetens et al. 1997) for grating-based imaging with a phase grating.

Furthermore, the use of commercially available sandpaper is very cost-effective and enables access to the refraction information in the horizontal as well as the vertical directions without the need for elaborate fabrication of 2D structures with small, high-precision period.

From a resources point of view, the reconstruction in real space is more computationally expensive than Fourier-based algorithms. Fast processing can, however, be achieved by GPU (graphics processing unit)-based computation.

For many applications, the advantages of the speckle-based technique outweigh its challenges. In particular, some of the limitations of both speckle- and grating-based imaging can be overcome simultaneously by the recently proposed advanced operational approach developed within the scope of this Ph.D. project, coined unified modulated pattern analysis (UMPA), which can be applied to both random and periodic reference patterns (Zdora et al. 2017b), see Sect. 5.4.3.2 and Chap. 6.

5.4 Experimental Implementations

Several operational modes have been developed for X-ray speckle-based imaging to quantify the modulations of the interference pattern. The most suitable mode for a certain application depends on the desired speed of data acquisition, spatial resolution and signal sensitivity.

5.4.1 *Single-Shot X-ray Speckle-Tracking Mode (XST)*

The first implementations of X-ray speckle-based imaging were demonstrated in single-shot mode, known as X-ray speckle-tracking (XST), which only requires one reference image with the diffuser but without the sample in the beam and one sample image with both the diffuser and the sample in the beam (Bérujon et al. 2012b; Morgan et al. 2012). As mentioned in the previous section, this approach can be seen as a generalisation of single-shot 2D grid-based methods (Morgan et al. 2011a, 2013) to a random interference pattern.

A schematic of the setup for this approach is shown in Fig. 5.3a. One single sample image is acquired, as illustrated in Fig. 5.3b, and one reference image without the sample, see Fig. 5.3c. As explained in Sect. 5.3.2, the X-ray absorption, refraction and small-angle scattering properties of the specimen lead to local changes of the mean intensity, position and visibility of the speckle pattern. Examples of subsets around a pixel of interest in the sample and reference image, respectively, can be seen in Fig. 5.3d, e. The speckle pattern is shifted in Fig. 5.3d compared to the reference in Fig. 5.3e and the intensity is reduced due to the presence of the sample.

In the XST implementation, the local displacement of the pattern is analysed using a windowed zero-normalised cross-correlation (Pan et al. 2009) in real space (Morgan et al. 2012; Bérujon et al. 2012a, b). This means that the refraction signal in each pixel of the image is reconstructed by selecting an analysis window around

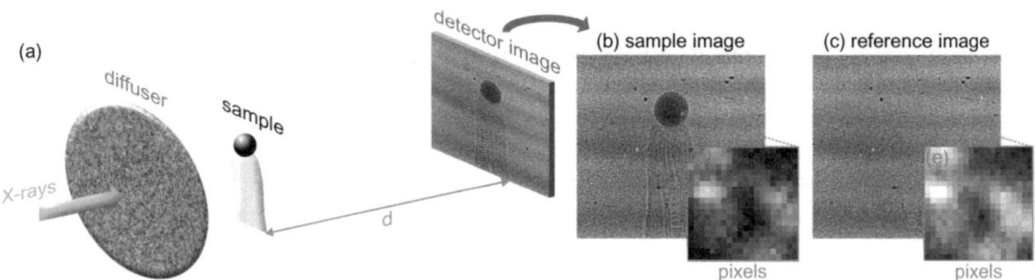

Fig. 5.3 Principle of single-shot XST imaging. **a** Setup for XST imaging. **b** One sample image and **c** one reference image (without the sample) are acquired. **d, e** A subset window larger than the speckle size is selected around each pixel in the image for reconstruction. Figure reprinted with permission from Zdora (2018), licensed under CC BY 4.0

this pixel, as shown in Fig. 5.3d, e, and performing a normalised cross-correlation between the reference and the corresponding sample window.

The cross-correlation coefficient γ between reference and sample windows is given by (Bérujon et al. 2012b):

$$\gamma = \sum_{i=-M}^{M} \sum_{j=-M}^{M} \left[\frac{[I_0(x_i, y_j) - \bar{I}_0][I(x_i', y_j') - \bar{I}]}{\Delta I_0 \Delta I} \right]. \tag{5.5}$$

Here, $I_0(x_i, y_j)$ describes the value in a pixel of the subset of the reference speckle pattern centred at (x_0, y_0) and $I(x_i', y_j')$ the value in a corresponding subset of the sample speckle pattern centred at (x_0', y_0'). The sums run over all pixels in the analysis window of size $2M + 1$. $\bar{I}$ and $\bar{I}_0$ are the mean values and ΔI and ΔI_0 the standard deviations of the sample and reference patterns in the window. If only a rigid translation of the subset is considered, we can write $x_i' = x_i + u_x$ and $y_i' = y_i + u_y$ and the location of the cross-correlation peak $\gamma^{\max}$ corresponds to the local displacement (u_x, u_y) of the speckle pattern in the two orthogonal directions. This can then be converted to a refraction angle signal (α_x, α_y) by geometrical considerations (in small-angle approximation):

$$\begin{aligned} \alpha_x &= \frac{u_x \cdot p_{\text{eff}}}{d} \\ \alpha_y &= \frac{u_y \cdot p_{\text{eff}}}{d}, \end{aligned} \tag{5.6}$$

where p_{eff} is the effective pixel size in the detector plane and d the propagation distance.[1] The analysis window slides across the whole image and refraction, transmission and dark-field signals are obtained locally for each pixel.

The transmission signal can be calculated from the ratio of the mean intensities in the sample and reference windows:

$$T = \bar{I}/\bar{I}_0. \tag{5.7}$$

The dark-field image is typically retrieved as the ratio of the sample and reference visibilities, which can be quantified for each pixel as the ratio of the standard deviation divided by the mean intensity in the respective sample and reference analysis windows (Bérujon et al. 2012a):

$$D = \frac{\Delta I/\bar{I}}{\Delta I_0/\bar{I}_0} = \frac{1}{T}\frac{\Delta I}{\Delta I_0}. \tag{5.8}$$

[1]The propagation distance d corresponds to the sample-detector distance for the configuration in Fig. 5.3a where the diffuser is placed upstream of the sample, but to the diffuser-detector distance if the diffuser is placed downstream of the sample.

It has also been proposed that alternatively the reduction of the cross-correlation peak value can be taken as a measure for the dark-field signal (Wang et al. 2015a), see Sect. 5.5.

A different approach for the image reconstruction of XST data was introduced a bit later (Zanette et al. 2014). The idea is based on a physical model of the speckle interference pattern in the detector plane that takes into account the modulations of the pattern by the presence of the sample. For a certain pixel (x, y), the sample interference pattern I can be described in terms of the reference interference pattern I_0, but modulated in intensity, amplitude and position by the properties of the sample:

$$I(x, y) = T(x, y)\left[\bar{I}_0 + D(x, y)\left(I_0(x + u_x, y + u_y) - \bar{I}_0\right)\right]. \tag{5.9}$$

Here, $\bar{I}_0$ is the mean intensity of the reference pattern and $T(x, y)$ the local transmission through the sample reducing the intensity of the speckle pattern. The amplitude $\left(I_0(x + u_x, y + u_y) - \bar{I}_0\right)$ of the reference pattern is reduced by the factor $D(x, y)$ corresponding to the local dark-field signal. The refraction in the sample is taken into account by the quantities u_x, u_y, describing the displacement of the interference pattern in the horizontal and vertical directions, respectively. For the reconstruction of the multimodal signals, a windowed least-squares minimisation between the model in Eq. (5.9) and the measured sample speckle pattern I is performed. The minimisation procedure is conducted pixel-by-pixel using the sum over the pixels in an analysis window w around the pixel of interest (x_0, y_0):

$$\begin{aligned}\mathcal{L} = \sum_{i=-M}^{M} \sum_{j=-M}^{M} w(x_i, y_j) \big\{ I(x_i, y_j) - T(x_i, y_j) & \\ \left[\bar{I}_0 + D(x_i, y_j)\left(I_0(x_i + u_x, y_j + u_y) - \bar{I}_0\right)\right]\big\}^2 &\end{aligned} \tag{5.10}$$

Minimisation of the function $\mathcal{L}$ delivers the multimodal image signals u_x, u_y, T and D. From the speckle displacement (u_x, u_y), the refraction angle (α_x, α_y) can be obtained via Eq. (5.6).

The extent of the analysis window w should be larger than the average speckle size to achieve a good reconstruction result. Different window types can be used from a simple square window with equal weighting for all pixels to Hamming or Tukey (tapered cosine) windows that give less weight to pixels at the edges. The latter can often lead to improved results with reduced artefacts.

Commonly, in the XST analysis approach, as outlined above, only a rigid translation of the speckle pattern is considered and higher-order modulations of the sample subset compared to the reference subset are neglected. However, it has been shown that considering the distortions of the analysis subset can improve the robustness and accuracy of the reconstruction algorithm and, furthermore, delivers additional information e.g. on the local curvature of the X-ray wavefront (Bérujon et al. 2015). The coefficients of the higher-order distortions can be obtained from a minimisation approach after determining the rigid translation of the subset. Consideration of higher-order subset distortions can be beneficial e.g. for analysing focussing samples such as X-ray refractive lenses. The information from higher-order distortions can in this case help to reduce artefacts arising from the demagnification of the reference pattern in the lens.

The main advantage of the XST implementation is the fast image acquisition, which makes it suitable for dynamic imaging and in-vivo studies. It was demonstrated that a successful reconstruction can be achieved from a single image with sub-μs exposure time at a synchrotron source (Aloisio et al. 2015). Furthermore, XST does not require any special equipment, such as high-accuracy, high-precision scanning stages that are needed for the speckle-scanning method discussed in the next section. As it is essential that the position of the diffuser is identical for the reference and the sample scan, some stability of the setup is required. However, a slight displacement of the speckle pattern caused by drift or movement of the diffuser or beam instabilities can be corrected for by realigning the reference and sample images in the empty space background, e.g. via cross-correlation, as discussed in Sect. 5.3.3.

The main drawback of the single-shot approach is the limited spatial resolution that is given by twice the FWHM of the size of the analysis window, which needs to be larger than the speckle size to reliably track the speckle displacement from a single diffuser position. The ultimate limit for the resolution of this operational mode is hence the speckle size.

5.4.2 X-ray Speckle-Scanning Modes (XSS)

For applications where high resolution is more important than image acquisition speed, the speckle-scanning (XSS) mode, also called speckle-stepping mode, is more suitable. It was proposed shortly after the single-shot approach and can be considered as a generalised version of X-ray grating interferometry in phase-stepping mode (Bérujon et al. 2012a). However, the analysis of speckle-scanning data is performed in real space, as opposed to the Fourier analysis for X-ray grating interferometry. The speckle-scanning mode has been demonstrated in two experimental ways (Zdora et al. 2017a): 2D and 1D scanning, which are described in the following.

5.4.2.1 2D Scanning (2D XSS)

The first speckle-stepping implementation was reported for scanning of the diffuser in both the horizontal and the vertical direction in small equidistant steps (Bérujon et al. 2012a), as illustrated in Fig. 5.4a. This way, a signal is recorded at each diffuser position, with and without the sample in the beam, see Fig. 5.4b, c. A sample and a reference 2D array, which contain the intensities at each diffuser step with and without the sample in the beam, respectively, are obtained for each pixel in the detector plane. Examples of these arrays for one pixel are shown in Fig. 5.4d, e. The reconstruction can then be performed pixel-wise and effectively in the sample

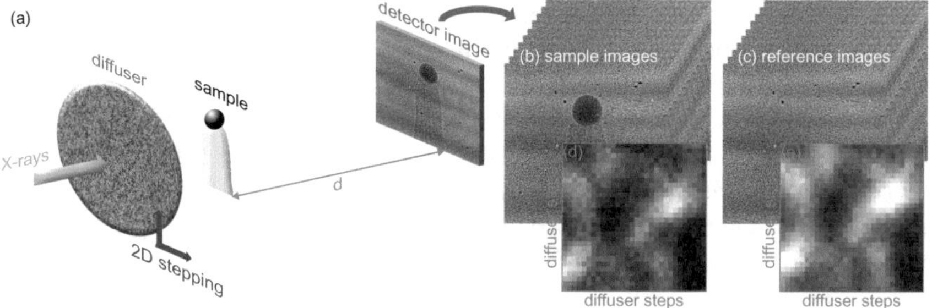

Fig. 5.4 Principle of multi-frame 2D XSS imaging. **a** Setup for 2D XSS imaging. The diffuser is stepped in two directions in small equidistant steps on a regular grid. **b** A sample image and **c** a reference image are acquired at each of the several hundred diffuser positions. **d, e** For each pixel, a 2D array with the signal at each diffuser position is obtained, enabling a high-resolution pixel-wise reconstruction. Figure reprinted with permission from Zdora (2018), licensed under CC BY 4.0

plane[2] by zero-normalised cross-correlation (see Eq. (5.5)) of these sample and reference arrays. The retrieval of the three complementary image signals—transmission, refraction and dark field—is conducted analogous to the single-shot case, but with the analysis arrays built from the signals at different diffuser positions rather than different pixels of an analysis window. The displacement (u_x, u_y) of the speckle pattern between reference and sample arrays is now given in units of diffuser steps and can be converted to a refraction angle signal in the horizontal and vertical direction separately:

$$\begin{aligned} \alpha_x &= \frac{u_x \cdot s}{d} \\ \alpha_y &= \frac{u_y \cdot s}{d}, \end{aligned} \tag{5.11}$$

where s is the diffuser step size[3] and d the propagation distance.

The sensitivity of the refraction angle measurement critically depends on the diffuser step size (see Sect. 5.4.4) and typically small steps in the range of the pixel size or smaller are chosen. For step sizes much smaller than the effective pixel size, it is important to ensure that the intensity variation between subsequent steps is sufficient. For a certain experimental arrangement, this sets the limit of the achievable angular sensitivity. Typically, the diffuser is scanned on a grid of several tens of steps across in each direction, adding up to a total number of hundreds of frames for the reconstruction of one image, which makes this approach unsuitable for fast imaging applications. Furthermore, due to the small regular step sizes, XSS requires delicate

[2]The reconstruction is effectively performed in the sample plane if the sample is placed downstream of the diffuser, but in the diffuser plane if the sample is placed upstream of the diffuser.

[3]To be precise: s is the diffuser step size in the sample plane in the case that the sample is placed downstream of the diffuser, but the step size in the diffuser plane in the case that the sample is placed upstream of the diffuser.

and costly high-accuracy, high-precision scanning stages, which should be aligned carefully with the beam direction to ensure equal step sizes in both directions.

Compared to the XST approach, XSS is significantly more sensitive to instabilities of the setup. The technique requires the speckle pattern to be shifted by a known constant step, as it defines the size of the pixels in the cross-correlated arrays that are built from the intensity values at the different diffuser positions, see Fig. 5.4d, e. Deviations from the desired position of the speckle pattern, caused by instabilities of the beam or setup (see Sect. 5.3.3), cannot be corrected for, as this would alter the effective step size.

In contrast to the XST approach, however, where several pixels in an analysis window contribute to the signal reconstruction of one pixel, the stepping mode allows a real pixel-wise analysis. This enables a much higher resolution down to the pixel size, which is the main advantage of the XSS technique. In practice, the point-spread function of the detector and other factors might deteriorate the resolution.

5.4.2.2 1D Scanning (1D XSS)

To reduce the number of acquired images, it was proposed that two orthogonal 1D scans could be performed instead of a full 2D grid scan in cases of small speckle displacement, i.e. for short propagation distances or moderately phase-shifting samples (Bérujon et al. 2012a).

This was simplified further by taking only one single 1D scan to obtain the 2D refraction information (Wang et al. 2016b). In this mode, here called 1D XSS, the diffuser is stepped only in one direction—horizontally or vertically—in equidistant steps that are much smaller than the average speckle size and on the order of the pixel size. This is done with and without the sample in the beam, see Fig. 5.5b, c. To be able to track the 2D speckles, a few nearby pixels are selected in the orthogonal direction

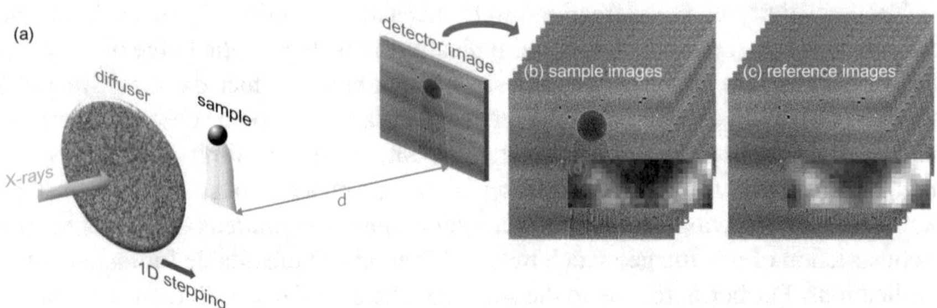

Fig. 5.5 Principle of multi-frame 1D XSS imaging. **a** Setup for 1D XSS imaging. The diffuser is stepped in only one direction in small equidistant steps. **b** A sample image and **c** a reference image are acquired at each of the several tens of diffuser positions. **d, e** For each pixel, a 2D array is built from the signal at each diffuser position in the pixel of interest and a few surrounding pixels in the direction that is not scanned. Figure reprinted with permission from Zdora (2018), licensed under CC BY 4.0

that is not scanned. For each pixel to be reconstructed, one obtains a signal at each diffuser step and takes a 1D window of a few pixels in the other direction, giving a 2D array per pixel. A cross-correlation is then performed between the sample and the reference arrays constructed this way. An example of the signal for one pixel for the case of scanning in the horizontal direction is shown with and without the sample in Fig. 5.5d, e, respectively. Typically, several tens of steps are taken in one direction and only a few pixels are selected in the orthogonal direction. Effectively, the 1D XSS approach can be considered a hybrid between the 2D XSS and the XST cases.

The reconstruction of the transmission and dark-field signals is performed the same way as in the 2D XSS and XST implementations by looking at the local changes in mean intensity and visibility within the analysis arrays for each pixel. The location of the cross-correlation peak gives the displacement (u_x, u_y) of the speckle pattern in the two directions. However, the two axes of the analysis array are not the same and the displacement is given in units of diffuser steps in the scanning direction and in units of effective pixel size in the orthogonal direction. For horizontal scanning, the conversion from the measured displacement to refraction angle signal is hence given by:

$$\begin{aligned} \alpha_x &= \frac{u_x \cdot s}{d} \\ \alpha_y &= \frac{u_y \cdot p_{\text{eff}}}{d}, \end{aligned} \tag{5.12}$$

where d is the propagation distance, s the diffuser step size and p_{eff} the effective pixel size.

The 1D XSS approach allows for a significantly faster image acquisition than 2D XSS. However, still several tens of frames have to be acquired for a successful image reconstruction. For most experimental realisations of 1D XSS found in the literature, a set of 60 diffuser steps was used; the minimum number reported is 40 steps (Wang et al. 2016a, b). The more steps are taken, the smaller the required number of pixels in the other direction and vice versa. This makes the approach more flexible than 2D XSS, while allowing a better spatial resolution than XST. On the other hand, the spatial resolution is reduced compared to 2D XSS as several surrounding pixels contribute to the reconstruction of the signal in one pixel. Furthermore, the sensitivity of the reconstructed images is not the same in the two orthogonal directions (Zdora et al. 2017a). The sensitivity is generally lower along the non-scanned direction as it is dependent on the effective pixel size, whereas it is determined by the step size for the scanning direction. 1D XSS could be used effectively in cases with a preferred direction of interest, for which the signals of the sample in the other direction are not crucial. The additional refraction information in the orthogonal direction can still be used to improve and reduce artefacts in the integrated phase that arise when the differential phase signal is only available in one direction.

5.4.2.3 Scanning with Self-correlation Analysis

For imaging purposes, 2D XSS or 1D XSS are typically performed in differential mode relying on the acquisition of reference and sample scans, as explained in the previous sections. However, for metrology applications (see Sect. 5.8.1), in particular for the characterisation of strongly focussing optics, often another mode is used, which is sometimes called self-correlation mode (Wang et al. 2017b). In this implementation, the local curvature, i.e. the second derivative of the wavefront, is measured as opposed to the first derivative obtained from the commonly employed differential mode (Bérujon et al. 2012a).

For the self-correlation mode, image acquisition is performed by scanning the diffuser following the same schemes as for the common XSS methods, but no reference images are taken. The correlation procedure is then applied to the signals recorded in two nearby pixels during the same image acquisition. The two pixels that are separated in the detector plane by a pixels in the x-direction and b pixels in the y-direction (i.e. by the absolute distances ap_{eff} and bp_{eff}, respectively) will see the same signal, but at different times depending on the diffuser step size s and the X-ray wavefront. Cross-correlation between the signals in the two pixels gives the delay $(\chi_x, \chi_y) = (u_x s, u_y s)$ for the observation of the same signal in the pixels, where (u_x, u_y) is the position of the maximum of the correlation coefficient. From geometrical considerations, one can approximate the local radius R of the wavefront in the x- and y-directions as follows (Bérujon et al. 2012a):

$$\begin{aligned} R_x &= \frac{d \cdot ap_{\text{eff}}}{ap_{\text{eff}} - \chi_x} = \frac{d \cdot ap_{\text{eff}}}{ap_{\text{eff}} - u_x s} \\ R_y &= \frac{d \cdot bp_{\text{eff}}}{bp_{\text{eff}} - \chi_y} = \frac{d \cdot bp_{\text{eff}}}{bp_{\text{eff}} - u_y s}. \end{aligned} \tag{5.13}$$

Here, d is the propagation distance and p_{eff} the effective pixel size. For small angles, the local radius R of the wavefront W is directly related to its local curvature or second derivative, which is in turn proportional to the second derivative of the beam phase Φ:

$$\begin{aligned} \frac{1}{R_x} &\approx \frac{\partial^2 W}{\partial x^2} = \frac{\lambda}{2\pi} \frac{\partial^2 \Phi}{\partial x^2} \\ \frac{1}{R_y} &\approx \frac{\partial^2 W}{\partial y^2} = \frac{\lambda}{2\pi} \frac{\partial^2 \Phi}{\partial y^2}, \end{aligned} \tag{5.14}$$

where λ is the wavelength.

The self-correlation analysis approach can deliver different kinds of information, depending on the location of the diffuser: It gives the wavefront distortions induced by the object under study, if the diffuser is mounted upstream of the sample, or the wavefront distortions caused by all optics and components in the beam upstream, if the diffuser is mounted downstream of the sample.

The self-correlation analysis is often used for metrology applications, in particular for the characterisation of X-ray mirrors, see Sect. 5.8.1. Self-correlation analysis of 1D scanning data can be applied to obtain the 1D slope of mirrors and in this case not the signal delay between two different single pixels but between two rows[4] of the detector image is considered. A 2D array is built for each row by stacking the signals in that row at each diffuser position. The correlation procedure is applied to the arrays of two neighbouring rows delivering the delay signal that can then be converted to the local wavefront curvature along one direction (Bérujon et al. 2014). The retrieval of the 2D wavefront curvature can be achieved by scanning the diffuser in two separate orthogonal 1D scans along the vertical and the horizontal directions. Furthermore, it has been shown that the 2D information on the wavefront curvature can also be accessed from only a single 1D scan by looking at the signal delay in two neighbouring pixels along the scanning direction and at the same time noting the displacement of the speckle pattern in the direction that is not scanned (Wang et al. 2015b; Kashyap et al. 2016c). One should be aware that for this approach, the displacement of the speckles along the non-scanned axis needs to be small and the sensitivity in this direction is typically lower than in the scanning direction, similar to the conventional 1D XSS mode. The technique can be useful to reduce artefacts in the reconstructed slope profile of a mirror arising when only the 1D speckle displacement is considered.

5.4.2.4 2D Scanning with Sparse Sampling

The 1D XSS approach requires significantly less diffuser steps than 2D XSS, which, however, comes at the cost of a reduced sensitivity in the direction orthogonal to the scanning axis. Moreover, still several dozens of diffuser positions are needed for a successful reconstruction. Recently, a stepping scheme has been proposed that uses the concept of 2D XSS, but with a sparse sampling for the sample scan (Bérujon and Ziegler 2017). The acquisition of the reference patterns without the specimen in the beam is performed with a 2D raster scan of the diffuser as for classic 2D XSS, while sample images are taken only at every n-th point of the diffuser scanning grid used for the reference scan. The missing sample images are then obtained via interpolation. It was demonstrated that a coarse scanning grid of only 5×5 steps for the sample acquisition and subsequent interpolation to 25×25 step arrays is sufficient to obtain images of good quality comparable to the full 2D XSS data. This significantly reduces the scan time and dose to the sample. However, a four-fold reduction of the sensitivity has been reported for the 5×5-step sparse scanning scheme compared to a conventional 25×25-step 2D XSS scan (Bérujon and Ziegler 2017).

[4]The signal delay between two rows is analysed if the diffuser is scanned vertically. For the case of scanning the diffuser horizontally, the signal delay between two columns is determined.

5.4.2.5 Analysis of the Scattering Distribution

A conceptually different approach for retrieving the information about the sample from 2D diffuser scanning data is the recovery of the ultrasmall-angle scattering distribution of the sample (Bérujon and Ziegler 2015). This was inspired by an analogue reconstruction process introduced for 1D and 2D grating interferometry (Modregger et al. 2012, 2014). In general, the sample interference pattern can be expressed as the convolution of the reference signal without the sample in the beam and the optical transfer function of the specimen, equivalent to the sample scattering distribution. The reconstruction approach relies on recovering the optical transfer function using iterative methods like the Richardson-Lucy deconvolution (Richardson 1972; Lucy 1974). Modregger et al. (2012, 2014) claim that the various moments of the scattering distribution carry different information about the sample. The zeroth moment is equivalent to the transmission through the specimen, while the first order moments can be interpreted as the differential phase signals in the two directions. The authors suggest that the second moments give directional information about the small-angle scattering strength, equivalent to the common dark-field signal, and the third and fourth moments quantify the skewness and kurtosis of the distribution, respectively.[5]

Although this approach gives a large number of contrast channels, some of which are not accessible with the other reconstruction methods, and has been shown to have improved angular sensitivity (Modregger et al. 2012, 2014), the cumbersome and computationally expensive reconstruction procedure has so far impeded its wider implementation for speckle-based imaging.

5.4.3 Acquisition with Random Diffuser Positions

As discussed in the previous sections, there are some crucial limitations of the classic implementations of speckle-based imaging in the single-shot XST and XSS modes. While the XST approach is quite limited in spatial resolution, the XSS modes require a large number of acquired frames and small equidistant steps. The 1D XSS scheme results in different sensitivities for the horizontal and vertical directions and a reduced resolution. Recently, efforts have been made to develop experimental implementations that provide a trade-off between the advantages and drawbacks of the two classic modes XST and XSS.

Three approaches for this purpose have been proposed, namely the speckle-vector tracking technique (XSVT) (Bérujon and Ziegler 2015), the mixed XSVT approaches (Bérujon and Ziegler 2016, 2017) and the unified modulated pattern analysis (UMPA)

[5]The kurtosis describes the weight of the tails compared to the central peak of the sample's scattering distribution. The skewness is a measure of the asymmetry of the distribution. The physical meanings of the two quantities are not yet identified so far. It has been suggested that in the case of powder-like samples the kurtosis of the small-angle scattering distribution relates to the typical structure sizes (Modregger et al. 2017).

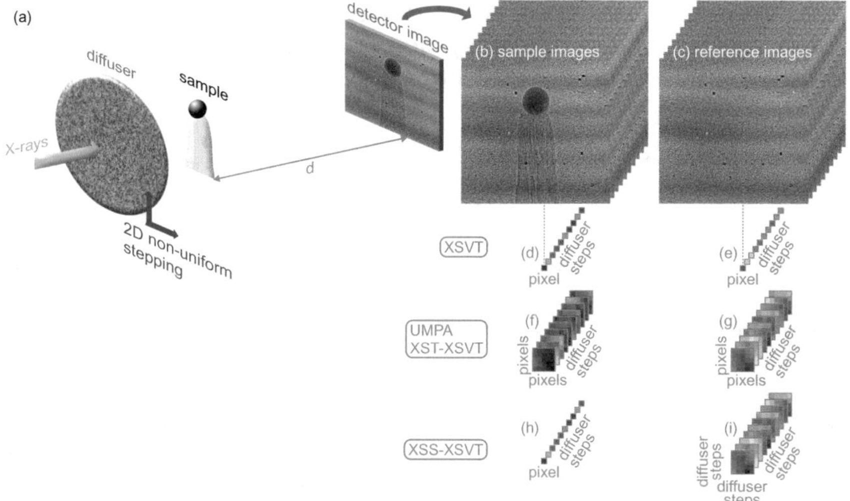

Fig. 5.6 Principle of advanced modes for multi-frame speckle imaging. **a** Setup for speckle imaging with a random scan pattern of non-equidistant, large steps (XSVT and UMPA). The diffuser is stepped in two directions in a few large random steps. **b** A sample image and **c** a reference image are acquired at each of the diffuser positions. For XSVT, the analysis is performed for each pixel by comparing the **d** sample and **e** reference vectors built from the intensity in a single pixel at each diffuser step. For the UMPA and XST-XSVT approaches, a small subset window is chosen around the pixel under consideration in each of the **f** sample and **g** reference images at the different diffuser positions, allowing one to reduce the number of steps and improving the reconstruction result. For the XSS-XSVT reconstruction, a **h** sample vector is built as for the XSVT case, while for **i** the reference vector, the diffuser is scanned in small equidistant steps around each of the initial diffuser positions. Figure reprinted with permission from Zdora (2018), licensed under CC BY 4.0

(Zdora et al. 2017b). They all rely on taking sample and reference scans at several different diffuser positions, as shown in Fig. 5.6a. In contrast to the XSS mode, in the case of the advanced methods, the diffuser positions can be randomly chosen and step sizes should be significantly larger than the speckle size. This allows for the use of less accurate, less costly stepping stages. They, however, still need to be precise and repeatable to ensure that sample and reference images are taken at the same diffuser positions. The number of required steps is much lower than for the XSS case, leading to shorter scan times.

5.4.3.1 X-ray Speckle-Vector Tracking (XSVT) and Mixed XSVT Approaches

The XSVT method considers the signal in each pixel to be a vector made up from the measured intensities at all N diffuser positions. A sample vector $\mathbf{i_r} = (i_{\mathbf{r},1}, \ldots, i_{\mathbf{r},N})$ (see Fig. 5.6d) and a reference vector $\mathbf{o_r} = (o_{\mathbf{r},1}, \ldots, o_{\mathbf{r},N})$ (see Fig. 5.6e) can be created for each pixel $\mathbf{r} = (x, y)$. For the reconstruction of the multimodal images, a zero-normalised cross-correlation is performed between the reference and sample speckle vectors (Bérujon and Ziegler 2015):

$$\gamma(\mathbf{i}_\mathbf{r}, \mathbf{o}_{\mathbf{r+h}}) = \frac{\sum_{k=1}^{N}(i_{\mathbf{r},k} - \bar{i}_\mathbf{r})(o_{\mathbf{r+h},k} - \bar{o}_{\mathbf{r+h}})}{\sqrt{\sum_{k=1}^{N}(i_{\mathbf{r},k} - \bar{i}_\mathbf{r})^2 \sum_{k=1}^{N}(o_{\mathbf{r+h},k} - \bar{o}_{\mathbf{r+h}})^2}}, \quad (5.15)$$

where $\mathbf{h}$ is a small displacement and $\bar{i}_\mathbf{r}$ and $\bar{o}_{\mathbf{r+h}}$ are the mean values of the sample and reference vectors, respectively, for the pixel $\mathbf{r}$. The position of the correlation peak $\mathbf{u} = \arg\max_\mathbf{h} \ \gamma(\mathbf{i}_\mathbf{r}, \mathbf{o}_{\mathbf{r+h}})$ gives the displacement $\mathbf{u} = (u_x, u_y)$ of the speckle pattern due to refraction in the sample, which can then be converted into a refraction angle signal using Eq. (5.6). Here, arg max stands for "arguments of the maxima" and $\arg\max_\mathbf{h} \ \gamma(\mathbf{i}_\mathbf{r}, \mathbf{o}_{\mathbf{r+h}})$ corresponds to the displacement $\mathbf{h}$ for which $\gamma(\mathbf{i}_\mathbf{r}, \mathbf{o}_{\mathbf{r+h}})$ reaches its maximum value. The transmission signal of a pixel can be obtained from the ratio of the mean intensities of the sample and reference speckle vectors. The dark-field signal is retrieved from the ratio of the standard deviations of the sample and reference speckle vectors normalised by the transmission.

To allow reducing the number of diffuser steps further and to make the method more flexible, a mixed XST-XSVT approach was proposed (Bérujon and Ziegler 2016). The principle of image reconstruction based on the correlation of speckle vectors is the same. However, for each of the acquired images, at the same time, a small analysis window is chosen around the pixel under consideration (similar to XST), as illustrated in Fig. 5.6f, g. The information from the surrounding pixels in the analysis window at the different diffuser positions contributes to the speckle vector of each pixel and is included in the correlation analysis to obtain the image signals. This is done analogous to Eq. (5.15), but the sum now runs not only over all diffuser positions, but also over all pixels in the window.

Recently, also a mixed XSS-XSVT approach has been proposed (Bérujon and Ziegler 2017). The acquisition of the sample interference patterns follows the normal XSVT scheme and sample images are taken at several random diffuser positions building up a sample vector for each pixel to be reconstructed, see Fig. 5.6h. For the reference images, the diffuser is additionally scanned in small regular steps around each of the diffuser positions of the sample scan. The recorded signals for all of the positions can be arranged in a reference vector for each pixel, see Fig. 5.6i. The reconstruction of the multimodal images is performed by the correlation of sample and reference vectors. The displacement of the speckles is obtained from the location of the cross-correlation peak and the refraction angle can be calculated via Eq. (5.11).

5.4.3.2 Unified Modulated Pattern Analysis (UMPA)

The same acquisition scheme as for the combined XST-XSVT approach is used for the recently proposed UMPA method, which is presented in more detail in Chap. 6. Images at a few different random diffuser positions are recorded and a small analysis window around the pixel of interest is applied, as shown in Fig. 5.6f, g. However, UMPA proposes a different concept for data analysis that is based on a least-squares minimisation between a model and the measurement of the sample interference pattern summed over all diffuser positions (Zdora et al. 2017b). This model was first proposed for the XST mode (Zanette et al. 2014), see Eq. (5.9). For the UMPA approach, the model in Eq. (5.9) holds for each interference pattern at diffuser position n. In the least-squares minimisation process (see Eq. (5.10)) of the function $\mathcal{L}$, the sum now runs not only over all pixels in the analysis window, but also over all diffuser positions n:

$$\mathcal{L} = \sum_{n} \sum_{i=-M}^{M} \sum_{j=-M}^{M} w(x_i, y_j) \left\{ I_n(x_i, y_j) - T(x_i, y_j) \left[\bar{I}_0 + D(x_i, y_j) \left(I_{0n}(x_i + u_x, y_j + u_y) - \bar{I}_0 \right) \right] \right\}^2 . \tag{5.16}$$

Here, $w(x_i, y_j)$ is the window function, which typically has a much smaller extent than for the XST case; $I_n(x_i, y_j)$ and $I_{0n}(x_i, y_j)$ are the intensities in pixel (x_i, y_j) at step n of the diffuser with and without sample, respectively; and $\bar{I}_0$ is the mean intensity of the reference pattern over all diffuser positions. The local speckle displacement (u_x, u_y), transmission T and dark-field signal D are obtained directly from the reconstruction and the refraction angle can be calculated from the displacement using Eq. (5.6).

For both the XST-XSVT and the UMPA approaches, the use of an analysis window around the pixel to be reconstructed allows one to significantly reduce the number of acquired frames by adding information from the surrounding pixels. The size of the analysis window is typically only a few pixels across, resulting in a moderate reduction in spatial resolution. However, the choice of the number of steps and window size are always coupled. The exact parameter combinations depend on the focus of the specific experiment, in particular on the desired spatial resolution and sensitivity. Larger window sizes generally allow for the use of fewer diffuser steps, but lead to a reduced spatial resolution, while a larger number of diffuser positions enables high-resolution imaging with a small analysis window at the cost of long acquisition times and a high dose to the sample. Therefore, these approaches can be seen as a trade-off between the XST and XSS modes and they allow flexible tuning of the reconstruction result. This will aid the straightforward implementation of speckle-based technique for a wider range of applications with different requirements on scan times, spatial resolution and signal sensitivity, also at laboratory sources.

Furthermore, it has been demonstrated that UMPA can be successfully applied not only to random speckle patterns, but also periodic reference patterns such as the Talbot self-image created by a beam-splitter phase grating (Zdora et al. 2017b), see also Sect. 6.3.5, Chap. 6. Advantages of the UMPA approach over other phase-sensitive imaging methods making use of periodic wavefront markers, such as grating interferometry, include the simplicity of the experimental setup and data acquisition, the absence of phase-wrapping artefacts thanks to analysis in real space and the compatibility with non-sinusoidal interference patterns. Hence, UMPA will facilitate the implementation of flexible and tunable phase-contrast and dark-field imaging at most existing X-ray phase-contrast imaging setups without the need for significant modifications.

5.4.4 Angular Sensitivity and Spatial Resolution

Analogous to X-ray grating interferometry, the two main criteria for assessing the quality of the reconstructed phase-contrast images are the spatial resolution and the angular sensitivity.

The spatial resolution strongly depends on the experimental implementation and processing method. For XST (Sect. 5.4.1), it is determined by the size of the subset window chosen in the reconstruction process and is ultimately limited by the speckle size. For 2D XSS (Sect. 5.4.2.1), it can go down to the effective detector pixel size as a pixel-wise reconstruction is performed. In practice, the point-spread function of the detector and other factors might deteriorate the resolution. For 1D XSS (Sect. 5.4.2.2), the spatial resolution is reduced compared to the 2D scanning case, as a few pixels taken along the axis orthogonal to the scanning direction contribute to the signal formation. For the sparse sampling variation of 2D XSS, a resolution equivalent to the resolution of the detector system could in principle be realised, but due to the interpolation step used for the sample image, it might effectively be lower. The XSVT as well as the mixed XSS-XSVT approaches (Sect. 5.4.3.1) can also achieve a resolution down to the pixel size. On the other hand, the mixed XST-XSVT (Sect. 5.4.3.1) and UMPA (Sect. 5.4.3.2) approaches show a lower spatial resolution that is determined by the extent of the subset window taken around the pixel of interest. Typically, the window sizes are much smaller than for XST and hence a higher resolution can be achieved with UMPA and XST-XSVT. The resolution limit can be quantified as twice the FWHM of the window extent (Zdora et al. 2017b), see also Chap. 6, in particular Sect. 6.3.7.3.

The second property commonly used to evaluate the quality of the reconstructed phase-contrast images is the angular sensitivity, which is a measure of the smallest refraction angle or differential phase shift that can be measured with a certain setup and acquisition scheme. The sensitivity is typically quantified as the standard deviation of the reconstructed refraction angle signal in a small region of interest in a homogeneous background region without sample (e.g. air), analogous to the X-ray grating interferometry case, see Sect. 4.2.3 in Chap. 4. As for the spatial resolution, it also strongly depends on the processing scheme. In general, it is inversely proportional to the propagation distance and dependent on the accuracy of the reconstruction algorithm and the photon noise, amongst other factors. A detailed study on the noise properties (which are directly related to the angular sensitivity) in the differential phase signals of XST measurements based on simulations and experimental validation can be found by Zhou et al. (2016). For the XST, XSVT, mixed XST-XSVT and

UMPA methods that perform the reconstruction in the detector plane, the angular sensitivity is, furthermore, directly proportional to the effective pixel size p_{eff}. For the approaches that operate in the sample (diffuser) plane,[6] such as 2D XSS and mixed XSS-XSVT, it is proportional to the diffuser step size s in the sample (diffuser) plane instead. This means that these operational modes can achieve a better sensitivity for a given setup, as s is typically smaller than p_{eff}. An up to 100-fold improvement of the sensitivity for 2D XSS compared to XST has been reported (Bérujon and Ziegler 2016). For the 1D XSS analysis, a high sensitivity dependent on the step size can be achieved in the scanning direction, whereas the sensitivity in the other direction that is not scanned is proportional to the pixel size.

Further quantities influencing the angular sensitivity are the number N of diffuser steps for the reconstruction of the image and the extent w of the analysis subset. It can be shown that the angular sensitivity is inversely related to w and $\sqrt{N}$ (Zdora et al. 2017b), see also Sect. 6.3.4 in Chap. 6. This relationship makes the UMPA and XST-XSVT approaches very attractive since the angular sensitivity can be controlled by changing N and w. As mentioned above, the choice of w also determines the spatial resolution of the reconstructed images. Hence, the UMPA and XST-XSVT modes allow flexible tuning of the resolution and sensitivity that can be adjusted to specific experimental requirements. In a practical implementation, also the constraints in scan time and dose, which inherently increase with N, might play a role. The choice of N and w ultimately depends on the focus and the desired outcome of the experiment.

5.5 Speckle-Based X-ray Dark-Field Imaging Approaches

In the very first demonstrations of X-ray speckle-based imaging (Morgan et al. 2012; Bérujon et al. 2012b), the focus was solely on the phase-contrast signal. However, it was soon recognised that complementary dark-field information can be obtained simultaneously from a speckle imaging data set (Bérujon et al. 2012a). Although the capabilities of the dark-field image have not yet been extensively exploited for X-ray speckle-based imaging applications, there is great potential in particular for dark-field tomography (see Sect. 5.7) for medical and materials science applications.

The dark-field signal gives information about small-angle scattering in the sample (Yashiro et al. 2010; Pfeiffer et al. 2008). For speckle imaging, it is related to the loss in visibility of the speckle pattern caused by a decrease in the coherence of the X-rays after undergoing scattering in the specimen. Different models have been developed to measure the dark-field signal from the acquired speckle data. The first proposed method is analogous to the treatment in X-ray grating interferometry, where the dark field is defined as the ratio of the amplitudes of the sample and reference phase-stepping curves normalised by the transmission (Pfeiffer et al. 2008). For speckle-based imaging, the same concept can be used and the standard deviation of

[6]Depending on the mounting of the diffuser upstream or downstream of the sample, the reconstruction is effectively performed in the sample or the diffuser plane, respectively.

the interference pattern can be taken as a measure for the amplitude. The equivalent description of the dark-field signal D for speckle imaging is then given by the ratio of the sample and reference standard deviations, σ_{sam} and σ_{ref}, normalised by the transmission T (Bérujon et al. 2012a). For a pixel (x, y), this can be expressed as:

$$D(x, y) = \frac{1}{T(x, y)} \frac{\sigma_{\text{sam}}(x, y)}{\sigma_{\text{ref}}(x + u_x, y + u_y)}, \tag{5.17}$$

where u_x and u_y are the displacements of the sample interference pattern in the two orthogonal directions due to refraction and σ is the standard deviation operator over all diffuser positions for the scanning-based modes or all pixels in the subset window for XST. Although first derived for 2D XSS, the same procedure for the calculation of the dark-field signal can also be applied to the other operational modes such as XSVT (Bérujon and Ziegler 2015) and mixed XSVT approaches (Bérujon and Ziegler 2016).

Another way to extract the dark-field signal was first proposed for the single-shot XST method (Zanette et al. 2014) and was later extended for the UMPA mode (Zdora et al. 2017b). As outlined in Sect. 5.4.1, the reduction of amplitude due to small-angle scattering can be included in a model that expresses the sample speckle pattern as a modulated version of the reference speckle pattern, see Eq. (5.9). A windowed least-squares minimisation procedure delivers here directly the dark-field signal D. The same model is used in the UMPA approach (see Sect. 5.4.3), but the signal from several diffuser positions is combined, which allows for a higher sensitivity and spatial resolution, also for the dark-field signal (Zdora et al. 2018b).

It should be noted that, although the reconstruction approaches in the previous two paragraphs differ, the physical principle of the dark-field contrast generation that they are based on is the same, as pointed out by Bérujon (2015).

An alternative view on the dark-field signal was presented by Wang et al. for the XSS technique (Wang et al. 2015c). Here, the dark field is extracted by taking a normalised cross-correlation of the interference patterns in neighbouring pixels. This is done separately for the sample and for the reference pattern. The normalised maximum correlation coefficient in a pixel (x, y) is defined as the ratio of the maximum sample and the maximum reference correlation coefficients (Wang et al. 2015c):

$$M(x, y) = M_{\text{sam}}(x, y)/M_{\text{ref}}(x, y). \tag{5.18}$$

The change in $M(x, y)$ is taken as a measure for the small-angle scattering in the sample and a reduction of $M(x, y)$ from one pixel to a neighbouring pixel is interpreted as an increased dark-field signal. The absolute dark-field signal in this approach is defined as (Wang et al. 2015c):

$$D(x, y) = -2 \ln M(x, y). \tag{5.19}$$

Using 1D scanning, Wang et al. report that this approach can deliver directional dark-field images that include contributions from small-angle scattering as well as the second derivative of the wavefront phase (Wang et al. 2015c). However, Bérujon claims that the normalised maximum correlation coefficient cannot accurately describe the scattering behaviour of a sample (Bérujon 2015). He, furthermore, argues that, rather than the second derivative of the phase, the method senses optical phase discontinuities at pixel boundaries that are larger than the pixel size and hence cannot be regarded as a dark-field signal (Bérujon 2015).

In another implementation, Wang et al. used an approach based on the reduction of the peak value of the local cross-correlation coefficient between reference and sample signals to calculate the dark-field image for the XST single-shot analysis (Wang et al. 2015a). The dark-field signal is here defined as $D = 1 - \gamma^{\max}$, where $\gamma^{\max}$ is the maximum (peak) value of the cross-correlation coefficient, which is obtained for each pixel from the zero-normalised cross-correlation of the sample and reference subset windows.

The correlation coefficient was also used in a dark-field approach proposed for 1D XSS dark-field tomography (Wang et al. 2016a, b, c). The sample speckle pattern is modelled as the convolution of the reference speckle pattern and the optical transfer function of the specimen. The latter can be approximated by taking into account the phase shift as well as the scattering in the sample, where the scattering is modelled as Gaussian and isotropic. Cross-correlation is performed between the sample arrays and the reference arrays. From these considerations and with some further approximations, the maximum of the cross-correlation coefficient can be expressed as (Wang et al. 2016b):

$$\gamma^{\max} = \exp\left(-\frac{8\pi^4 d^2 \sigma^2}{\zeta^2_{\text{speckle}}}\right), \tag{5.20}$$

where d is the propagation distance, σ^2 the second moment of the scattering angle distribution and ζ_{speckle} the average speckle size, which can be estimated from the position of the maximum of the power spectrum of the speckle pattern. This can be rearranged to obtain the dark-field signal D:

$$D = \sigma^2 = \frac{-\zeta^2_{\text{speckle}}}{8\pi^4 d^2} \ln\left(\gamma^{\max}\right). \tag{5.21}$$

The model of the sample interference pattern as a convolution of the sample scattering distribution and the reference pattern is also used by Bérujon and Ziegler (2015). However, here the full scattering distribution is analysed using iterative methods (see Sect. 5.4.2.5) and the transmission, differential phase and dark-field signals are interpreted as its different moments. In this framework, the two second normalised moments can be seen as the characteristic scattering width in the two orthogonal directions, equivalent to the dark-field signal. Further complementary scattering signals corresponding to various phenomena can also be obtained with this approach.

Although there is great potential for X-ray dark-field imaging using UMPA, which delivers this image modality in an efficient, straightforward manner, it has not been studied extensively to this date. More work and further development of UMPA dark-field imaging will be performed in the future, in particular exploring advanced analysis methods, e.g. directional dark-field imaging, and new fields of applications.

5.6 Translation to Laboratory Sources and High X-ray Energies

The relatively low requirements on the temporal and spatial coherence of the X-ray beam (Zdora et al. 2015) make speckle-based imaging an ideal candidate for the application at laboratory-based systems with conventional X-ray tubes. The translation of the speckle-based technique to a laboratory source was first demonstrated with the single-shot XST technique (Zanette et al. 2014) at a liquid-metal-jet source (Excillum) (Hemberg et al. 2003). Transmission, differential phase and dark-field images were successfully reconstructed for several samples, such as the plastic flower shown in Fig. 5.7.

Shortly after, the implementation of the 2D XSS method at the same laboratory source was also reported (Zhou et al. 2015).

The liquid-metal-jet source used by Zanette et al. (2014), Zhou et al. (2015) and in Chap. 9, Sect. 9.4 is a laboratory source with relatively high flux and a small spot size that has a polychromatic spectrum dominated by the gallium, indium and tin emission lines of the liquid anode material (Hemberg et al. 2003). However, it has been shown that also conventional micro-focus sources with lower flux and a broader spectrum can be used for speckle-based imaging (Wang et al. 2016a, b). At a liquid-metal-jet source, typically smaller spot sizes can be used as the liquid anode material significantly relaxes the power limit, allowing for a higher flux at small spot sizes. At conventional sources, the spot size is often much larger and hence the transverse coherence conditions are inferior, making it more challenging to produce

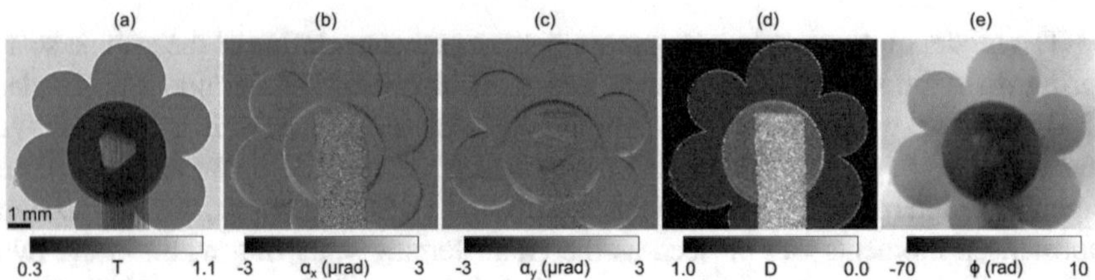

Fig. 5.7 First demonstration of X-ray speckle-based imaging at a laboratory source. **a** Transmission, **b** differential phase in the horizontal and **c** the vertical direction and **d** the dark-field signal of a plastic flower on a wooden support could be successfully retrieved. **e** Wavefront phase obtained from integration of (**b**) and (**c**). Figure reprinted with permission from Zanette et al. (2014). Copyright (2014) by the American Physical Society

X-ray near-field speckle, which relies on scattering and interference effects. To tackle this, an alternative approach has been explored that relies on creating a high-visibility reference pattern by exploiting the absorption of small random structures, e.g. from coarse sandpaper (Wang et al. 2016b) or a "random absorption mask" such as steel wool (Wang et al. 2016a). The "absorption-speckle" method can be easily applied to a large range of laboratory sources. However, one should be aware that this approach is not speckle imaging in its original definition as the reference pattern is not a speckle pattern based on interference effects. It has been shown that absorption speckle allows for the use of high-energy X-rays, for which the contrast of a conventional near-field speckle pattern created by a piece of sandpaper is usually low (Wang et al. 2016a). On the other hand, recently, near-field speckle-based imaging was demonstrated also using conventional phase speckle with high-energy X-rays from a filtered synchrotron bending magnet beam with a mean energy of 65 keV and 25% bandwidth of the detected spectrum (Bérujon and Ziegler 2017). A speckle pattern of high visibility was achieved in this setup by stacking several sheets of sandpaper (see also Chap. 9, Sect. 9.3).

For a practical implementation of the speckle-based technique with polychromatic X-rays, one should be aware that, as for other imaging methods, artefacts may arise from beam hardening in the specimen, in particular for high-density samples. This has been investigated in detail in a simulation study (Zdora et al. 2015) and the effect has been observed experimentally in the dark-field signal of XST measurements conducted at a micro-focus laboratory X-ray source (Vittoria et al. 2017).

5.7 Speckle-Based X-ray Phase-Contrast and Dark-Field Tomography

For many applications, the 2D data alone are not sufficient and it is essential to obtain quantitative 3D information of the inner density distribution in a specimen. Often, also the 3D scattering distribution is of interest and can give complementary information. As illustrated in the previous sections, speckle-based imaging can provide quantitative phase-contrast signals as well as transmission and dark-field images from a single data set. The extension from 2D projection imaging to 3D tomography is straightforward. Projections with the sample in the beam are taken at typically a few hundred or thousand different viewing angles of the specimen between 0° and 180° (or 360°).[7] Depending on the operational mode (see Sect. 5.4), images are acquired at one or several diffuser positions. References without the sample do not need to be taken for each projection and in principle it is sufficient to have one reference image at each diffuser position. However, commonly, a few sets of references are recorded to account for beam instabilities. For each of the projections, the multimodal image signals are then reconstructed from the acquired raw data. Subsequently, a tomographic reconstruction algorithm, e.g. filtered back-projection

[7]For fan- or cone-beam geometry, the angular range must be extended by the fan-/cone-beam angle.

(Kak and Slaney 2001), is applied to obtain the phase, transmission and dark-field tomograms.

Phase tomography using the speckle-based technique has been demonstrated both at highly brilliant synchrotron sources as well as in the laboratory. In a first report, the phase and transmission tomograms of a human artery obtained with the single-shot XST mode (see Sect. 5.4.1) were shown (Wang et al. 2015a). The superior sensitivity to density differences of the phase signal over the transmission signal, here between the artery lumen and walls, was observed in a qualitative way, see Fig. 5.10I in Sect. 5.8.2.

At around the same time, a quantitative analysis of speckle tomography data was presented from XST measurements at a liquid-metal-jet laboratory source (Zanette et al. 2015). Here, it was shown that the complementary quantitative absorption and refraction information from transmission and phase tomograms, respectively, can be combined for identifying and characterising different materials with similar refraction and absorption properties in a sample, see Fig. 5.10V in Sect. 5.8.2.

Furthermore, quantitative phase and dark-field tomographies of a phantom sample were successfully demonstrated using the 1D XSS method (see Sect. 5.4.2.2) (Wang et al. 2016c). However, it should be noted that an object with features oriented mainly along the axis orthogonal to the scanning direction was chosen and only the refraction signal in the scanning direction was considered in the tomography reconstruction. The sensitivity along the axis opposite the scanning direction is typically lower for the 1D XSS method.

Also, the XSVT and the mixed XST-XSVT approaches (see Sect. 5.4.3) have been implemented in tomographic mode and it was shown that complementary absorption, phase and dark-field tomograms of berry samples could successfully be reconstructed (Bérujon and Ziegler 2016). An intelligent interlaced acquisition scheme, similar to the one proposed for grating interferometry (Zanette et al. 2011), was used here to reduce the number of required diffuser steps for the tomography scan even further by including the information from several subsequent projections. This way, as few as five diffuser positions per projection could be used for the interlaced XSVT tomography when combining the information from additional two projections preceding and following the projection of interest (Bérujon and Ziegler 2016). With an otherwise identical acquisition scheme, this was reduced to only one diffuser position per projection for the mixed XST-XSVT approach, which in the analysis includes the information from neighbouring pixels in a small window, here 3×3 pixels, centred around the pixel under consideration (Bérujon and Ziegler 2016).

A similar interlaced system was applied to the acquisition scheme of sparsely sampled XSS (see Sect. 5.4.2.4) (Bérujon and Ziegler 2017). Considering the information from two preceding and two following projections in the analysis of one projection allowed for stepping of the diffuser effectively on a 5×5 grid, while only taking five diffuser steps per projection. For projection $p + 5k$ with $k = 0, 1, \ldots, (N - 5)/5$; $p = 1, 2, \ldots, 5$, five images were acquired only at the diffuser positions in row p of the 5×5 grid, where N is the total number of projections of the tomography scan. The reference pattern was scanned on a denser grid following the conventional 2D XSS scheme and the corresponding missing sample frames were obtained by interpolation. High-quality absorption and phase volumes were retrieved with this approach, while the exposure time could be significantly reduced thanks to the sparse sampling and interlaced acquisition scheme.

The translation of UMPA (see Sect. 5.4.3) from 2D projection to 3D tomographic mode has just recently been achieved and first results are shown in some of the following chapters for biomedical imaging (see Chap. 8), geology (see Chap. 9, Sect. 9.3.4) and materials science (see Chap. 9, Sect. 9.3.5) applications. Most of these tomographies were performed with 20–25 diffuser steps per projection. However, it is demonstrated that good image quality can still be achieved when reducing the number of steps down to five or possibly less per projection without the need for interpolation or other computationally expensive preparation of the raw data.

5.8 Applications of the X-ray Speckle-Based Technique

As X-ray near-field speckle imaging is a versatile, robust and easily implemented technique, it can be expected to find applications in a wide range of fields. Being a relatively young method, a lot of the tremendous potential of speckle-based X-ray imaging has yet to be explored. The main applications of the technique that have been demonstrated so far are illustrated in the following.

5.8.1 Metrology and Wavefront Sensing

A focus of applications has been the use of X-ray near-field speckle for metrology, optics characterisation and beam phase sensing. The simple and robust experimental arrangement and high angular sensitivity make the speckle-based technique an ideal candidate for metrology. The idea of applying speckle imaging to wavefront measurements and optics characterisation was presented early on in the first publications on the technique (Bérujon et al. 2012a, b). In the following years, increasing use of near-field speckle was reported for the characterisation of refractive lenses (Bérujon et al. 2013b; Wang et al. 2014, 2015d, 2017b; Zdora et al. 2017b, 2018c) and X-ray mirrors (Sawhney et al. 2013; Bérujon et al. 2014; Wang et al. 2014, 2015b, e; Kashyap et al. 2016a, c; Wang et al. 2017a, b), as well as analysing the local beam wavefront (Bérujon et al. 2012b, a; Wang et al. 2014) and measuring the transverse coherence length of the X-ray beam (Kashyap et al. 2015; Wang et al. 2017b).

For metrology, speckle-based phase-sensing is commonly operated in one of the two modes (Bérujon et al. 2012a; Wang et al. 2017b): the differential mode or the self-correlation mode, see Sect. 5.4.

Moderately refracting optical elements such as single compound refractive lens (CRL) elements can be analysed using the common differential mode, which is based on acquiring one or more reference interference patterns and one or more sample patterns with the optics in the beam and subsequent reconstruction of the differential phase using one of the available analysis methods (see Sect. 5.4). The wavefront phase Φ downstream of the optical element can then be obtained from the differential phase (refraction angle) signal via integration.

The characterisation of a 2D CRL element was first demonstrated using XST (Bérujon et al. 2013b; Wang et al. 2014) and 2D XSS (Bérujon et al. 2012a) in differential mode, see Fig. 5.8I. Furthermore, a 1D parabolic lens made from beryllium was analysed with 1D XSS, as shown in Fig. 5.8II, and the aberrations from the expected wavefront downstream of the lens were retrieved (Wang et al. 2015d). Recently, parabolic 1D and 2D CRL elements made from SU-8 polymer material (Nazmov et al. 2004) were inspected using the UMPA approach implemented in two configurations, with either a piece of random sandpaper or a periodic phase grating as a phase modulator (Zdora et al. 2017b, 2018c), see Fig. 5.8III. The analysis allowed for the sensitive identification and quantification of deviations from the expected refraction behaviour caused by beam damage and shape errors of the lenses, as shown in Fig. 5.8IV.

On the other hand, the self-correlation mode (Sect. 5.4.2.3) can be employed to directly measure the absolute effective local wavefront curvature, i.e. the second derivative of the wavefront after passing through the lens (Bérujon et al. 2012a).

For strongly focussing (or defocussing) optics such as X-ray mirrors, it is essential to use the self-correlation mode (see Sect. 5.4.2.3), as in this case the X-ray beam is significantly (de-)magnified by the optical element and the conventional correlation procedure between sample and reference scans will not succeed in accurately measuring the wavefront distortions. The self-correlation analysis has been applied for the characterisation of mechanically bent and piezo bimorph X-ray mirrors. A single 1D scan of the diffuser allows retrieving the 1D mirror slope (Bérujon et al. 2014) and it has been shown that also the 2D slope can be accessed from 1D scanning (Wang et al. 2015b).

As mentioned in Sect. 5.4.2.3, measurements can be performed with the diffuser upstream or downstream of the optical element under consideration (Kashyap et al. 2016c). The two configurations are illustrated in Fig. 5.9I.

The first configuration gives information about the wavefront distortions caused by the optical element downstream of the diffuser only. This approach is suitable for the characterisation of mirror surfaces in order to detect slope errors. The reflective surface of the mirror and its errors can be analysed directly, as shown in Fig. 5.9II, III. The separation in the plane of the incident wavefront of two rays that are adjacent in the detector plane is obtained from the signal delay given by the correlation procedure. Subsequent integration allows for calculating the position of the rays in the diffuser plane and an iteration process delivers the mirror slope (Bérujon et al. 2014). It was demonstrated that this way slope errors can be accurately determined, allowing precise optimisation of the mirror (Bérujon et al. 2014; Wang et al. 2015b, 2017b; Kashyap et al. 2016c).

The downstream configuration (see Fig. 5.9I bottom) senses the total beam wavefront modulations by all optics in the beam upstream of the diffuser, rather than the properties of an optical element itself. Typically, the local radius of curvature of the

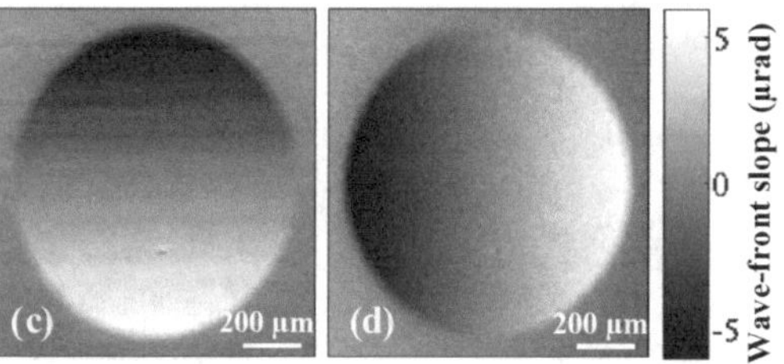

(**I**) Wavefront slope for a 2D beryllium CRL element measured with 2D XSS.

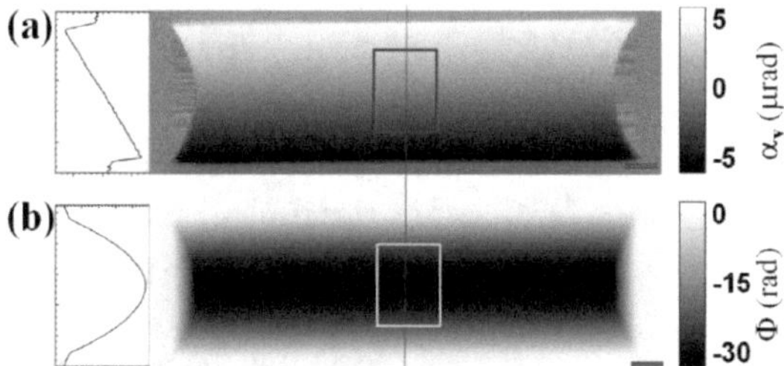

(**II**) Vertical refraction angle, i.e. wavefront slope (top) and reconstructed phase (bottom) for a 1D beryllium refractive lens measured with 1D XSS. Scale bar corresponds to 0.2 mm.

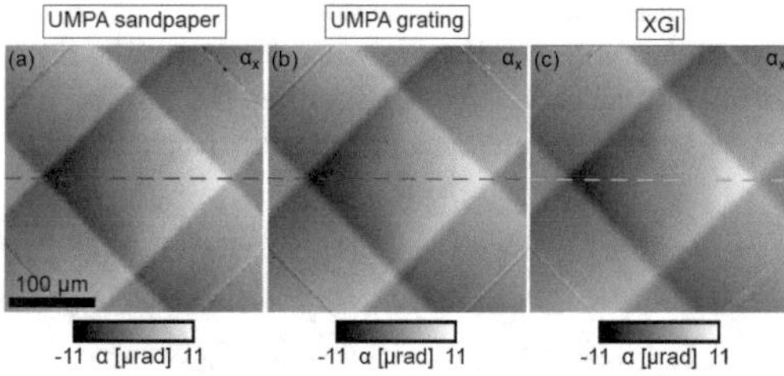

(**III**) Horizontal refraction angle for a 2D polymer CRL element measured with UMPA using a random (left) and a periodic (centre) phase modulator to create a reference pattern and comparison with grating interferometry (right).

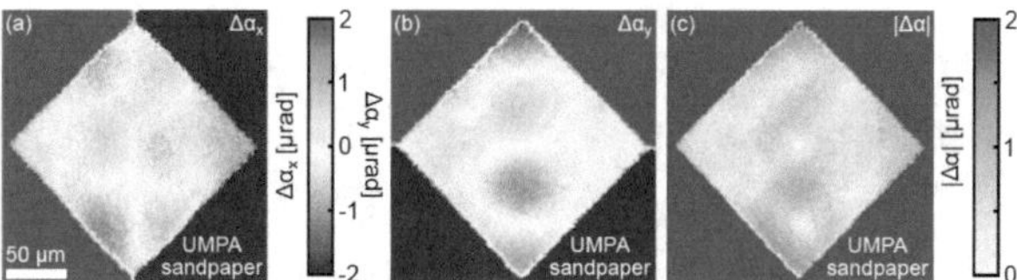

(**IV**) Deviation from the expected horizontal (left) and vertical (centre) refraction angle and absolute deviation (right) in the aperture of the 2D polymer CRL in (III), measured with UMPA using a speckle pattern as a wavefront marker.

Fig. 5.8 Examples of applications for the characterisation of X-ray refractive lenses. Figures reprinted with permission: **I** from Bérujon et al. (2012a), Copyright (2012) by the American Physical Society; **II** from Wang et al. (2015d) © The Optical Society; **III**, **IV** from Zdora et al. (2018c), licensed under CC BY 4.0. Composite figure reprinted and rearranged with permission from Zdora (2018), licensed under CC BY 4.0

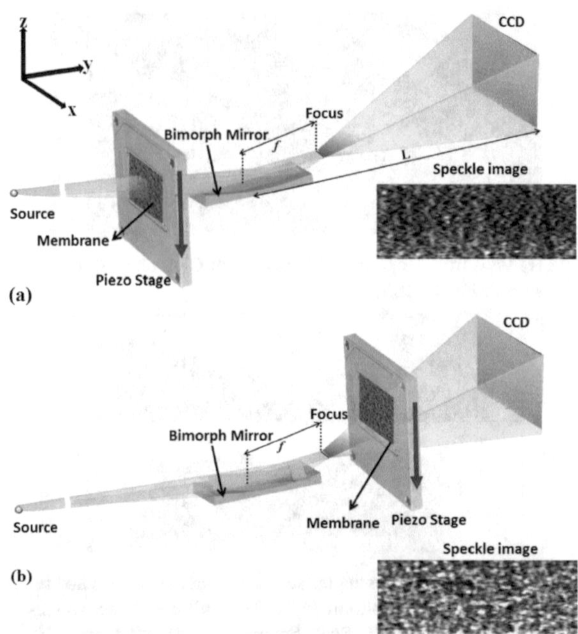

(I) Two different configurations for the optimisation of an X-ray mirror by 1D speckle-scanning with a self-correlation analysis. The diffuser can be placed upstream (top) or downstream (bottom) of the mirror for mirror-slope measurement or beam wavefront characterisation, respectively.

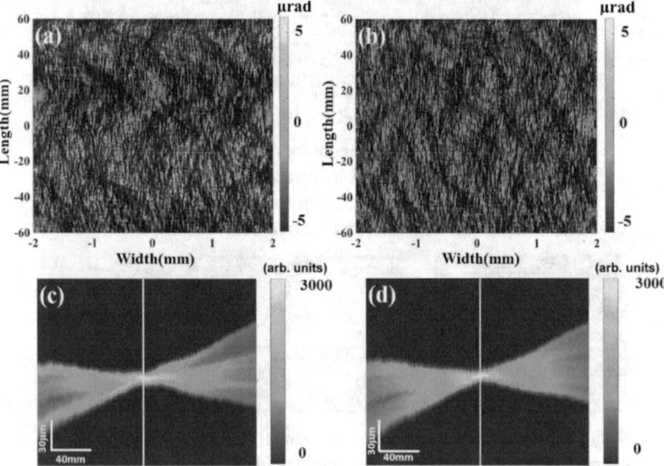

(II) 2D slope errors and intensity profiles along the propagation direction before (left) and after (right) optimisation of an X-ray bimorph mirror, measured with the upstream configuration in (I) (top).

Fig. 5.9 Examples of applications for X-ray mirror characterisation. Figures reprinted with permission: **I**, **II** from Kashyap et al. (2016c), licensed under CC BY 4.0; **III** from Bérujon et al. (2014) © The Optical Society; **IV** from Wang et al. (2015e) © The Optical Society. Composite figure reprinted and rearranged with permission from Zdora (2018), licensed under CC BY 4.0

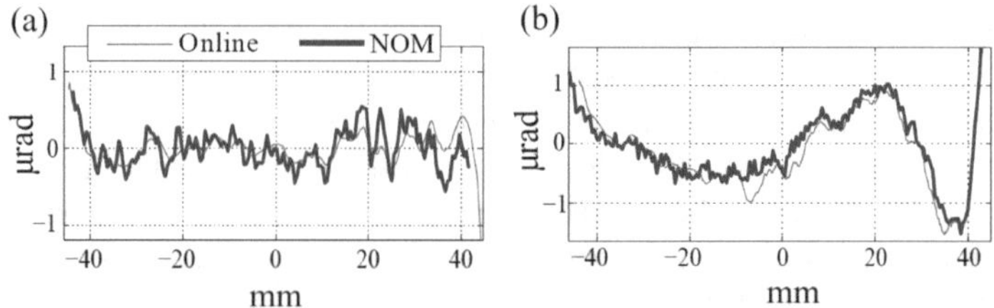

(III) 1D slope errors of a vertically (left) and a horizontally (right) focussing X-ray mirror measured with the diffuser upstream of the mirrors and comparison with NOM (nanometer optical metrology) results.

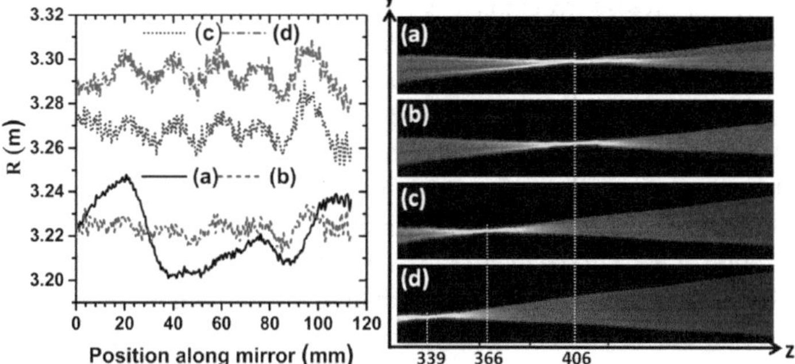

(IV) Wavefront radius of curvature at the detector measured with the downstream configuration in (I) (bottom) and corresponding intensity profiles along the propagation direction (distance z in mm from the mirror) (a) before optimisation, (b) after optimisation (focussed) and (c,d) after optimisation (defocussed) of a bimorph mirror used for beam shaping.

Fig. 5.9 (continued)

wavefront is reconstructed, which is directly related to the second derivative of the wavefront phase Φ (see Eq. (5.14)) (Bérujon et al. 2012a, 2014; Wang et al. 2015e; Kashyap et al. 2016c). This approach has been applied successfully to the fast, precise optimisation of bimorph mirrors with the aim to obtain a desired beam size and shape (Wang et al. 2015e; Kashyap et al. 2016a, c; Wang et al. 2017a, b), see Fig. 5.9IV. Sensitivities down to 2 nrad have been reported for these applications (Wang et al. 2015e; Kashyap et al. 2016c). An in-situ portable metrology device based on this concept has been developed at Diamond Light Source (Kashyap et al. 2016a). Its use for the characterisation of elliptical mirrors, the optimisation of bimorph mirrors and mirror alignment has been demonstrated (Wang et al. 2017a).

In addition to the investigation of optical elements such as X-ray refractive lenses and mirrors, it can also be of interest to characterise the absolute beam phase without additional beam-shaping optics. This was shown with the XST method in absolute mode (Bérujon et al. 2012b; Wang et al. 2014). In this configuration, images are recorded at two different detector positions along the beam path and the cross-correlation is performed between these two images without the use of a reference speckle pattern. The recovered local displacement of the speckle pattern relates to the refraction angle, i.e. the first derivative of the wavefront, which can then be integrated to obtain the beam phase. This method, however, requires the beam to remain stable over the course of the two image acquisitions. For cases where the X-ray beam is fluctuating in time, it is more appropriate to record the two images simultaneously and a different variation of this type of measurement was demonstrated for this purpose (Bérujon et al. 2015). The setup consists of a diffuser and a first camera with a semi-transparent mirror and scintillator that records an image, but at the same time transmits part of the X-rays, which is recorded by a second camera further downstream. Cross-correlation according to the XST approach between the two simultaneously recorded images delivers the first derivative of the beam phase. Subset distortions can also be taken into account to gather additional information on the second derivative of the beam phase. This approach is particularly suitable for pulsed wavefronts found e.g. at X-ray free-electron lasers, where the beam profile changes from shot to shot. A similar setup relying on two cameras with semi-transparent scintillators and optical mirrors with holes has been developed recently for beam characterisation at the European X-ray free electron laser (Bérujon et al. 2017).

Apart from the beam phase, also information about the transverse coherence of the X-ray beam can be obtained using X-ray near-field speckle. This had first been shown in the early days of X-ray near-field speckle by using a colloidal suspension as a diffuser (Alaimo et al. 2009) and later with speckle from a filter membrane (Kashyap et al. 2015; Wang et al. 2017b). Only a single exposure of the diffuser is necessary for the analysis and the transverse coherence length can be retrieved by looking at the Fourier power spectrum of the flat-field corrected speckle interference pattern. As demonstrated by Cerbino et al. (2008), the power spectrum can be decomposed into the 2D scattered intensity distribution and a transfer function. The latter contains contributions from the Talbot effect, the detector response and the partial coherence of the beam. The detector response can be measured and the Talbot contribution is a known function. When a Gaussian intensity distribution is assumed, the partial coherence term is a function of the transverse coherence lengths in the two orthogonal directions. The known contributions from the detector response and Talbot effect as well as the model of the partial coherence term can be included in a fit function of the angular power spectrum. By fitting of the measured angular power spectrum to the function, using e.g. a least-squares minimisation procedure, the transverse coherence lengths of the X-ray beam can be determined (Kashyap et al. 2015).

5.8.2 Imaging for Biomedical and Materials Science Applications

Other important and promising areas of applications of the speckle-based technique are X-ray phase-contrast and dark-field imaging, in particular for biological, biomedical and pre-clinical research as well as materials science.

For biomedical and biological soft-tissue specimens, the phase-contrast signal is of particular interest, as it shows a much higher sensitivity to small density differences than the absorption image for this type of sample. Speckle-based phase-contrast tomography of biomedical and biological specimens has been explored with XST and the phase tomogram of a human artery was successfully obtained, showing superior contrast compared to the transmission signal (Wang et al. 2015a), see Fig. 5.10I. Furthermore, 1D XSS was employed to measure a whole fish and the multimodal images shown in Fig. 5.10III illustrate the complementary character of the different signals, which allows revealing different parts of the sample (Wang et al. 2016b). The multi-contrast signals of a chicken wing were measured with 1D XSS at a microfocus laboratory source (Wang et al. 2016b), which is a promising step towards the large-scale and accessible implementation of speckle-based imaging for biomedical applications. The results on the fish and chicken wing (Wang et al. 2016b) as well as the high-contrast scans of different kinds of berries using variations of the XSVT technique (Bérujon and Ziegler 2015, 2016, 2017) (see Fig. 5.10II) indicate possible potential in the area of food inspection for quality control and foreign body detection. Furthermore, the multimodal UMPA method has proven to be suitable for the investigation of biological samples, see Fig. 5.10IV. It allows for flexible tuning of the reconstruction result essential for optimising the trade-off between dose on the specimen and image quality, which is of great importance for biomedical specimens.

The complementary character of the various contrast modalities provided by speckle-based imaging can also be exploited for materials science applications. It was demonstrated that XST tomography implemented at laboratory sources can be employed for the identification of different materials in a sample. Several types of plastic were successfully distinguished using the combined information from phase and absorption tomograms (Zanette et al. 2015), as shown in Fig. 5.10V. In another publication, the chip of a computer memory card was imaged using 1D XSS at a laboratory source and the complementary phase and dark-field signals enabled identifying different components of the chip (Wang et al. 2016a), see Fig. 5.10VI. These examples suggest the promising potential of the speckle-based technique for the inspection and quality control of electronics as well as identification of different materials in a sample, which can be performed with a simple setup at widely available laboratory sources.

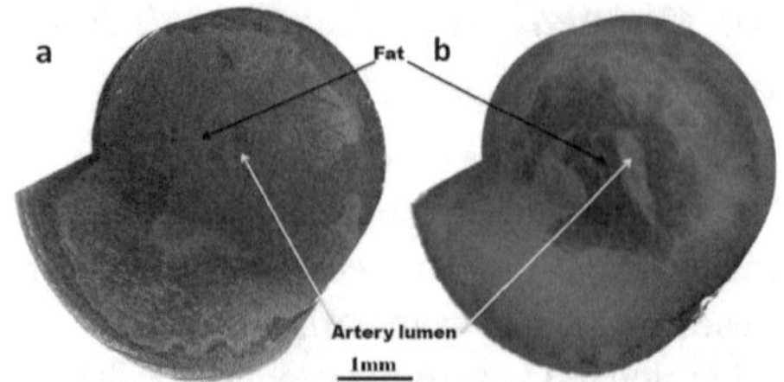

(I) Volume rendering of the transmission (left) and phase (right) volumes of a human carotid artery obtained with XST tomography.

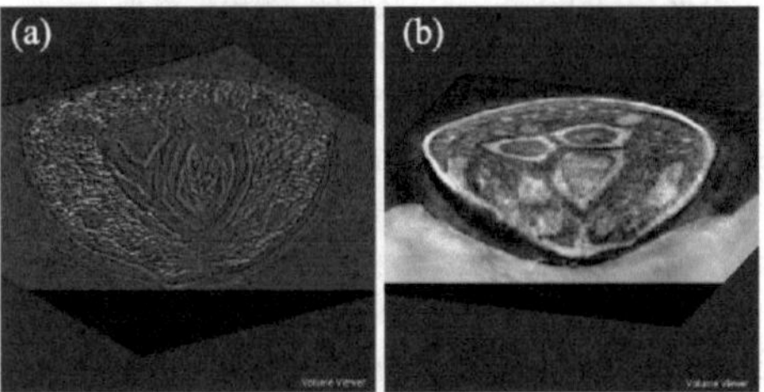

(II) Volume rendering of the dark-field (left) and phase (right) volumes of a juniper berry obtained with interlaced XSVT tomography.

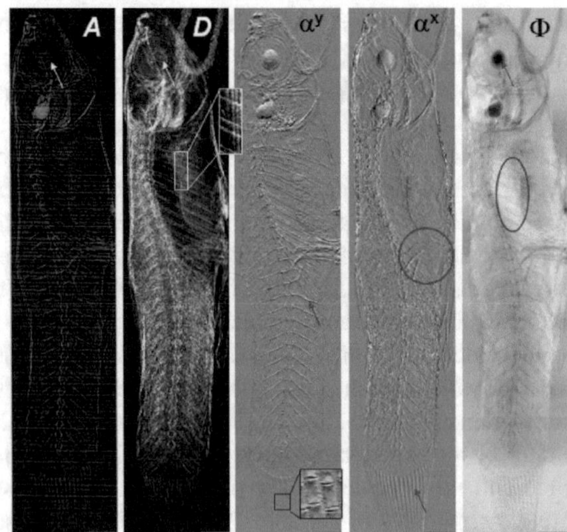

(III) Reconstructed absorption, dark-field, differential phase in the vertical and horizontal directions and integrated phase signal (left to right) of a fish obtained with 1D XSS projection imaging (vertical scanning). Scale bar corresponds to 1 mm.

Fig. 5.10 Examples of applications for biomedical and biological imaging and materials science. Figures reprinted with permission: **I** from Wang et al. (2015a), licensed under CC BY 4.0; **II** from Bérujon and Ziegler (2016), Copyright (2016) by the American Physical Society; **III** from Wang et al. (2016b), licensed under CC BY 4.0; **IV** from Zdora et al. (2017b), licensed under CC BY 4.0; **V** from Zanette et al. (2015); **VI** from Wang et al. (2016a), licensed under CC BY 4.0. Composite figure reprinted and rearranged with permission from Zdora (2018), licensed under CC BY 4.0

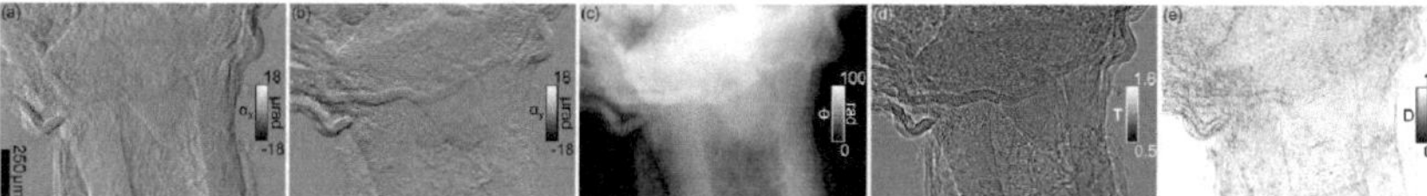

(IV) Horizontal, vertical refraction angle, integrated phase, transmission and dark-field (left to right) projections of a small flower bud obtained with UMPA.

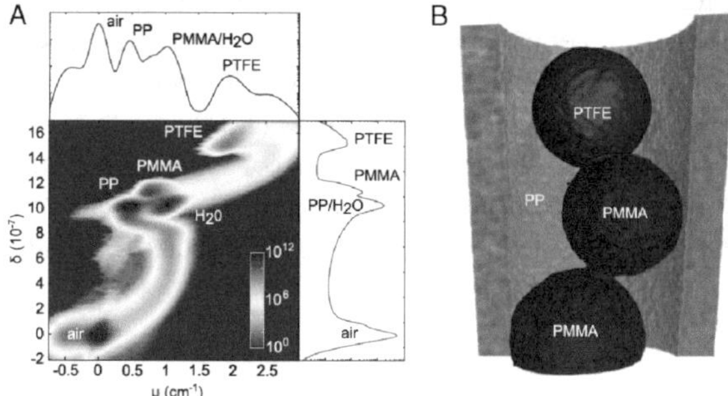

(V) Material characterisation using XST tomography at a liquid-metal-jet laboratory source. Different kinds of plastic in a phantom sample can be distinguished by combining attenuation and refraction information. The diameter of the spheres is 1.5 mm.

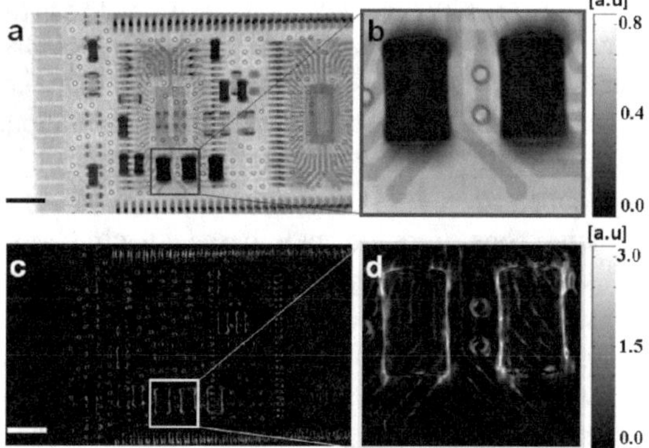

(VI) Reconstructed transmission (top) and dark-field (bottom) images of a microchip imaged at a laboratory X-ray micro-focus source using 1D XSS with an absorption speckle pattern. Scale bar corresponds to 2 mm.

Fig. 5.10 (continued)

5.8.3 *Other Applications*

A different application of X-ray near-field speckle is the use for capturing dynamic processes such as blood flow (Kim and Lee 2006, 2009; Irvine et al. 2008; Park et al. 2016; Izadifar et al. 2018) and the movement of mouse lungs (Murrie et al. 2015, 2016). Here, one makes use of speckle created directly by the specimen under study and no diffuser is needed. The speckle from blood or from the alveoli of the lung can be used as a marker to track the dynamic processes of blood flow and breathing, respectively. A windowed cross-correlation analogous to the XST imaging approach is performed between the speckle images recorded at different points in time. From the displacement of the speckle pattern, the speed of the particles can be estimated.

The recent advances in acquisition schemes and reconstruction approaches allow for short scan times, flexible tuning of the signal sensitivity and spatial resolution and a straightforward implementation at laboratory sources. Following these developments, it can be expected that the range of applications of the speckle-based technique will increase further in the next few years and the method will be employed widely at synchrotron and laboratory sources. In particular in the field of biomedical and pre-clinical imaging, speckle-based phase-contrast as well as dark-field imaging have great potential due to their robustness, cost- and dose-effectiveness, amongst others.

5.9 Conclusions and Outlook

Despite being developed just a few years ago, X-ray near-field speckle-based imaging has already seen rapid development and is receiving rising interest in the X-ray imaging community. The method has been demonstrated in various acquisition and reconstruction modes, catering to different demands on the spatial resolution, angular sensitivity and scan time. The latest advances in the operational modes of the technique offer the opportunity to flexibly tune these properties by adjusting reconstruction and scan parameters.

Developed at synchrotrons, X-ray speckle-based imaging was soon translated to laboratory sources with reduced temporal and spatial coherence without major efforts, making the method available for a wide range of users.

Most of the applications of the speckle-based technique to date have been focussed on metrology, optics characterisation and beam phase sensing, for which extremely high sensitivities down to a few nanoradians were achieved. The results obtained with X-ray phase-contrast and dark-field tomography for biomedical applications and materials science indicate the high potential of speckle imaging in these fields. Further applications of speckle-based tomography for multimodal quantitative visualisation of the inner structure of samples are anticipated and some presented in this thesis.

Future work on improving the existing speckle imaging implementations might include the development of alternative diffuser materials that can be adapted to specific experimental setups as well as further optimisation and acceleration of reconstruction algorithms. Some collaborative work on the first topic was carried out during this Ph.D. project, see Sect. 9.2 in Chap. 9.

The robustness and ease of implementation of X-ray speckle-based imaging has attracted increased interest and extensive research on the technique in the last few years. Following the recent developments, the widespread use of X-ray speckle-based imaging and metrology can be expected for applications in an expanding range of fields.

5.10 Most Recent Advances of X-ray Speckle-Based Imaging

Since the publication of the review article (Zdora 2018) on the X-ray speckle-based technique, which was the basis for this chapter, the increasing interest in the method has led to a number of more recent papers. They mainly focus on broadening the range of applications of speckle-based imaging via algorithmic and experimental optimisation, but also highlight the areas of applications that have drawn most in the technique. These developments include:

- optimising existing and developing new data acquisition and processing approaches that make the technique more flexible, robust and time-efficient as well as compatible with X-ray laboratory sources (Zdora et al. 2018a, b; Labriet et al. 2018, 2019; Paganin et al. 2018; Wang et al. 2019; Pavlov et al. 2019; Morgan and Paganin 2019; Paganin and Morgan 2019),
- work on speckle-based dark-field imaging with omni-directional sensitivity (Zhou et al. 2018c),
- pushing speckle imaging to higher X-ray energies, broader X-ray spectra and X-ray laboratory sources (Zdora et al. 2018; Zhou et al. 2018; Wang et al. 2018), see also Sect. 9.3 in Chap. 9, and
- optimising and automating speckle-based metrology for routine use at synchrotron and XFEL beamlines (Zhou et al. 2018, 2019; Bérujon et al. 2019; Jiang et al. 2019b; Seaberg et al. 2019).

It should be noted that some further ongoing developments, which are not published yet, were carried out as part of this Ph.D. project. These include the first demonstrations of 3D virtual histology using speckle-based phase tomography, see Chap. 8, work on the development of customised phase modulators for optimised speckle-based imaging, see Chap. 9, Sect. 9.2, the implementation of UMPA phase tomography at high X-ray energies for imaging of geological and materials science samples, see Chap. 9, Sect. 9.3, and the first report on UMPA speckle-based imaging at a laboratory source, see Chap. 9, Sect. 9.4.

5.10.1 Optimisation and New Development of Operational Modes

Regarding the first point in above summary, the optimisation and application of the existing UMPA (Zdora et al. 2018a, b) (see also following chapters) and XSS methods (Labriet et al. 2018) has been studied recently.

Moreover, a new processing scheme was developed that is suitable for fast imaging applications and promising for high-energy and polychromatic laboratory setups (Paganin et al. 2018; Labriet et al. 2019). This method called "single-image geometric-flow x-ray speckle tracking" relies on the concepts of optical energy conservation and geometric flow and delivers the speckle displacement field as well as the deflection angles and the phase shift of the X-rays in the specimen (Paganin et al. 2018). The method has been demonstrated on a phantom sample and a cicade in 2D projection and 3D tomographic mode, respectively. Quasi-monochromatic high-energy X-rays of 52 keV and a polychromatic (filtered) white beam with a 20 keV bandwidth were used, suggesting the compatibility with higher X-ray energies and laboratory X-ray sources. Recently, a first experimental demonstration of the approach at a laboratory source was reported and its potential for biomedical imaging applications was explored using synchrotron radiation (Labriet et al. 2019).

The geometric flow method was very recently extended by lifting the assumption of a low-absorbing sample and making it applicable to the case of an absorbing single-material object (Pavlov et al. 2019). The potential of the new approach was demonstrated on the phase tomogram of a mouse knee. The method can be seen as a merging of TIE-based reconstruction approaches for propagation-based phase-contrast imaging and speckle-based reconstruction methods.

The most recently proposed algorithmic approach for signal extraction in X-ray phase-contrast imaging is based on the Fokker-Planck equation (Morgan and Paganin 2019; Paganin and Morgan 2019). It is demonstrated that it can be applied to different multimodal imaging techniques such as X-ray grating interferometry, coded aperture imaging and also speckle-based imaging. It allows for the extraction of the transmission, phase and dark-field signals using a mathematical model and is promising for decoupling and understanding cross-talk between the different image modalities. Experimental validations of the proposed approach are anticipated for the near future.

An optimisation of the experimental procedure of speckle-based phase tomography using the 1D XSS mode was recently reported for fast imaging at high X-ray energies (Wang et al. 2019). The data acquisition speed was increased compared to previous tomography implementations of 1D XSS by using fly-scan tomography. The method was applied to several volcanic rock samples at an X-ray of 53 keV and a battery sample at 120 keV, see also Sect. 5.10.3. The total scan time for tomography with this 1D XSS tomography approach is, however, still almost two hours.

Measurements of volcanic rock samples and a mortar specimen at high X-ray energies were also part of this Ph.D. project and are presented in Chap. 9, Sect. 9.3. They were acquired prior to the publication by Wang et al. (2019) and with a similar setup at the same synchrotron beamline. Using the UMPA implementation of speckle-based imaging for phase tomography lead to an improved angular sensitivity compared to the 1D XSS mode. Furthermore, the recent advancements and optimisations of UMPA led to the demonstration of UMPA phase tomography for 3D virtual histology of unstained biomedical soft-tissue specimens shown in Chap. 8 and for the implementation of UMPA at laboratory X-ray sources, see Chap. 9, Sect. 9.4.

5.10.2 Development of Omni-Directional X-ray Speckle-Based Dark-Field Imaging

Although it has so far not been investigated extensively for speckle-based imaging, the dark-field signal has shown promising potential for applications in the medical field and materials science (see Sect. 4.5 in Chap. 4). An advantage of speckle-based over grating-based dark-field imaging is the simplicity of the setup that does not require any specialised optics. Furthermore, the 2D character of the speckles makes multi-directional dark-field imaging feasible. However, previous speckle-based dark-field methods (see Sect. 5.5) only extracted one- or two-directional information. Recently, a method for omni-directional dark-field imaging using single-shot speckle imaging was proposed (Zhou et al. 2018c). The idea is to transform the coordinates in each analysis subset window into polar coordinates via interpolation. The method was successfully demonstrated with mono- as well as polychromatic X-rays and also at high energies up to 52 keV.

5.10.3 Implementation and Optimisation at High X-ray Energies

High-energy speckle imaging using the existing XST and 1D XSS processing schemes was recently investigated for materials science and geology applications and is demonstrated in this thesis also for UMPA.

Multimodal XST phase tomography was reported at a laboratory source using an average energy of 40 keV (Zhou et al. 2018). For this demonstration, a stack of coarse sandpaper was used as a diffuser.

In a different publication Wang et al. investigated different types of engineered porous materials for use as random absorption masks (Wang et al. 2018). Speckle visibilities of 6–7% were obtained with these masks at an X-ray energy of 53 keV. For 1D XSS imaging with custom-made alloy absorption masks, angular sensitivities in the range of (0.21–0.45) μrad were reported (using 50 diffuser steps of 2 μm each at an effective detector pixel size of 3.8 μm). For the most recent high-energy tomography measurements, steel wool strands were used to produce a high-contrast reference pattern by absorption effects (Wang et al. 2019), which enabled speckle-based tomography at energies up to 120 keV with a speckle visibility of 14%.

Another publication aiming at improving speckle-based imaging at higher X-ray energies and with a broader spectrum focussed on the optimisation of the detection system. It was proposed to place the scintillation screen of the detector system in a flowing inert-gas environment to avoid the accumulation of dust (Zdora et al. 2018). It was demonstrated that with this setup high-energy speckle imaging with the 1D XSS method could be performed with a filtered white beam of peak energy 57.5 keV without any degradation of the scintillator.

5.10.4 *Further Development for X-ray Optics and Beam Characterisation*

In the last years, the speckle-based technique has seen increasing uptake for X-ray optics and beam characterisation thanks to its simplicity and high sensitivity, see Sect. 5.8.1. A recent publication reviews the principles of X-ray speckle-based beam, lens and mirror characterisation and its optimisations over the past years (Bérujon et al. 2019). Furthermore, latest experimental results on absolute beam state and beam stability measurements as well as the characterisation of 2D mirror surfaces, multilayer interference fringes and X-ray lenses and the optimisation of adaptive optics are reported in this publication.

The latter has been subject of intense efforts in the last years and it has been shown that the speckle-based technique offers a convenient and simple way for online mirror alignment and optimisation. An approach that can be almost fully automated was demonstrated for the case of elliptical mirrors that are commonly used at synchrotron beamlines (Zhou et al. 2018). The idea is to determine the longitudinal ray aberration to quantify the misalignment of the mirrors, which results in a different focus-detector distance for different pixels. This concept for mirror alignment was shown to be efficient and easy to implement at synchrotron beamlines and recently a user-friendly software featuring a graphical user interface was developed for this purpose (Zhou et al. 2019).

As one of the most recent developments, the translation of speckle-based beam-phase sensing and optics characterisation to XFEL sources should be mentioned. A wavefront sensor based on the X-ray speckle-based technique was proposed for this purpose (Bérujon et al. 2017), which has now been implemented and tested at the European XFEL (Seaberg et al. 2019). The potential of the instrument was demonstrated on the characterisation of a phase plate.

For optics characterisation at the synchrotron, other recently published reports on speckle-based metrology apply the scanning approach to the in-situ characterisation of multilayers (Jiang et al. 2017, 2019b) and the iterative optimisation of a piezo-electric deformable mirror (Jiang et al. 2019a). The implementation of the setup for the first was similar to the one proposed by Bérujon et al. (2014) using 1D scanning of the diffuser. From the speckle measurements, medium-spatial-frequency (spatial wavelength in the range of 1 mm to 1 μm) of inner structural features of the multilayers were extracted. This allowed for the quantitative determination of the intrinsic roughness, replication factor and figure errors of the multilayers. Furthermore, the complementary dark-field signal reconstructed from the same data set was shown to carry information on fluctuations of the inner layer density and interfacial roughness (Jiang et al. 2019b). For the second case, speckle-scanning as proposed by Wang et al. (2015b, e) was applied to perform an iterative optimisation of the mirror shape to minimise wavefront errors.

5.11 Concluding Remarks

In this chapter a comprehensive review of the X-ray speckle-based technique was given, including the different available data acquisition and analysis schemes, developments of the method and various applications that have been demonstrated in the last years. Furthermore, the most recent research and advances of the speckle-based method for imaging and metrology purposes were discussed in the last section.

The chapter provides the fundamental basis for the following content of this thesis, which includes work on the development, optimisation and applications of X-ray speckle-based imaging and optics characterisation that was conducted during this Ph.D. project.

References

Aizu Y, Asakura T (1991) Bio-speckle phenomena and their application to the evaluation of blood flow. Opt Laser Technol 23(4):205–219

Aizu Y, Asakura T (1996) Bio-speckles, chapter 2, pages 27–49, Lasers and optical engineering. Academic Press, San Diego, CA, USA

Alaimo MD, Potenza MAC, Manfredda M, Geloni G, Sztucki M, Narayanan T, Giglio M (2009) Probing the transverse coherence of an undulator x-ray beam using Brownian particles. Phys Rev Lett 103(19):194805

Alexander TL, Harvey JE, Weeks AR (1994) Average speckle size as a function of intensity threshold level: comparison of experimental measurements with theory. Appl Opt 33(35):8240–8250

Aloisio IA, Paganin DM, Wright CA, Morgan KS (2015) Exploring experimental parameter choice for rapid speckle-tracking phase-contrast X-ray imaging with a paper analyzer. J Synchrotron Radiat 22(5):1279–1288

Ando M, Sunaguchi N, Shimao D, Pan A, Yuasa T, Mori K, Suzuki Y, Jin G, Kim J-K, Lim J-H, Seo S-J, Ichihara S, Ohura N, Gupta R (2016) Dark-Field imaging: recent developments and potential clinical applications. Phys Med 32(12):1801–1812

Bates RHT (1982) Astronomical speckle imaging. Phys Rep 90(4):203–297

Baum T, Eggl E, Malecki A, Schaff F, Potdevin G, Gordijenko O, Garcia EG, Burgkart R, Rummeny EJ, Noël PB, Bauer JS, Pfeiffer F (2015) X-ray dark-field vector radiography—a novel technique for osteoporosis imaging. J Comput Assist Tomogr 39(2):286–289

Bennett EE, Kopace R, Stein AF, Wen H (2010) A grating-based single-shot x-ray phase contrast and diffraction method for in vivo imaging. Med Phys 37(11):6047–6054

Bérujon S (2015) Comment on hard-x-ray directional dark-field imaging using the speckle scanning technique. Researchgate. https://doi.org/10.13140/rg.2.1.3975.9608

Bérujon S, Cojocaru R, Piault P, Celestre R, Roth T, Barrett R, Ziegler E (2019) X-ray optics and beam characterization using random modulation. arXiv:1902.09418

Bérujon S, Wang H, Alcock S, Sawhney K (2014) At-wavelength metrology of hard X-ray mirror using near field speckle. Opt Express 22(6):6438–6446

Bérujon S, Wang H, Pape I, Sawhney K (2013a) X-ray phase microscopy using the speckle tracking technique. Appl Phys Lett 102(15):154105

Bérujon S, Wang H, Sawhney K (2012a) X-ray multimodal imaging using a random-phase object. Phys Rev A 86(6):063813

Bérujon S, Wang H, Sawhney KJS (2013b) At-wavelength metrology using the X-ray speckle tracking technique: case study of a X-ray compound refractive lens. J Phys Conf Ser 425(5):052020

Bérujon S, Ziegler E (2015) Near-field speckle-scanning-based x-ray imaging. Phys Rev A 92(1):013837

Bérujon S, Ziegler E (2016) X-ray multimodal tomography using Speckle-Vector tracking. Phys Rev Appl 5(4):044014

Bérujon S, Ziegler E (2017) Near-field speckle-scanning-based x-ray tomography. Phys Rev A 95(6):063822

Bérujon S, Ziegler E, Cerbino R, Peverini L (2012b) Two-Dimensional X-ray beam phase sensing. Phys Rev Lett 108(15):158102

Bérujon S, Ziegler E, Cloetens P (2015) X-ray pulse wavefront metrology using speckle tracking. J Synchrotron Radiat 22(4):886–894

Bérujon S, Ziegler E, Cojocaru R, Martin T (2017) Development of a hard x-ray wavefront sensor for the EuXFEL. Proc SPIE 10237:102370K

Boas DA, Dunn AK (2010) Laser speckle contrast imaging in biomedical optics. J Biomed Opt 15(1):011109

Brogioli D, Vailati A, Giglio M (2002) Heterodyne near-field scattering. Appl Phys Lett 81(22):4109–4111

Carnibella RP, Kitchen MJ, Fouras A (2012) Determining particle size distributions from a single projection image. Opt Express 20(14):15962–15968

Carnibella RP, Kitchen MJ, Fouras A (2013) Decoding the structure of granular and porous materials from speckled phase contrast X-ray images. Opt Express 21(16):19153–19162

Cerbino R (2007) Correlations of light in the deep Fresnel region: an extended Van Cittert and Zernike theorem. Phys Rev A 75(5):053815

Cerbino R, Peverini L, Potenza MAC, Robert A, Bösecke P, Giglio M (2008) X-ray-scattering information obtained from near-field speckle. Nat Phys 4(3):238–243

Chellappan KV, Erden E, Urey H (2010) Laser-based displays: a review. Appl Opt 49(25):F79–F98

Cloetens P, Guigay JP, De Martino C, Baruchel J, Schlenker M (1997) Fractional Talbot imaging of phase gratings with hard x rays. Opt Lett 22(14):1059–1061

Dainty J (ed) (1984a) Laser speckle and related phenomena. Topics in applied physics. Springer, Berlin, Germany

Dainty JC (1984b) Stellar speckle interferometry. In: Dainty J (ed) Laser speckle and related phenomena, vol 9 of Topics in applied physics. Springer, Berlin, Germany

David C, Nöhammer B, Solak HH, Ziegler E (2002) Differential x-ray phase contrast imaging using a shearing interferometer. Appl Phys Lett 81(17):3287–3289

Ennos AE (1975) Speckle interferometry, vol 9. Topics in applied physics, chapter 6. Springer, Englewood, CO, UK

Federation of European Producers of Abrasives (2019) FEPA P-grit sizes coated abrasives. https://www.fepa-abrasives.com/abrasive-products/grains. Accessed 21 June 2019

Frankot RT, Chellappa R (1988) A method for enforcing integrability in shape from shading algorithms. IEEE Trans Pattern Anal Mach Intell 10(4):439–451

Fujii H, Asakura T, Nohira K, Shintomi Y, Ohura T (1985) Blood flow observed by time-varying laser speckle. Opt Lett 10(3):104–106

Gatti A, Magatti D, Ferri F (2008) Three-dimensional coherence of light speckles: theory. Phys Rev A 78(6):063806

Giglio M, Brogiol D, Potenza MAC, Vailati A (2004) Near field scattering. Phys Chem Chem Phys 6(7):1547–1550

Giglio M, Carpineti M, Vailati A (2000) Space intensity correlations in the near field of the scattered light: a direct measurement of the density correlation function $g(r)$. Phys Rev Lett 85(7):1416–1419

Giglio M, Carpineti M, Vailati A, Brogioli D (2001) Near-field intensity correlations of scattered light. Appl Opt 40(24):4036–4040

Goodman JW (1984) Statistical properties of laser speckle patterns. In: Dainty J (ed) Laser speckle and related phenomena, vol 9 Topics in applied physics. Springer, Berlin, Germany

Goodman JW (2007) Speckle phenomena in optics: theory and applications. Roberts & Company Publishers, Englewood, CO, US

Gromann LB, De Marco F, Willer K, Noël PB, Scherer K, Renger B, Gleich B, Achterhold K, Fingerle AA, Münzel D, Auweter S, Hellbach K, Reiser M, Bähr A, Dmochewitz M, Schröter TJ, Koch FJ, Meyer P, Kunka D, Mohr J et al (2017) In-vivo X-ray dark-field chest radiography of a pig. Sci Rep 7(1):4807

Hamed A (2013) Recognition of direction of new apertures from the Eloganted Speckle images: simulation. Opt Photonics J 3(3):250–258

Harker M, O'Leary P (2008) Least squares surface reconstruction from measured gradient fields. In: Proceedings of the IEEE conference on computer vision and pattern recognition, Anchorage, USA, 23–28 June 2008. IEEE, Piscataway, NJ, US, pp 1–7

He C, Sun W, MacDonald CA, Petruccelli JC (2019) The application of harmonic techniques to enhance resolution in mesh-based x-ray phase imaging. J Appl Phys 125(23):233101

Hellbach K, Yaroshenko A, Willer K, Pritzke T, Baumann A, Hesse N, Auweter S, Reiser MF, Eickelberg O, Pfeiffer F, Hilgendorff A, Meinel FG (2016) Facilitated diagnosis of Pneumothoraces in newborn mice using X-ray dark-field radiography. Invest Radiol 49(10):597–601

Hemberg O, Otendal M, Hertz HM (2003) Liquid-metal-jet anode electron-impact x-ray source. Appl Phys Lett 83(7):1483–1485

Høgmoen K, Pedersen HM (1977) Measurement of small vibrations using electronic speckle pattern interferometry: theory. J Opt Soc Am 67(11):1578–1583

Horch E (1995) Speckle imaging in astronomy. Int J Imaging Syst Technol 6(4):401–417

Irvine SC, Paganin DM, Dubsky S, Lewis RA, Fouras A (2008) Phase retrieval for improved three-dimensional velocimetry of dynamic x-ray blood speckle. Appl Phys Lett 93(15):153901

Izadifar M, Kelly ME, Peeling L (2018) Synchrotron speckle-based x-ray phase-contrast imaging for mapping intra-aneurysmal blood flow without contrast agent. Biomed Phys Eng Express 4(1):015011

Jiang H, Tian N, Liang D, Du G, Yan S (2019a) A piezoelectric deformable X-ray mirror for phase compensation based on global optimization. J Synchrotron Rad 26(3):729–736

Jiang H, Yan S, Liang D, Tian N, Wang H, Li A (2017) X-ray multilayer mid-frequency characterizations using speckle scanning techniques. Proc SPIE 10385:103850Q

Jiang H, Yan S, Tian N, Liang D, Dong Z, Zheng (2019b) Extraction of medium-spatial-frequency interfacial waviness and inner structure from X-ray multilayers using the speckle scanning technique. Opt Mater Express 9(7):2878–2891

Jiang M, Wyatt CL, Wang G (2008) X-ray phase-contrast imaging with three 2D gratings. Int J Biomed Imaging 2008:827152

Jones R, Wykes C (1989) Holographic and speckle interferometry, Cambridge studies in modern optics. Cambridge University Press, Cambridge, UK

Kak AC, Slaney M (2001) Principles of computerized tomographic imaging. Society for Industrial and Applied Mathematics, Philadelphia, PA, USA

Kashyap Y, Wang H, Sawhney K (2015) Two-dimensional transverse coherence measurement of hard-x-ray beams using near-field speckle. Phys Rev A 92(3):033842

Kashyap Y, Wang H, Sawhney K (2016a) Development of a speckle-based portable device for *in* situ metrology of synchrotron X-ray mirrors. J Synchrotron Radiat 23(5):1131–1136

Kashyap Y, Wang H, Sawhney K (2016b) Experimental comparison between speckle and grating-based imaging technique using synchrotron radiation X-rays. Opt Express 24(16):18664–18673

Kashyap Y, Wang H, Sawhney K (2016c) Speckle-based at-wavelength metrology of X-ray mirrors with super accuracy. Rev Sci Instrum 87(5):052001

Kim GB, Lee SJ (2006) X-ray PIV measurements of blood flows without tracer particles. Exp Fluids 41(2):195–200

Kim GB, Lee SJ (2009) Contrast enhancement of speckle patterns from blood in synchrotron X-ray imaging. J Biomech 42(4):449–454

Kitchen MJ, Paganin D, Lewis RA, Yagi N, Uesugi K, Mudie ST (2004) On the origin of speckle in x-ray phase contrast images of lung tissue. Phys Med Biol 49(18):4335

Kottler C, David C, Pfeiffer F, Bunk O (2007) A two-directional approach for grating-based differential phase contrast-imaging using hard x-rays. Opt Express 15(3):1175–1181
Kuratomi Y, Sekiya K, Satoh H, Tomiyama T, Kawakami T, Katagiri B, Suzuki Y, Uchida T (2010) Speckle reduction mechanism in laser rear projection displays using a small moving diffuser. J Opt Soc Am A 27(8):1812–1817
Labriet H, Bérujon S, Broche L, Fayard B, Bohic S, Stephanov O, Paganin DM, Lhuissier P, Salvo L, Bayat S, Brun E (2019) 3D histopathology speckle phase contrast imaging: from synchrotron to conventional sources. Proc SPIE 10948:109481S
Labriet H, Brun E, Fayard B, Bohic S, Bérujon S (2018) Performance of x-ray speckle tracking phase retrieval algorithms towards a dose improvement. Microsc Microanal 24(S2):36–37
Lauridsen T, Willner M, Bech M, Pfeiffer F, Feidenhans'l R (2015) Detection of sub-pixel fractures in X-ray dark-field tomography. Appl Phys A 121(3):1243–1250
Leong AFT, Asare E, Rex R, Xiao XH, Ramesh KT, Hufnagel TC (2019) Determination of size distributions of non-spherical pores or particles from single x-ray phase contrast images. Opt Express 27(12):17322–17347
Leong AFT, Paganin DM, Hooper SB, Siew ML, Kitchen MJ (2013) Measurement of absolute regional lung air volumes from near-field x-ray speckles. Opt Express 21(23):27905–27923
Liba O, Lew MD, SoRelle ED, Dutta R, Sen D, Moshfeghi DM, Chu S, de la Zerda A (2017) Speckle-modulating optical coherence tomography in living mice and humans. Nat. Commun 8:15845
Lucy LB (1974) An iterative technique for the rectification of observed distributions. Astron J 79(6):745
Magatti D, Gatti A, Ferri F (2009) Three-dimensional coherence of light speckles: experiment. Phys Rev A 79(5):053831
Meinel FG, Yaroshenko A, Hellbach K, Bech M, Müller M, Velroyen A, Bamberg F, Eickelberg O, Nikolaou K, Reiser MF, Pfeiffer F, Yildirim AÖ (2014) Improved diagnosis of pulmonary emphysema using In Vivo Dark-Field radiography. Invest Radiol 51(10):653–658
Modregger P, Kagias M, Irvine SC, Brönnimann R, Jefimovs K, Endrizzi M, Olivo A (2017) Interpretation and utility of the moments of small-angle x-ray scattering distributions. Phys Rev Lett 118(26):265501
Modregger P, Rutishauser S, Meiser J, David C, Stampanoni M (2014) Two-dimensional ultra-small angle X-ray scattering with grating interferometry. Appl Phys Lett 105(2):024102
Modregger P, Scattarella F, Pinzer BR, David C, Bellotti R, Stampanoni M (2012) Imaging the ultrasmall-angle x-ray scattering distribution with grating interferometry. Phys Rev Lett 108(4):048101
Mohon N, Rodemann A (1973) Laser speckle for determining ametropia and accommodation response of the eye. Appl Opt 12(4):783–787
Momose A, Kawamoto S, Koyama I, Hamaishi Y, Takai K, Suzuki Y (2003) Demonstration of x-ray talbot interferometry. Jpn J Appl Phys 42:L866–L868
Morgan KS, Modregger P, Irvine SC, Rutishauser S, Guzenko VA, Stampanoni M, David C (2013) A sensitive x-ray phase contrast technique for rapid imaging using a single phase grid analyzer. Opt Lett 38(22):4605–4608
Morgan KS, Paganin DM (2019) Applying the Fokker-Planck equation to grating-based x-ray phase and dark-field imaging. Sci Rep 9(1):17465
Morgan KS, Paganin DM, Siu KKW (2011a) Quantitative single-exposure x-ray phase contrast imaging using a single attenuation grid. Opt Express 19(20):19781–19789
Morgan KS, Paganin DM, Siu KKW (2011b) Quantitative x-ray phase-contrast imaging using a single grating of comparable pitch to sample feature size. Opt Lett 36(1):55–57
Morgan KS, Paganin DM, Siu KKW (2012) X-ray phase imaging with a paper analyzer. Appl Phys Lett 100(12):124102
Murrie RP, Morgan KS, Maksimenko A, Fouras A, Paganin DM, Hall C, Siu KKW, Parsons DW, Donnelley M (2015) Live small-animal X-ray lung velocimetry and lung micro-tomography at the Australian Synchrotron Imaging and Medical Beamline. J Synchrotron Rad 22(4):1049–1055

Murrie RP, Paganin DM, Fouras A, Morgan KS (2016) Phase contrast x-ray velocimetry of small animal lungs: optimising imaging rates. Biomed Opt Express 7(1):79–92

Nazmov V, Reznikova E, Mohr J, Snigirev A, Snigireva I, Achenbach S, Saile V (2004) Fabrication and preliminary testing of X-ray lenses in thick SU-8 resist layers. Microsys Technol 10(10):716–721

Nesterets YI (2008) On the origins of decoherence and extinction contrast in phase-contrast imaging. Opt Commun 281(4):533–542

Noël PB, Willer K, Fingerle AA, Gromann LB, Marco FD, Scherer KH, Herzen J, Achterhold K, Gleich B, Münzel D, Renz M, Renger BC, Fischer F, Braun C, Auweter S, Hellbach K, Reiser MF, Schröter T, Mohr J, Yaroshenko A et al (2017) First experience with x-ray dark-field radiography for human chest imaging (Conference Presentation). Proc SPIE 10132:1013215

Paganin DM, Labriet H, Brun E, Bérujon S (2018) Single-image geometric-flow x-ray speckle tracking. Phys Rev A 98(5):053813

Paganin DM, Morgan KS (2019) X-ray Fokker-Planck equation for paraxial imaging. Sci Rep 9(1):17537

Pagot E, Cloetens P, Fiedler S, Bravin A, Coan P, Baruchel J, Härtwig J, Thomlinson W (2003) A method to extract quantitative information in analyzer-based x-ray phase contrast imaging. Appl Phys Lett 82(20):3421–3423

Pan B, Qian K, Xie H, Asundi A (2009) Two-dimensional digital image correlation for in-plane displacement and strain measurement: a review. Meas Sci Technol 20(6):062001

Pan J-W, Shih C-H (2014) Speckle reduction and maintaining contrast in a LASER pico-projector using a vibrating symmetric diffuser. Opt Express 22(6):6464–6477

Park H, Yeom E, Lee SJ (2016) X-ray PIV measurement of blood flow in deep vessels of a rat: an in vivo feasibility study. Sci Rep 6:19194

Pavlov KM, Li H, Paganin DM, Bérujon S, Rougé-Labriet H, Brun E (2019) Single-shot x-ray speckle-based imaging of a single-material object. arXiv:1908.00411

Pfeiffer F, Bech M, Bunk O, Kraft P, Eikenberry EF, Brönnimann C, Grünzweig C, David C (2008) Hard-X-ray dark-field imaging using a grating interferometer. Nat Mater 7:134–137

Prade F, Chabior M, Malm F, Grosse CU, Pfeiffer F (2015) Observing the setting and hardening of cementitious materials by X-ray dark-field radiography. Cem Concr Res 74:19–25

Prade F, Fischer K, Heinz D, Meyer P, Mohr J, Pfeiffer F (2016) Time resolved X-ray dark-field tomography revealing water transport in a fresh cement sample. Sci Rep 6:29108

Rabal H, Braga R (eds) (2008) Dynamic laser speckle and applications, optical science and engineering. CRC Press, Boca Raton, FL, US

Revol V, Jerjen I, Kottler C, Schütz P, Kaufmann R, Lüthi T, Sennhauser U, Straumann U, Urban C (2011) Sub-pixel porosity revealed by x-ray scatter dark field imaging. J Appl Phys 110(4):044912

Revol V, Plank B, Kaufmann R, Kastner J, Kottler C, Neels A (2013) Laminate fibre structure characterisation of carbon fibre-reinforced polymers by X-ray scatter dark field imaging with a grating interferometer. NDT E Int 58:64–71

Richardson WH (1972) Bayesian-Based iterative method of image restoration∗. J Opt Soc Am 62(1):55–59

Sawhney K, Alcock S, Sutter J, Bérujon S, Wang H, Signorato R (2013) Characterisation of a novel super-polished bimorph mirror. J Phys Conf Ser 425(5):052026

Schaff F, Bachmann A, Zens A, Zäh MF, Pfeiffer F, Herzen J (2017) Grating-based X-ray dark-field computed tomography for the characterization of friction stir welds: a feasibility study. Mater Charact 129:143–148

Schleede S, Meinel FG, Bech M, Herzen J, Achterhold K, Potdevin G, Malecki A, Adam-Neumair S, Thieme SF, Bamberg F, Nikolaou K, Bohla A, Yildirim AÖ, Loewen R, Gifford M, Ruth R, Eickelberg O, Reiser M, Pfeiffer F (2012) Emphysema diagnosis using X-ray dark-field imaging at a laser-driven compact synchrotron light source. Proc Natl Acad Sci 109(44):17880–17885

Schmitt JM, Xiang SH, Yung KM (1999) Speckle in optical coherence tomography. J Biomed Opt 4(1):95–105

Seaberg M, Cojocaru R, Bérujon S, Ziegler E, Jaggi A, Krempasky J, Seiboth F, Aquila A, Liu Y, Sakdinawat A, Lee HJ, Flechsig U, Patthey L, Koch F, Seniutinas G, David C, Zhu D, Mikeš L, Makita M, Koyama T et al (2019) Wavefront sensing at X-ray free-electron lasers. J Synchrotron Rad 26(4)

Sharp B (1989) Electronic speckle pattern interferometry (ESPI). Opt Laser Eng 11(4):241–255

Sun W, MacDonald CA, Petruccelli JC (2019) Propagation-based and mesh-based x-ray quantitative phase imaging with conventional sources. Proc SPIE 10990:109900U

Suzuki Y, Yagi N, Uesugi K (2002) X-ray refraction-enhanced imaging and a method for phase retrieval for a simple object. J Synchrotron Rad 9(3):160–165

Vittoria FA, Endrizzi M, Olivo A (2017) Retrieving the ultrasmall-angle x-ray scattering signal with polychromatic radiation in speckle-tracking and beam-tracking phase-contrast imaging. Phys Rev Appl 7(3):034024

Wang F, Wang Y, Wei G, Du G, Xue Y, Hu T, Li K, Deng B, Xie H, Xiao T (2017) Speckle-tracking X-ray phase-contrast imaging for samples with obvious edge-enhancement effect. Appl Phys Lett 111(17):174101

Wang H, Atwood RC, Pankhurst MJ, Kashyap Y, Cai B, Zhou T, Lee PD, Drakopoulos M, Sawhney K (2019) High-energy, high-resolution, fly-scan X-ray phase tomography. Sci Rep 9(1):8913

Wang H, Bérujon S, Herzen J, Atwood R, Laundy D, Hipp A, Sawhney K (2015a) X-ray phase contrast tomography by tracking near field speckle. Sci Rep 5: 8762

Wang H, Bérujon S, Sutter J, Alcock SG, Sawhney K (2014) At-wavelength metrology of x-ray optics at Diamond Light Source. Proc SPIE 9206:920608

Wang H, Cai B, Pankhurst MJ, Zhou T, Kashyap Y, Atwood R, Le Gall N, Lee P, Drakopoulos M, Sawhney K (2018) X-ray phase-contrast imaging with engineered porous materials over 50 keV. J Synchrotron Rad 25(4):1182–1188

Wang H, Kashyap Y, Cai B, Sawhney K (2016a) High energy X-ray phase and dark-field imaging using a random absorption mask. Sci Rep 6:30581

Wang H, Kashyap Y, Laundy D, Sawhney K (2015b) Two-dimensional in situ metrology of X-ray mirrors using the speckle scanning technique. J Synchrotron Radiat 22(4):925–929

Wang H, Kashyap Y, Sawhney K (2015c) Hard-X-Ray Directional Dark-Field Imaging Using the Speckle Scanning Technique. Phys Rev Lett 114(10):103901

Wang H, Kashyap Y, Sawhney K (2015d) Speckle based X-ray wavefront sensing with nanoradian angular sensitivity. Opt Express 23(18):23310–23317

Wang H, Kashyap Y, Sawhney K (2016b) From synchrotron radiation to lab source: advanced speckle-based X-ray imaging using abrasive paper. Sci Rep 6:20476

Wang H, Kashyap Y, Sawhney K (2016c) Quantitative X-ray dark-field and phase tomography using single directional speckle scanning technique. Appl Phys Lett 108(12):124102

Wang H, Kashyap Y, Zhou T, Sawhney K (2017a) Speckle-based portable device for in-situ metrology of x-ray mirrors at diamond light source. Proc SPIE 10385:1038504

Wang H, Sutter J, Sawhney K (2015e) Advanced in situ metrology for x-ray beam shaping with super precision. Opt Express 23(2):1605–1614

Wang H, Zhou T, Kashyap Y, Sawhney K (2017b) Speckle-based at-wavelength metrology of x-ray optics at diamond light source. Proc SPIE 10388:103880I

Weitkamp T, Diaz A, David C, Pfeiffer F, Stampanoni M, Cloetens P, Ziegler E (2005) X-ray phase imaging with a grating interferometer. Opt Express 13(16):6296–6304

Wen HH, Bennett EE, Kopace R, Stein AF, Pai V (2010) Single-shot x-ray differential phase-contrast and diffraction imaging using two-dimensional transmission gratings. Opt Lett 35(12):1932–1934

Willer K, Fingerle AA, Gromann LB, De Marco F, Herzen J, Achterhold K, Gleich B, Muenzel D, Scherer K, Renz M, Renger B, Kopp F, Kriner F, Fischer F, Braun C, Auweter S, Hellbach K, Reiser MF, Schroeter T, Mohr J et al (2018) X-ray dark-field imaging of the human lung-A feasibility study on a deceased body. PLoS ONE 13(9):1–12

Yang F, Prade F, Griffa M, Jerjen I, Di Bella C, Herzen J, Sarapata A, Pfeiffer F, Lura P (2014) Dark-field X-ray imaging of unsaturated water transport in porous materials. Appl Phys Lett 105(15):154105

Yang L, Xie X, Zhu L, Wu S, Wang Y (2014) Review of electronic speckle pattern interferometry (ESPI) for three dimensional displacement measurement. Chinese J Mech Eng En 27(1):1–13

Yaroshenko A, Hellbach K, Yildirim AÖ, Conlon TM, Fernandez IE, Bech M, Velroyen A, Meinel FG, Auweter S, Reiser M, Eickelberg O, Pfeiffer F (2015) Improved In vivo assessment of pulmonary fibrosis in mice using x-ray dark-field radiography. Sci Rep 5:17492

Yaroshenko A, Pritzke T, Koschlig M, Kamgari N, Willer K, Gromann L, Auweter S, Hellbach K, Reiser M, Eickelberg O, Pfeiffer F, Hilgendorff A (2016) Visualization of neonatal lung injury associated with mechanical ventilation using x-ray dark-field radiography. Sci Rep 6:24269

Yashiro W, Terui Y, Kawabata K, Momose A (2010) On the origin of visibility contrast in x-ray Talbot interferometry. Opt Express 18(16):16890–16901

Zanette I, Bech M, Pfeiffer F, Weitkamp T (2011) Interlaced phase stepping in phase-contrast x-ray tomography. Appl Phys Lett 98(9):094101

Zanette I, Weitkamp T, Donath T, Rutishauser S, David C (2010) Two-Dimensional x-ray grating interferometer. Phys Rev Lett 105(24):248102

Zanette I, Zdora M-C, Zhou T, Burvall A, Larsson DH, Thibault P, Hertz HM, Pfeiffer F (2015) X-ray microtomography using correlation of near-field speckles for material characterization. Proc Natl Acad Sci USA 112(41):12569–12573

Zanette I, Zhou T, Burvall A, Lundström U, Larsson DH, Zdora M-C, Thibault P, Pfeiffer F, Hertz HM (2014) Speckle-Based x-ray phase-contrast and dark-field imaging with a laboratory source. Phys Rev Lett 112(25):253903

Zdora M-C (2018) State of the art of x-ray speckle-based phase-contrast and dark-field imaging. J Imaging 4(5):60

Zdora M-C, Thibault P, Deyhle H, Vila-Comamala J, Kuo W, Rau C, Zanette I (2018a) Advanced X-ray phase-contrast and dark-field imaging with the unified modulated pattern analysis (UMPA). Microsc Microanal 24(S2):20–21

Zdora M-C, Thibault P, Deyhle H, Vila-Comamala J, Rau C, Zanette I (2018b) Tunable X-ray speckle-based phase-contrast and dark-field imaging using the unified modulated pattern analysis approach. J Instrum 13(05):C05005

Zdora M-C, Thibault P, Pfeiffer F, Zanette I (2015) Simulations of x-ray speckle-based dark-field and phase-contrast imaging with a polychromatic beam. J Appl Phys 118(11):113105

Zdora M-C, Thibault P, Rau C, Zanette I (2017a) Characterisation of speckle-based X-ray phase-contrast imaging. J Phys Conf Ser 849(1):012024

Zdora M-C, Thibault P, Zhou T, Koch FJ, Romell J, Sala S, Last A, Rau C, Zanette I (2017b) X-ray phase-contrast imaging and metrology through unified modulated pattern analysis. Phys Rev Lett 118(20):203903

Zdora M-C, Zanette I, Zhou T, Koch FJ, Romell J, Sala S, Last A, Ohishi Y, Hirao N, Rau C, Thibault P (2018c) At-wavelength optics characterisation via X-ray speckle- and grating-based unified modulated pattern analysis. Opt Express 26(4):4989–5004

Zheng B, Pleass CM, Ih CS (1994) Feature information extraction from dynamic biospeckle. Appl Opt 33(2):231–237

Zhou T, Wang H, Connolley T, Scott S, Baker N, Sawhney K (2018a) Development of an X-ray imaging system to prevent scintillator degradation for white synchrotron radiation. J Synchrotron Rad 25(3):801–807

Zhou T, Wang H, Fox O, Sawhney K (2018b) Auto-alignment of X-ray focusing mirrors with speckle-based at-wavelength metrology. Opt Express 26(21):26961–26970

Zhou T, Wang H, Fox OJL, Sawhney KJS (2019) Optimized alignment of X-ray mirrors with an automated speckle-based metrology tool. Rev Sci Instrum 90(2):021706

Zhou T, Wang H, Sawhney K (2018c) Single-shot X-ray dark-field imaging with omnidirectional sensitivity using random-pattern wavefront modulator. Appl Phys Lett 113(9):091102

Zhou T, Yang F, Kaufmann R, Wang H (2018d) Applications of laboratory-based phase-contrast imaging using speckle tracking technique towards high energy x-rays. J Imaging 4(5)

Zhou T, Zanette I, Zdora M-C, Lundström U, Larsson DH, Hertz HM, Pfeiffer F, Burvall A (2015) Speckle-based x-ray phase-contrast imaging with a laboratory source and the scanning technique. Opt Lett 40(12):2822–2825

Zhou T, Zdora M-C, Zanette I, Romell J, Hertz HM, Burvall A (2016) Noise analysis of speckle-based x-ray phase-contrast imaging. Opt Lett 41(23):5490–5493

Chapter 6
The Unified Modulated Pattern Analysis

6.1 Introductory Remarks

In this chapter an advanced processing scheme for X-ray speckle-based imaging is presented, which was developed during this Ph.D. project, the unified modulated pattern analysis (UMPA). The name of the method stems from its fundamental idea to unify X-ray grating—and speckle-based imaging and their various operational modes in a single approach. The proposed technique can be applied to existing grating—and speckle-imaging setups and is adaptable to different experimental conditions through tuning of the scan and reconstruction parameters. It provides a flexible solution that fills the gap between the existing single-shot and scanning modes of X-ray speckle—and grating-based imaging.

For the case of speckle-based imaging, it was discussed in the previous chapter that originally two main operational modes were proposed, single-shot XST and multi-frame XSS. Although first demonstrations were successfully conducted with these modes, their wider use for scientifically relevant applications has been impeded by some of their inherent limitations, in particular the relatively low spatial resolution for XST and the long scan times and high requirements on the diffuser scanning stages for XSS. The UMPA acquisition and processing scheme was developed to overcome these limiting factors and bridge the gap between the XST and XSS modes, as illustrated in the diagram in Fig. 6.1. UMPA encompasses the extreme cases of the single-shot and scanning approaches, but also allows for freely choosing the number of diffuser steps and reconstruction parameters to achieve sensitivities and spatial resolutions in between the ones accessible by XST and XSS. This makes UMPA a flexible technique for speckle—and grating-based based imaging that, furthermore, does not require specialised equipment such as high-precision scanning stages and can be operated with a simple, robust setup.

This chapter begins with a section discussing the limiting factors of the XST and XSS techniques for speckle-based imaging, which is an adapted version of an original conference proceedings article published as (Zdora et al. 2017b): M.-C. Zdora,

M.-C. Zdora, *X-ray Phase-Contrast Imaging Using Near-Field Speckles*,
Springer Theses, https://doi.org/10.1007/978-3-030-66329-2_6

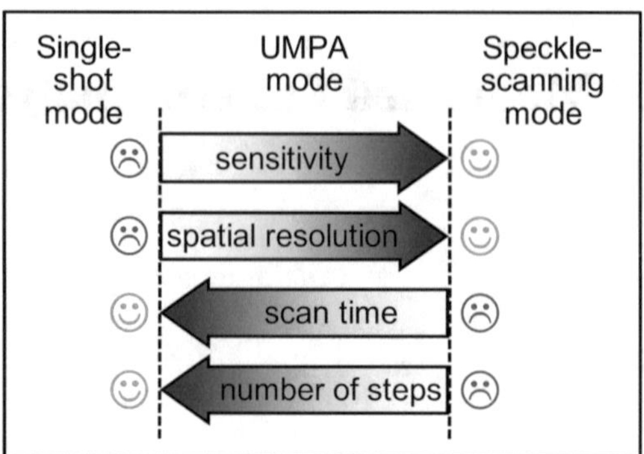

Fig. 6.1 Relationship of the single-shot XST, the speckle-scanning XSS and the new UMPA operational modes. XST suffers from limited spatial resolution and sensitivity, while XSS requires a long scan time with a large number of small, equidistant phase modulator steps. UMPA encompasses both and closes the gap between the two methods by allowing for flexible tuning of the angular sensitivity and spatial resolution

P. Thibault, C. Rau, and I. Zanette, "Characterisation of speckle-based X-ray phase-contrast imaging," J. Phys. Conf. Ser. **849**(1), 012024 (2017), licensed under CC BY 3.0. The effects of the scan and reconstruction parameters on the image quality of the differential phase signal obtained with XST, 2D XSS and 1D XSS are discussed.

The second and main section of the chapter explains in detail the data acquisition and reconstruction concept of UMPA and how it tackles the limitations of the previous implementations of speckle imaging. First demonstrations of the method for multimodal imaging and optics characterisation are presented and it is shown that UMPA is compatible with random and periodic phase modulators. This section is an adapted reprint of the paper published as (Zdora et al. 2017b): M.-C. Zdora, P. Thibault, T. Zhou, F. J. Koch, J. Romell, S. Sala, A. Last, C. Rau, and I. Zanette, "X-ray Phase-Contrast Imaging and Metrology through Unified Modulated Pattern Analysis," Phys. Rev. Lett. **118**(20), 203903 (2017), licensed under CC BY 4.0.

In the last section of the chapter, the tunable character of the UMPA approach is illustrated in more detail on a demonstration sample reconstructed with different parameter combinations. This section contains excerpts of a proceedings article published as (Zdora et al. 2018): M.-C. Zdora, P. Thibault, H. Deyhle, J. Vila-Comamala, C. Rau, and I. Zanette, "Tunable X-ray speckle-based phase-contrast and dark-field imaging using the unified modulated pattern analysis approach," J. Instrum. **13**(5), C05005–C05005 (2018), licensed under CC BY 3.0.

The rest of this thesis following this chapter is based on the UMPA approach and various applications and further extensions of UMPA are presented.

6.2 Performance of the X-ray Speckle-Tracking and Speckle-Scanning Methods

In this section, it is studied how different scan and reconstruction parameters influence the image quality of the differential phase signals obtained from speckle-based X-ray phase-contrast imaging measurements in the single-shot as well as the 2D and 1D speckle-stepping modes. In particular, the effects of the analysis window size and the number of diffuser steps on the spatial resolution and signal sensitivity are investigated using a phantom sample and discussed. It is shown that the trade-off between spatial resolution, angular sensitivity, scan time and simplicity of the setup has to carefully be addressed.

6.2.1 Background and Motivation

As mentioned previously, X-ray speckle imaging is typically performed in two different operational modes, speckle tracking (XST) (Bérujon et al. 2012b; Morgan et al. 2012) and speckle scanning (XSS) (Bérujon et al. 2012a). While single-shot XST is fast and dose efficient, it suffers from a low spatial resolution of the images. The 2D XSS mode can achieve significantly higher resolution, but requires several hundreds of projections, an extremely stable setup and high-precision scanning motors. The 1D stepping approach (Wang et al. 2016) that was proposed a bit later and other advanced operational modes (Bérujon and Ziegler 2015, 2016, 2017), including the UMPA approach discussed in this chapter (Zdora et al. 2017b), allowed for a reduction of the number of diffuser steps and a simplification of the stepping scheme. In the following, the XST, 2D XSS and 1D XSS operational modes are characterised experimentally on a phantom sample and the effects of different scan and reconstruction parameters, in particular the analysis window size and the number of diffuser steps, are explored.

6.2.2 Experimental Setup and Signal Reconstruction

Measurements were performed at the I13-2 Diamond-Manchester Imaging Beamline at Diamond Light Source, UK using a monochromatic X-ray beam of energy 19 keV. The experimental setup is shown in Fig. 6.2a. The sample—silicon spheres (diameter: 480 µm) and quartz spheres (diameter: 350 µm) in a polypropylene pipette tip filled with water—was mounted on a hanging rotation stage and immersed in a water tank. The diffuser, here a piece of P800 sandpaper, was placed on a piezo xy-translation stage at a distance of approximately 10 cm downstream of the specimen. The detector system (150 µm-thick CdWO4 scintillation screen, optical microscope with

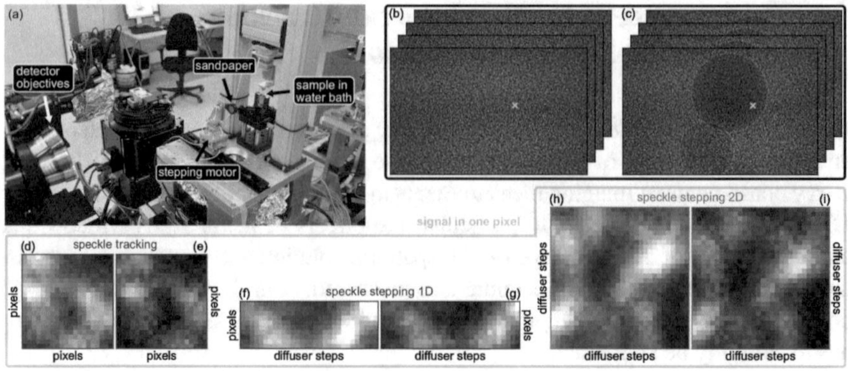

Fig. 6.2 Experimental setup and reconstruction principles for the XST, 1D XSS and 2D XSS modes of X-ray speckle-based imaging. **a** Photograph of the experimental setup at Diamond I13 beamline. **b** Reference images and **c** sample images are collected for different diffuser steps. The visibility of the pattern averaged over the whole field of view is 42%. The signal in one pixel (cyan cross) is shown for **d**, **e** the single-shot XST analysis, **f**, **g** the 1D XSS mode and **h**, **i** the 2D XSS mode, without (left) and with (right) the sample in the beam. Figure reprinted with permission from Zdora et al. (2017b), licensed under CC BY 3.0

20× magnification, pco.4000 CCD camera with 9 µm pixel size) giving an effective pixel size of $p_{\text{eff}} = 0.45\,\mu\text{m}$ was positioned 20 cm downstream of the diffuser.[1]

Reference and sample images, see Fig. 6.2b, c, respectively, were acquired for 30×30 translational steps of the diffuser (step size: $d_{\text{step}} = 0.45\,\mu\text{m}$) with an exposure time of 2 s per projection.

The reconstruction of the image signals for the XST and the 2D and 1D XSS cases is based on cross-correlation in real space (Zanette et al. 2014), delivering the displacement s of the speckle pattern caused by the refraction of X-rays in the sample, which can be converted into a refraction angle $\alpha = s\zeta/z$, where z is the diffuser-detector distance. Here, $\zeta = p_{\text{eff}}$ for the single-shot XST mode and $\zeta = d_{\text{step}}$ for the 2D speckle-stepping XSS mode. In case of 1D XSS, $\zeta = d_{\text{step}}$ for the signal in the scan direction and $\zeta = p_{\text{eff}}$ for the signal perpendicular to the scan direction.

For the XST analysis, a windowed cross-correlation was applied to a single reference and sample image, see Fig. 6.2d, e, while the analysis of the 2D XSS scan was performed pixel-wise in the diffuser plane over the sample and reference signals at the different diffuser steps, see Fig. 6.2h, i. The 1D XSS analysis was conducted considering only one row of 30 steps in the horizontal and a few pixels in the vertical direction for cross-correlation, see Fig. 6.2f, g. From the two differential phase signals, the phase shift was obtained via 2D Fourier integration (Kottler et al. 2007).

[1]The visibility of the speckle pattern is defined here as $(I_{\text{max}} - I_{\text{min}})/(I_{\text{max}} + I_{\text{min}})$, where I_{max} and I_{min} are the maximum and minimum intensities of the pattern in a 100 × 100 pixels region of interest in the background area.

6.2.3 Effects of Different Scan and Reconstruction Parameters

The different parameters used for the reconstructions using the XST, conventional 2D XSS and 1D horizontal XSS modes are summarised in Table 6.1.

For the XST analysis (see Chap. 5, Sect. 5.4.1) the analysis window size was varied from 5 × 5 pixels to 30 × 30 pixels. Figure 6.3a–d shows the corresponding refraction angle signals in the horizontal direction. A clear improvement in image quality with increasing window size can be observed. This is confirmed by the increase in the angular sensitivity, determined as the standard deviation of the refraction angle in a region of 100 × 100 pixels in the background area without sample, which can be found in Table 6.1. However, it can also be seen in Fig. 6.3a–d that a larger window results in a loss of spatial resolution. The trade-off between spatial resolution, which is for XST limited by the window size, and image quality has to carefully be considered.

As explained in Chap. 5, Sect. 5.4.2.1, a significant increase in spatial resolution can be achieved with the XSS mode by performing a 2D scan of the diffuser in steps smaller than the speckle size. However, this comes at the cost of a large number of acquisition frames and the need for very accurate and precise stepping motors to ensure equidistant step sizes. The refraction angle signals in the two directions were reconstructed for different numbers of steps keeping the step size constant at 0.45 μm, see Fig. 6.3e–h. The image quality significantly improves with more steps (see Table 6.1), however the immense increase in scan time and dose to the sample make 2D XSS impractical for most applications.

As a third operational mode that provides a trade-off between the two previous methods, 1D XSS can be applied, for which the diffuser is scanned in only one direction, which significantly reduces the number of acquired images, see Chap. 5, Sect. 5.4.2.2. However, for a 2D analysis of the speckle shift, a few additional pixels in the direction perpendicular to the scan direction have to be considered in the reconstruction process. Commonly, several tens of scan positions are required for 1D XSS in order to allow for a small window size of only several pixels (Wang et al. 2016). Here, a scan of 30 steps along the horizontal direction was used for the analysis and w_y pixels in the vertical direction at each scan position were included

Table 6.1 Different scan and analysis parameters and respective angular sensitivities of the reconstructed differential phase images of the phantom sphere sample. Table reprinted with permission from Zdora et al. (2017b), licensed under CC BY 3.0

XST	**5×5 px**	**10×10 px**	**20×20 px**	**30×30 px**
$\sigma_x \mid \sigma_y$ [nrad]	1982 | 1824	1309 | 1157	352 | 270	153 | 144
2D XSS	**5×5 steps**	**10×10 steps**	**20×20 steps**	**30×30 steps**
$\sigma_x \mid \sigma_y$ [nrad]	1979 | 1875	1346 | 1219	399 | 312	180 | 150
1D XSS	**30 steps, 5 px**	**30 steps, 10 px**	**30 steps, 20 px**	**30 steps, 30 px**
$\sigma_x \mid \sigma_y$ [nrad]	764 | 675	435 | 331	241 | 184	188 | 139

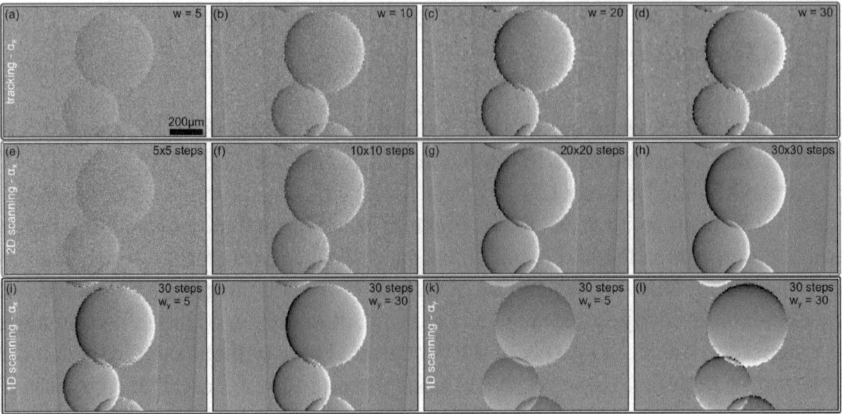

Fig. 6.3 Reconstructions of the phantom sample with different speckle imaging methods and parameters. Refraction angle signal α_x in the horizontal reconstructed with **a–d** XST analysis with different window sizes, **e–h** 2D XSS analysis with different numbers of steps and **i, j** 1D XSS analysis with 30 steps in the horizontal and different numbers of pixels in the vertical. **k, l** Refraction α_y in the vertical for 1D XSS. Grey value range: -6 µrad to 6 µrad. Figure reprinted with permission from Zdora et al. (2017b), licensed under CC BY 3.0

for the signal reconstruction. From the refraction angle signals in the horizontal and the vertical directions in Fig. 6.3i–l, respectively, it can be seen that taking a larger window w_y improves the angular sensitivity, see also Table 6.1, but at the cost of spatial resolution in the vertical direction.

The observations from the differential phase signals translate to the integrated phase shift, which is shown in Fig. 6.4a for XST with a window size of 30 × 30 pixels and in Fig. 6.4b for 2D XSS with 30 × 30 steps. The line plots through the centre of the silicon sphere in Fig. 6.4c illustrate that the phase shift values agree for the two methods, but the XST reconstruction appears more noisy, as expected from the lower angular sensitivity of the differential phase signals. Furthermore, the spatial resolution of the 2D XSS scan is superior to the XST case. Artefacts at the edges of the spheres can be observed for the XST mode, which are not present for the XSS case. They are the result of failure of the reconstruction algorithm in areas with too little diversity in information of the acquired data.

6.2.4 Conclusions

In this section, a study of the effects of the scan and reconstruction parameters on the image quality of the differential phase signals obtained with the originally proposed operational modes—XST, 2D XSS and 1D XSS—was presented.

For XST, the size of the analysis window for the given experimental conditions needed to be on the order of 20 × 20 pixels for an acceptable angular sensitivity,

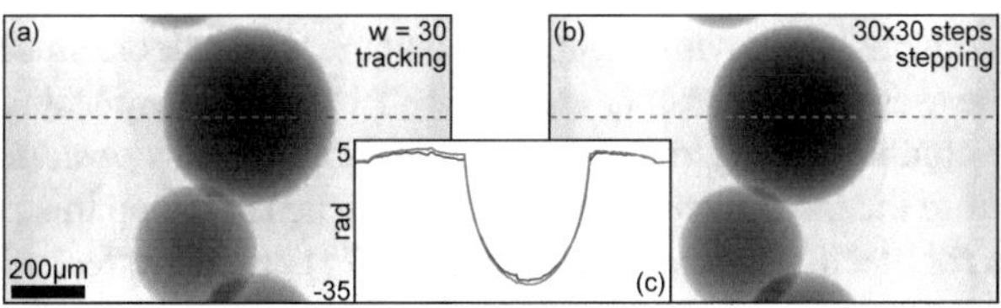

Fig. 6.4 Integrated phase shift signal of the phantom for **a** the XST (window of 30 × 30 pixels) and **b** the 2D XSS case (30 × 30 steps). Grey value window: −33 rad to 4 rad. **c** Horizontal line profiles through the centre of the silicon sphere in panels (**a**) and (**b**). Figure reprinted with permission from Zdora et al. (2017b), licensed under CC BY 3.0

which, however, led to a significant loss in spatial resolution, determined by the extent of the window. A comparable angular sensitivity at much higher spatial resolution was achieved with 2D XSS over at least 20 × 20 steps, which, furthermore, led to the elimination of artefacts at the edges of features where the XST analysis failed. 1D XSS required less diffuser steps than 2D XSS, but led to a reduced spatial resolution and angular sensitivity in the direction perpendicular to the scan direction.

When choosing the preferred operational mode for a speckle imaging experiment, there is an inevitable trade-off between spatial resolution, scan time, angular sensitivity and simplicity of the setup. To overcome the limitations of the XST and XSS techniques and combine their advantages, various advanced approaches have been proposed. One of them is the UMPA mode, which was developed during this Ph.D. project and is introduced in the following section.

6.3 Principle and First Demonstration of the Unified Modulated Pattern Analysis

In this section, a solution for tackling the main limitations and drawbacks of the previous implementations of X-ray speckle—as well as grating-based imaging techniques is presented. The proposed unified modulated pattern analysis (UMPA) is a versatile approach and allows for tuning of the signal sensitivity, spatial resolution and scan time. In the following, the method is characterised and it is demonstrated that it has promising potential for high-sensitivity, quantitative phase imaging and metrology.

6.3.1 Background and Motivation

In the last decades, phase-contrast imaging and phase sensing have found numerous applications in a broad range of fields from the visible light regime to the X-ray and electron domains. Improving computing resources and algorithms have changed profoundly the way measurements are made in these fields, pushing the trend to take multiple complementary measurements to disentangle *a posteriori* the useful information from the instrument response. Approaches based on this principle were used early on for electron microscopy (Misell 1973) and more recently for a range of applications such as laser wavefront sensing (Almoro and Hanson 2008), phase sensing and imaging through in-line holography and ptychography in the visible light (Zhang et al. 2003; Rodenburg et al. 2007a) and the X-ray regime (Wilkins et al. 1996; Cloetens et al. 1999; Rodenburg et al. 2007b; Thibault et al. 2008; Stockmar et al. 2013), and even as a mean to align space telescopes using near-infrared radiation (Acton et al. 2004).

The same evolution can be found in X-ray grating-based imaging (GBI) (David et al. 2002; Momose et al. 2003; Weitkamp et al. 2005), see Chap. 4, and, more recently, speckle-based imaging (SBI) (Bérujon et al. 2012b; Morgan et al. 2012; Bérujon et al. 2012a), see Chap. 5.

As outlined in previous chapters, both GBI and SBI were initially used in single-shot mode, through moiré pattern analysis (Momose et al. 2003; Bevins et al. 2012) and real-space speckle-tracking cross-correlation analysis (Bérujon et al. 2012b; Morgan et al. 2012). The analysis on single frames allows for short acquisition times, but suffers from limited spatial resolution, as areas covering several pixels are required to extract useful information. A second mode of operation, also commonly used for both techniques, involves scanning the reference pattern in equidistant sub-feature steps. This approach, known as phase-stepping (Weitkamp et al. 2005; Miao et al. 2013) and speckle-scanning (Bérujon et al. 2012a; Wang et al. 2016) modes, allows for a pixel-wise reconstruction to be performed, thus improving resolution substantially.

Despite ongoing efforts to improve these techniques (Wang et al. 2016; Bérujon and Ziegler 2015; Hipp et al. 2016; Marschner et al. 2016), important drawbacks still impede their widespread implementation. GBI stepping analysis assumes that the interference pattern is sinusoidal and undistorted in amplitude and period. The step sizes are often as small as 100 nanometres. The common implementation of GBI with two gratings is cumbersome, sensitive to instabilities and dose-inefficient. For SBI, the scanning mode involves the difficult task of acquiring several hundreds of frames taken with step sizes of tens of nanometres, as demonstrated in Sect. 6.2.

Here, a single solution is proposed for the current limitations of both GBI and SBI, which will facilitate their wider implementation, even under sub-optimal experimental conditions, while preserving the achievable image quality. The method, called "Unified Modulated Pattern Analysis" (UMPA), is applicable to any type of reference pattern, periodic or random. In the following, the performance of the technique for imaging and metrology experiments is characterised and its flexibility to tune image quality and scan time is demonstrated.

6.3.2 Materials and Methods

6.3.2.1 Experimental Setup

A sketch of the experimental setup is shown in Fig. 6.5. Experiments were carried out at I13-1 beamline at Diamond Light Source (Rau et al. 2011) with a 19 keV X-ray beam selected by a horizontally deflecting monochromator.

A near-field interference pattern is created in the detector plane by a phase modulator (PM): a near-field speckle pattern produced by a piece of abrasive paper (combination of granularities P800 and P5000[2]) in Fig. 6.5a or a line interference pattern from a beam-splitter phase grating (material: SU-8 photoresist polymer, period: 5.4 μm, line height: 26.7 μm, fabricated by IMT-KIT, Germany) in Fig. 6.5b. Projections with and without the sample in the beam were recorded for different positions of the PM. Each projection was exposed for 4 s. A pco4000 CCD camera (pixel size:

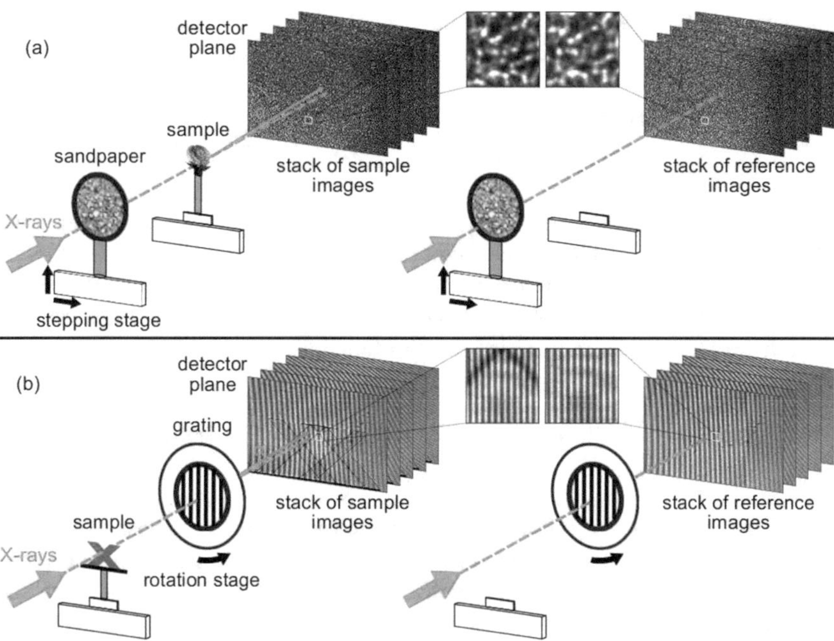

Fig. 6.5 Experimental setup for UMPA imaging. Setup using **a** a piece of abrasive paper and **b** a beam-splitter phase grating as a phase modulator to create a near-field interference pattern in the detector plane. The insets show 150 × 150 pixels regions of the patterns with (left) and without (right) the sample in the beam. Figure reprinted with permission from Zdora et al. (2017b), licensed under CC BY 4.0

[2]See FEPA P-grit classification for coated abrasives by Federation of European Producers of Abrasives (2019). The nominal mean grain sizes of the sandpaper are (21.8 ± 1) μm for the P800 sandpaper and 5 μm for the P5000 sandpaper.

9 µm) coupled to magnifying optics and a scintillation screen was used, providing an effective pixel size of $p_{\text{eff}} = 0.40\,\mu\text{m}$. The sandpaper was placed 0.17 m upstream of the sample and the detector 0.34 m downstream of the sample. The grating, on the other hand, was located 0.16 m downstream of the sample[3] and the detector was placed at the same position as before, i.e. another 0.18 m from the grating, close to the first fractional Talbot distance (Cloetens et al. 1997). The sandpaper was scanned in the transverse plane following a spiral pattern and images were recorded at 24 PM positions with step sizes of about 28 µm—larger than the average speckle size. The grating was rotated around the optical axis from 0 to 90 degrees in increments of 2 degrees between each of the 46 projections.

6.3.2.2 Analytical Model and Reconstruction of the Signals

A mathematical model of the modulations of the reference pattern caused by the presence of the sample has been described previously by Zanette et al. (2014). According to this model, the intensity $I_j(\mathbf{r})$ of the jth out of N interference patterns can be expressed in terms of the undisturbed pattern $I_{0j}(\mathbf{r})$ as follows:

$$\begin{aligned} I_j(\mathbf{r}) &= T(\mathbf{r})\left[\langle I\rangle + v(\mathbf{r})\left(I_{0j}(\mathbf{r}+\mathbf{u}) - \langle I\rangle\right)\right] \\ &= \beta(\mathbf{r})\langle I\rangle + \kappa(\mathbf{r}) I_{0j}(\mathbf{r}+\mathbf{u}), \end{aligned} \tag{6.1}$$

where $T(\mathbf{r})$ is the sample's transmission, $v(\mathbf{r})$ is the visibility, related to small-angle scattering, and $\mathbf{u}$ is a displacement vector proportional to the refraction angle. $\langle I\rangle$ is the mean intensity of the reference pattern and we define $\beta = T(1-v)$ and $\kappa = Tv$. A least-squares formulation of the problem allows to find these quantities from the set of measurements, by minimising the cost function:

$$\mathscr{L} = \sum_{\mathbf{r}} \sum_{j}^{N} \Gamma(\mathbf{r}-\mathbf{r_0}) \left| I_j(\mathbf{r}) - \beta(\mathbf{r})\langle I\rangle - \kappa(\mathbf{r}) I_{0j}(\mathbf{r}+\mathbf{u}) \right|^2 . \tag{6.2}$$

Here, Γ is a normalised window function centered at position $\mathbf{r}_0$, whose width influences both the sensitivity of the reconstructed refraction angle signal and the spatial resolution. Minimisation of Eq. (6.2) with respect to β, κ and $\mathbf{u}$ for all window centres $\mathbf{r}_0$ yields the four independent signal maps T, v and $\mathbf{u} = (u_x, u_y)$.[4] The refraction angle (α_x, α_y) and the phase gradient $(\partial\Phi/\partial x, \partial\Phi/\partial y)$ are obtained from $\mathbf{u}$ through simple geometric transformations (under small-angle approximation):

$$\left(\frac{\partial\Phi}{\partial x}, \frac{\partial\Phi}{\partial y}\right) = \frac{2\pi}{\lambda}\left(\alpha_x, \alpha_y\right) = \frac{2\pi}{\lambda}\left(u_x, u_y\right)\frac{p_{\text{eff}}}{d}, \tag{6.3}$$

[3]The two PMs were placed at different positions relative to the sample due to mechanical limitations in the setup, which, however, does not affect the quantitative results of the measurements.

[4]For a detailed mathematical description of the UMPA algorithm, see Appendix B.

where d is the propagation distance to the detector. The total phase shift Φ experienced by the X-rays passing through the sample is then recovered via 2D Fourier integration of the differential phase signals (Kottler et al. 2007; Morgan et al. 2012).

In a previous work, a similar method was used to obtain these quantities from a single speckle measurement (Zanette et al. 2014). However, for the problem to be well-behaved one had to set the width of Γ larger than the speckle size as smaller windows do not track speckles reliably. The approach with a single measurement also required diversity in both x and y direction, ruling out its application with linear gratings. Combining multiple measurements in a single reconstruction lifts these limitations. It can be shown that the model of Eq. (6.2) also encompasses the speckle-scanning (XSS) scheme, by letting the window reduce to a single-pixel box function.

6.3.3 Experimental Demonstration

As a first demonstration, the UMPA technique with a random PM was applied to a biological sample, a small flower bud (*Cotoneaster Dammeri Radicans*).

The visibility, defined as the ratio of standard deviation and mean intensity of the speckle pattern in a 150×150 pixels region without sample, was on average 43% in the left half of the field of view and 37% in the right part, with the difference due to slightly slanted mounting of the scintillation screen. The size of the speckles determined via 2D autocorrelation analysis of the speckle pattern, as described by Zdora et al. (2015), was approximately 34 pixels $\approx 13.7\,\mu\text{m}$ in the horizontal and 28 pixels $\approx 11.3\,\mu\text{m}$ in the vertical direction. This asymmetry is explained by a smaller transverse coherence in the horizontal direction.

The reconstruction of the differential phase, transmission and dark-field signals was performed using an analysis window of size $w \times w = 5 \times 5$ pixels (normalised Hamming window with a full width at half maximum, FWHM, of 2 pixels). The corresponding images are presented in Fig. 6.6. The differential phase signals in the horizontal and vertical directions in Fig. 6.6a, b, respectively, reveal the detailed inner structure of the flower bud. The petals in the upper part show a granular texture. In the bottom part, the vertical walls of the stalk are clearly visible, in particular in the horizontal refraction angle in Fig. 6.6a, and a complex tube network can be observed inside the stalk. This fine tubular composition and a strong phase shift in the upper part of the sample are visible in the integrated phase signal in Fig. 6.6c. While the transmission image in Fig. 6.6d mainly highlights the outlines of the features due to edge-enhancement effects occurring upon propagation, the dark-field in Fig. 6.6e reveals strong scattering from the small granular features, especially in the upper part of the specimen.

The spatial resolution of UMPA is not limited by the speckle size. Its ideal value is equal to twice the FWHM of the window, giving in the present case a resolution of 4 pixels $\approx 1.6\,\mu\text{m}$. For small window sizes, the achievable resolution reaches a limit imposed by experimental factors such as the point spread function of the detector. As

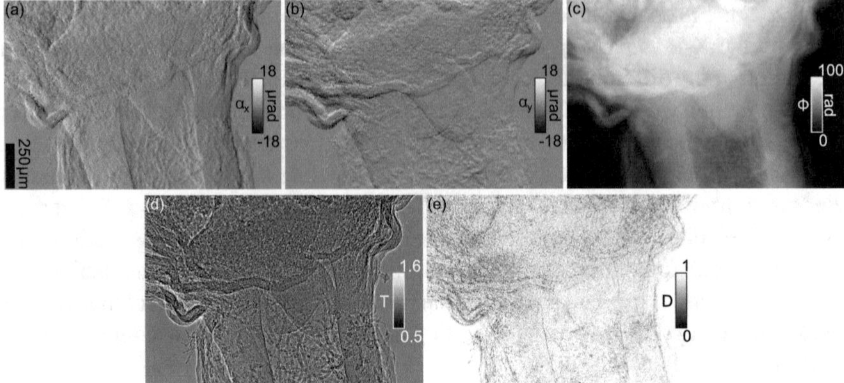

Fig. 6.6 Multimodal images of a small flower bud (*Cotoneaster Dammeri Radicans*) imaged with the speckle setup for UMPA and reconstructed with $N = 24$ and $w = 5$. **a** Refraction in the horizontal direction, **b** refraction in the vertical direction, **c** phase shift, **d** transmission, and **e** dark-field signal. Figure reprinted with permission from Zdora et al. (2017b), licensed under CC BY 4.0

a lower limit of the spatial resolution of the reconstructed phase image,[5] the FWHM of two of the smallest discernible features in Fig. 6.6c is considered, which gives values of 4.8 and 5.1 μm. This estimate is consistent with data from a test pattern, see Sect. 6.3.7. The angular sensitivity of the measurement—evaluated by computing the standard deviation of the refraction angle signal in a region of 150 × 150 pixels in the background area of the image—is 79 nrad in the horizontal and 66 nrad in the vertical direction. The sensitivity is better in the vertical due to the higher transverse coherence of the X-ray beam in this direction, leading to a better visibility and smaller size of the speckles along the vertical axis.

6.3.4 *Tuning of Angular Sensitivity, Spatial Resolution and Scan Time*

As in many other imaging systems, sensitivity and resolution are not independent quantities. Performing error propagation from the raw measurements to the reconstructed refraction maps, one finds that the variance σ^2 in the signal is inversely proportional to Nw^2, the number of independent contributions to the extraction of the signal in one pixel. Hence, the angular sensitivity σ, defined as the standard deviation of the refraction angle signal, can be expressed by the relation

$$\sigma = \frac{C}{w \times \sqrt{N}}, \tag{6.4}$$

[5]The spatial resolution of the phase image will in general be decreased compared to the differential phase signal due to the additional image processing step of 2D integration.

where C is a constant depending on the measurement conditions. This relation, already observed for photographic emulsions many decades ago (Selwyn 1935), generally applies to imaging systems whose noise is devoid of spatial correlations. While Eq. (6.4) does not encompass all experimental contributions to the signal error, such as photon-counting statistics, read-out noise and PM position errors as discussed e.g. by Yashiro et al. (2008) and Revol et al. (2010), it confirms the validity and tunable character of our reconstruction method.

Relation (6.4) explains how to beat the resolution limit imposed by the feature size of the reference pattern: to decrease w while keeping the same sensitivity σ, one simply needs to include more independent measurements N. Figure 6.7 illustrates this possibility with experimental data. A sample made from polymethyl methacrylate (PMMA) spheres of diameters up to 80 μm and polystyrene (PS) spheres of diameter 250–350 μm in a polyvinyl chloride (PVC) tube was imaged with the setup in Fig. 6.5 using a random PM. Figure 6.7a shows the refraction angle signal α_x with the parameters $N = 16$ and $w = 21$. A region of interest (ROI) of 370 × 370 pixels—as indicated by the red box—is shown in Fig. 6.7b. The relationship between the quantities in Eq. (6.4) is visualised for a few fixed w and N values in the line plots in Fig. 6.7i, j, respectively. The constant $C = 2236$ nrad was determined from a weighted least-square fit of the experimental σ values to Eq. (6.4), see Sect. 6.3.7.

The measured σ for selected parameter pairs N, w are plotted as coloured dots in Fig. 6.7i, j and they agree well with the theoretical model. For these parameters, the corresponding ROIs are presented in Fig. 6.7c–h. The gain in angular sensitivity can clearly be observed for increasing N, see Fig. 6.7c–e, and larger window sizes w, see Fig. 6.7f–h, as predicted by Eq. (6.4). The latter, however, also leads to a decrease in spatial resolution. For $N = 1$, equivalent to single-shot SBI (XST), in Fig. 6.7c, the sparsity of information gives rise to systematic errors at the edges of the sphere.

6.3.5 Experimental Realisation with a Periodic Wavefront Modulator

The resolution–sensitivity relation given by Eq. (6.4) holds true also when the reference pattern is periodic. Here, the grating setup described above is used to perform metrology on a polymer (SU-8 photoresist) point-focus compound refractive lens (CRL) (Nazmov et al. 2004) fabricated at IMT-KIT, Germany. The aim of this measurement was to evaluate beam damage from extreme X-ray exposure during previous use as an optics element.[6]

The period of the grating interference pattern, measured by fitting the line pattern to a sinusoidal curve, was approximately 13.5 pixels ≈ 5.4 μm. The mean visibility

[6]The lens was repeatedly exposed to X-rays from an undulator synchrotron source at SPring-8, BL10XU beamline with prefocus-lens (2 mm aperture beryllium lens) focussing the beam on a spot of approximately 50 μm. These extreme conditions led to heating and dose-induced damage of the polymer.

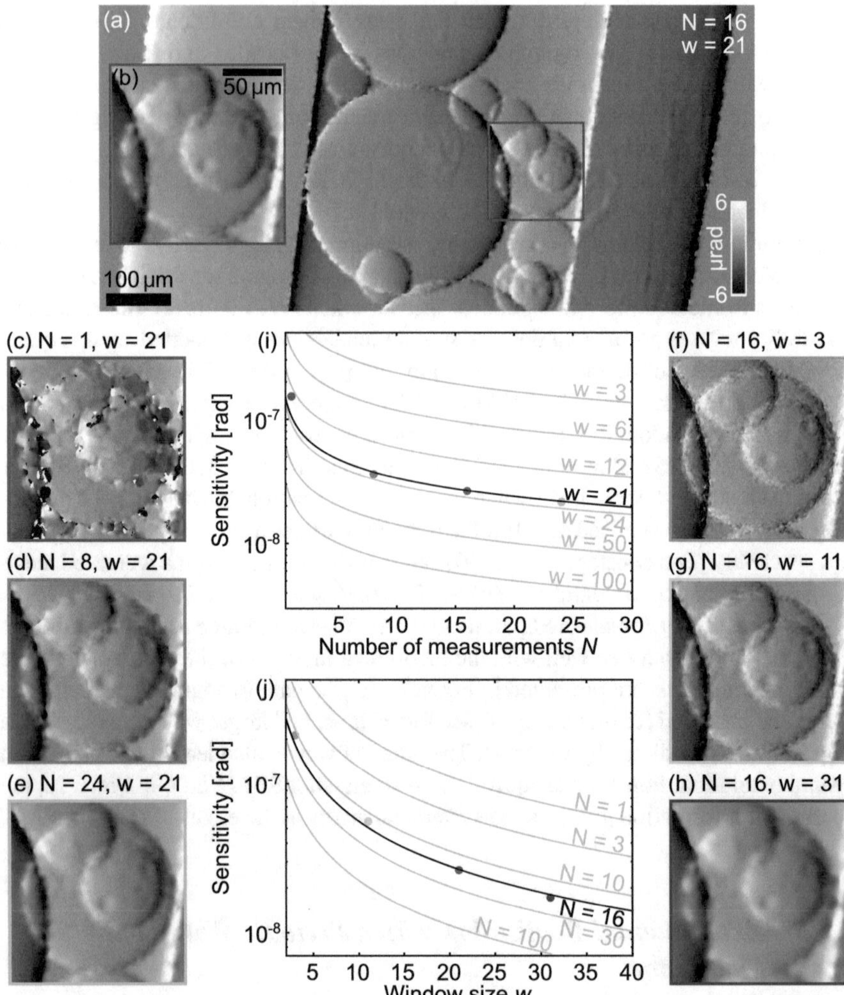

Fig. 6.7 Influence of the scan and reconstruction parameters on the angular sensitivity and spatial resolution. **a** Refraction in the horizontal direction α_x of a sphere sample imaged using a sandpaper diffuser and reconstructed with $N = 16$ and $w = 21$ and **b** region of interest (370×370 pixels). **c**–**h** Same regions reconstructed with different parameter combinations N, w. **i, j** Relationship between angular sensitivity σ and number of measurements N or analysis window size w, respectively, as described by Eq. (6.4). Experimentally measured values are plotted as coloured dots and agree well with the model. Figure reprinted with permission from Zdora et al. (2017b), licensed under CC BY 4.0

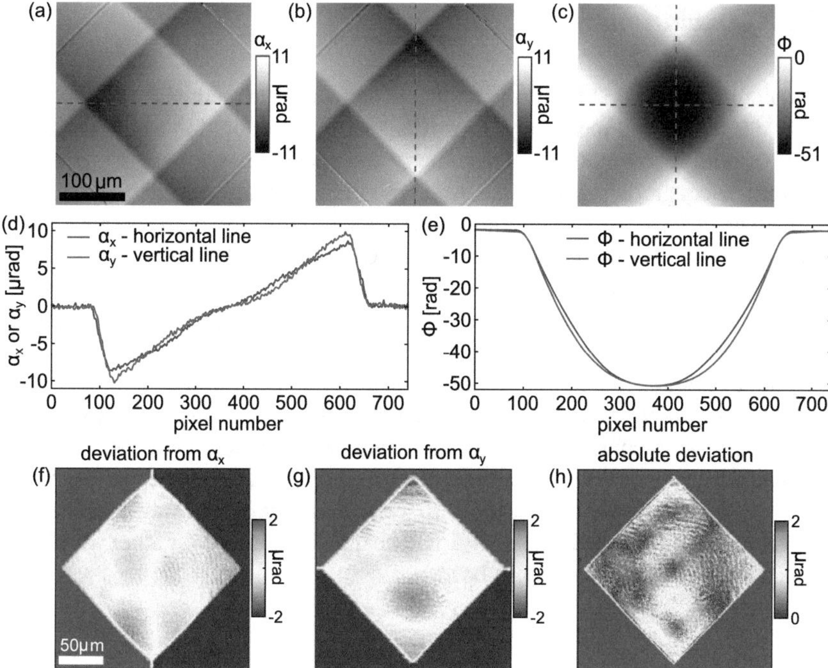

Fig. 6.8 Characterisation of a point-focus refractive lens using UMPA. Refraction signals in **a** the horizontal and **b** the vertical direction of a polymer compound refractive lens measured with a periodic line reference pattern and reconstructed with $N = 46$ and $w = 5$ and **c** integrated phase shift. Line profiles through the centre of the lens for **d** the refraction signal and **e** the phase signal. **f**–**g** Residuals from a linear fit to the refraction angle in the two directions and **h** absolute magnitude of the combined deviation vector. Figure reprinted with permission from Zdora et al. (2017b), licensed under CC BY 4.0

was determined from the same fit as $(I_{\max} - I_{\min})/(I_{\max} + I_{\min})$, where $I_{\min}$ and $I_{\max}$ are the minimum and maximum intensities of the fit curve, respectively. The visibility was on average 30.4% in the left half and 23.4% in the right half of the field of view. The angular sensitivities for this setup were $\sigma_x = 114$ nrad, $\sigma_y = 94$ nrad.

The refraction signals reconstructed with a window size $w = 5$ and the integrated phase are shown in Fig. 6.8a–c. The line profiles through the centre of the differential phase signals in the horizontal (blue) and the vertical direction (red) in Fig. 6.8d and through the integrated phase shift in Fig. 6.8e reveal an asymmetry in the refraction properties of the lens. This observation is confirmed by the residuals from a linear fit to the refraction angle (Koch et al. 2016) in the central focussing part of the lens shown in Fig. 6.8f, g. Figure 6.8h is the magnitude of the combined deviation vector visualising the absolute deviation from the expected refraction behaviour of the lens, which is about two orders of magnitude larger than for CRLs without beam damage (Koch et al. 2016).

6.3.6 Conclusions

It was demonstrated that UMPA is a tunable and versatile method for X-ray phase-sensitive imaging and metrology. It provides an elegant solution to the main limitations of both X-ray grating interferometry and speckle-based imaging. It lifts the need for perfect gratings and, when used with random phase modulators, decreases the number of frames by one order of magnitude with respect to previous implementations. By relaxing the requirements on the phase modulator structure and the scanning step size and precision, UMPA will accelerate the implementation of phase-contrast imaging setups, in particular at laboratory sources, thus also promoting the development of clinical applications (Tanaka et al. 2013). The algorithm, of which the original implementation is available for download,[7] allows to tune the spatial resolution and angular sensitivity and adapt them to the experimental constraints.

The proposed technique can be easily employed in tomographic mode and it is anticipated that it will find applications in a broad range of fields such as biomedical imaging and material characterisation, as well as metrology and wavefront sensing. Furthermore, it is expected that UMPA will be quickly adopted for other wavelength regimes such as laser and electron beams.

6.3.7 Supplementary Information

6.3.7.1 Validation of the UMPA Algorithm

The validity of the proposed reconstruction approach, can be confirmed by comparing the experimental data with Eq. (6.4). Equation (6.4) is a simple analytical model for the angular sensitivity, which does not consider other experimental contributions to the signal error, such as photon counting statistics, read-out noise and errors in the position of the PM, but adequately describes the experimental behaviour when the window size w and number of steps N are not too small.

A weighted least squares fit of the measured angular sensitivity values σ to the model function in Eq. (6.4) of the main text was performed, giving a proportionality constant $C = (2236 \pm 10)$ nrad. For this fit, data points with $(w\sqrt{N}) < 20$ were excluded, as for small N and w experimental contributions become more dominant and Eq. (6.4) does not hold anymore.

The experimental values (black markers) as well as the fit curve (red line) are shown in Fig. 6.9. Error bars of the data are smaller than the size of the symbols. It can be seen that Eq. (6.4) with $C = 2236$ nrad is a good fit through the experimental data. This confirms the validity of the the UMPA reconstruction algorithm and demonstrates that Eq. (6.4) can be used to tune the spatial resolution and contrast of the reconstructed images by adapting N and/or w.

[7] See https://github.com/pierrethibault/UMPA.

6.3.7.2 Demonstration of Quantitative Phase and Transmission Measurements

To demonstrate that the analysis performed with the UMPA technique yields quantitative results for the phase shift and transmission in the sample, the same sphere phantom as in Fig. 6.7 was used, measured with the sandpaper as well as the grating configuration with $N = 18$ steps of the phase modulator and an analysis window of $w = 7$. The refraction angle signals in the horizontal and the vertical directions and the transmission are shown in Fig. 6.10a–c for the sandpaper and Fig. 6.10d–f for the grating case. Note that the sample was dismounted between the two scans and therefore the orientation of the sample in the projections with sandpaper and with grating is not identical, but very similar. The quality of the sandpaper scan is better due to sample movement occurring during the grating scan, which has been partly corrected for, but could not completely be eliminated.

The line profiles through the centre of the large sphere (material: PS, diameter: $\approx 315\,\mu\text{m}$) show excellent agreement of the two measurements for the differential phase as well as the transmission signal of the sphere. Slight discrepancies outside the sphere are caused by the different viewing angles of the projections.

The phase shift integrated from the refraction angle signals in the two directions is shown in Fig. 6.10j, k for the sandpaper and grating case, respectively. The horizontal line profiles through the centre of the sphere in Fig. 6.10m and vertical line profiles in Fig. 6.10l confirm the consistence between the two setups.

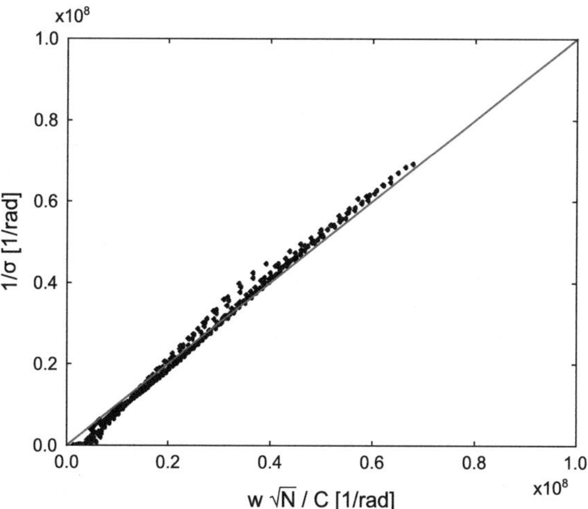

Fig. 6.9 Validation of the angular sensitivity model. Experimentally measured values $1/\sigma$ of the sphere phantom scan (markers) and curve resulting from a weighted least-square fit (red line) of the experimental σ-values to Eq. (6.4). Data points for $(w\sqrt{N}) < 20$ are excluded from the fitting procedure. The fit gives a constant of $C = 2236\,\text{nrad}$. The measured σ agree well with the fit function, confirming the validity of the reconstruction algorithm. Figure reprinted with permission from Supplementary Material of Zdora et al. (2017b), licensed under CC BY 4.0

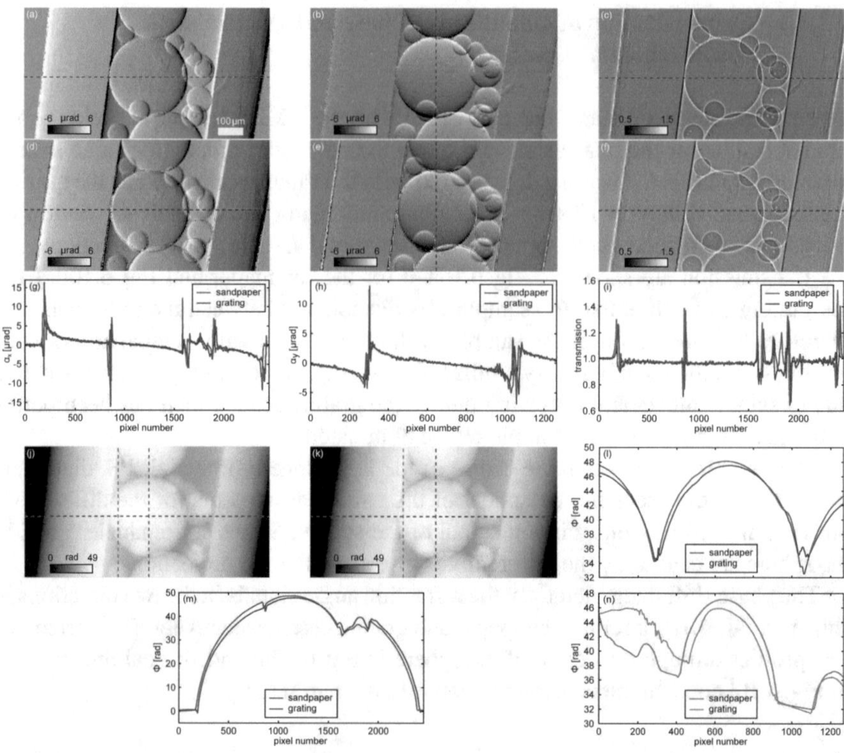

Fig. 6.10 Demonstration of quantitative phase and transmission measurements with UMPA. Multimodal signals of a plastic sphere phantom imaged using a piece of sandpaper as well as a periodic grating as a phase modulator. **a** Refraction angle in the horizontal, **b** refraction angle in the vertical, **c** transmission for the sandpaper case and **d** refraction in the horizontal, **e** refraction in the vertical, **f** transmission for the grating case. **g–i** Line profiles through the centre of the large polystyrene sphere (diameter: $\approx 315\,\mu m$). **j** Total phase shift integrated from (**a**) and (**b**) for the sandpaper case and **k** total phase shift integrated from (**d**) and (**e**) for the grating case. **l** Vertical and **m** horizontal line profiles through the polystyrene sphere. **n** Vertical line profiles through the centre of the smaller polymethyl methacrylate sphere. Discrepancies observed in the line profiles are due to the slightly different orientation of the sample for the two setups. Figure reprinted with permission from Supplementary Material of Zdora et al. (2017b), licensed under CC BY 4.0

For a quantitative analysis, the phase shift and transmission in the centre of the PS sphere is determined and corrected for the background contribution of the tube walls. The measured phase shifts are $\Phi_{\mathrm{SP,\,PS}} \approx 48.1$ rad for the sandpaper and $\Phi_{\mathrm{GR,\,PS}} \approx 47.6$ rad for the grating measurement. Subtracting the phase contribution of the tube $\Phi_{\mathrm{tube,\,SP,\,PS}} \approx 26.8$ rad and $\Phi_{\mathrm{tube,\,GR,\,PS}} \approx 25.9$ rad, estimated from the area below the sphere with only tube, a phase shift in the centre of the sphere of $\Phi_{\mathrm{sphere,\,SP,\,PS}} \approx 21.3$ rad and $\Phi_{\mathrm{sphere,\,GR,\,PS}} \approx 21.6$ rad is obtained.

A theoretical value for the phase shift Φ can be calculated using the relation

$$\Phi = \frac{2\pi}{\lambda}\delta t, \tag{6.5}$$

where λ is the X-ray wavelength, δ the refraction index decrement and t the thickness of the sample, here the sphere diameter. In our case, with $\delta_{PS,19\,keV} = 6.496 \times 10^{-7}$ for PS (chemical formula: C_8H_8, density: 1.05 g/cm^3) (Henke et al. 1993), $\lambda = 6.526 \times 10^{-11}$ m and $t = 315\,\mu m$, a theoretical phase shift of $\Phi_{lit,\,PS} = 19.7$ rad is expected in the centre of the PS sphere.

For the transmission, values of $T_{SP,\,PS} \approx 97.0\%$ and $T_{GR,\,PS} \approx 96.3\%$ are measured in the centre of the PS sphere. After normalisation with the background tube values of $T_{tube,\,SP,\,PS} \approx 98.5\%$ and $T_{tube,\,GR,\,PS} \approx 97.9\%$, transmissions of $T_{sphere,\,SP,\,PS} \approx 98.5\%$ and $T_{sphere,\,GR,\,PS} \approx 98.4\%$ are obtained, which compare well with the literature value $T_{lit,\,PS} \approx 98.7\%$ for 315 µm of PS (Henke et al. 1993).

The slight differences between literature and measured values can be attributed to inaccuracies in the centre position of the sphere and the background values for the contribution of the tube, which are only estimations as the thickness of the tube walls is unknown and not constant.

To confirm the quantitativeness of the measurements, one can also look at the phase signal of the small sphere (material: PMMA, diameter: $\approx 80\,\mu m$) at the bottom left part of the field of view. Vertical line profiles through the centre of the sphere are shown in Fig. 6.10n. Note that the difference in viewing angle of the sample is more evident at this position and hence the line profiles deviate, but are both going through the centre of the sphere. The phase shift values at this point are $\Phi_{SP,\,PMMA} \approx 37.2$ rad and $\Phi_{GR,\,PMMA} \approx 36.4$ rad for the sandpaper and the grating case, respectively. The tube background close to the small sphere shows a phase shift of $\Phi_{tube,\,SP,\,PMMA} \approx 31.5$ rad and $\Phi_{tube,\,GR,\,PMMA} \approx 30.8$ rad, respectively (the background phase shift differs from the values above as the tube does not have a constant wall thickness). After background subtraction with these values, $\Phi_{sphere,\,SP,\,PMMA} \approx 5.7$ rad and $\Phi_{sphere,\,GR,\,PMMA} \approx 5.6$ rad are obtained. This is in good agreement with the literature value $\Phi_{lit,\,PMMA} = 5.7$ rad using Eq. (6.5) with $\delta_{PMMA,19\,keV} = 7.386 \times 10^{-7}$ for PMMA (chemical formula: $C_5H_8O_2$, density: 1.19 g/cm^3) (Henke et al. 1993), $\lambda = 6.526 \times 10^{-11}$ m and $t = 80\,\mu m$.

6.3.7.3 Spatial Resolution of the Reconstructed Phase Images

The spatial resolution of the reconstructed refraction images is limited by the reconstruction algorithm to twice the FWHM of the analysis window w, but may in practice be dominated by experimental factors such as the point spread function of the detector. For $w = 5$ pixels, a resolution limit of 4 pixels $\approx 1.6\,\mu m$ is obtained when only considering the contribution from the reconstruction process. The resolution of the phase shift image after integration will in general be reduced by the additional image processing step. The spatial resolution of the phase signal is here estimated by looking at the smallest resolvable feature in the image and taking its full width at half

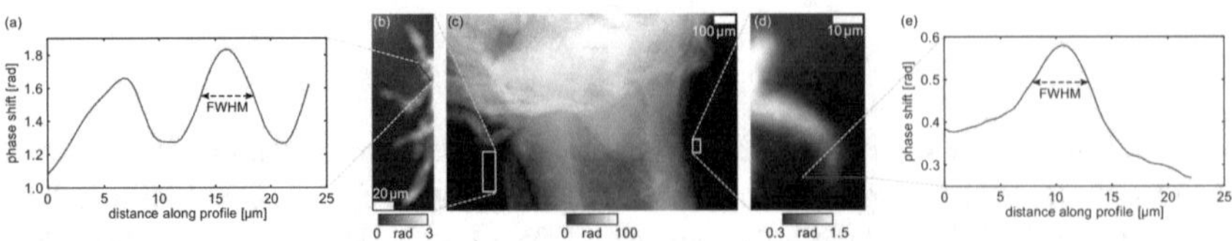

Fig. 6.11 Spatial resolution estimation. **c** Reconstructed phase signal of the flower bud presented in Sect. 6.3.3, recorded with a speckle pattern and parameters $w = 5$, $N = 24$. **b**, **d** Regions of interest and **a**, **e** corresponding line profiles through the smallest features in the regions. The FWHM is taken as an estimate for the spatial resolution of the phase image. Figure reprinted with permission from Supplementary Material of Zdora et al. (2017b), licensed under CC BY 4.0

maximum (FWHM) as a measure for the spatial resolution as shown by Zanette et al. (2010).

This analysis is performed for the phase image of the flower bud (*Cotoneaster Dammeri Radicans*) presented in Sect. 6.3.3, obtained with $w = 5$ and $N = 24$ using a speckle reference pattern. Figure 6.11c shows the phase signal with two regions of interest at the sides of the sample, from which the structures for the analysis were selected. They can be seen in Fig. 6.11b, d and the line profiles across the smallest features are plotted in Fig. 6.11a, e, respectively. The FWHMs of the structures are 4.8 µm for Fig. 6.11a and 5.1 µm for Fig. 6.11e. It should be noted that this analysis only provides an estimation of the resolution and should be seen as an upper limit.

Furthermore, a line test pattern was imaged with the same setup and parameters as the flower bud. The resulting phase shift, transmission and dark-field images are shown in Fig. 6.12. In the phase signal in Fig. 6.12a, the 5.0 µm lines can clearly be resolved, while the 2.5 µm lines are just at the limit of resolution. This is in agreement with the estimation from the FWHM of the smallest features in Fig. 6.11. A similar resolution is obtained for the transmission, for which, however, edge-enhancement effects dominate the signal. The dark-field is related to small-angle scattering and can commonly resolve features smaller than the resolution limit of the other modalities. In Fig. 6.12c the 2.5 µm lines can be resolved.

It should be noted that the setup was not fully optimised yet and higher resolution can be obtained in the future with improved setup components such as thinner scintillation screens and smaller effective pixel sizes, as well as by using reference patterns with smaller features. Further gain in spatial resolution could be realised by performing a deconvolution with the detector point-spread function.

6.4 Tunable Character and Parameter Choice

It was demonstrated in the previous section that the UMPA implementation of the X-ray speckle-based technique can overcome most of the limitations of the original XST and XSS methods and provides a flexible approach for multimodal imag-

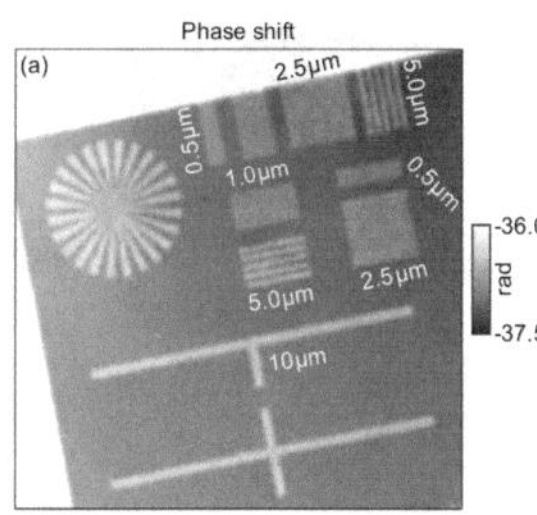

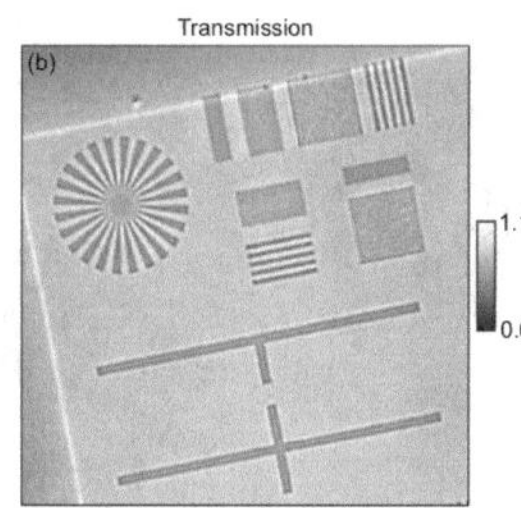

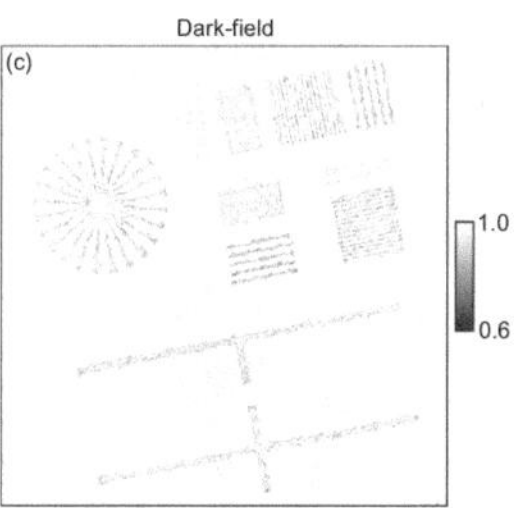

Fig. 6.12 Spatial resolution validation with a test pattern. **a** Phase shift, **b** transmission and **c** dark-field signals of a line test pattern, obtained with a speckle reference pattern and parameters $w = 5$, $N = 24$. Line widths are indicated in panel (**a**). The 5.0 μm lines can clearly be discerned in the phase and transmission images while the dark-field signal allows resolving features down to 2.5 μm size. Figure reprinted with permission from Supplementary Material of Zdora et al. (2017b), licensed under CC BY 4.0

ing. It can, moreover, be applied to periodic reference patterns and hence combines grating—and speckle-based imaging in a single approach.

In this section, the tunable character of UMPA phase-contrast and dark-field imaging with regard to the achievable spatial resolution, signal sensitivity and scan time is studied in more detail on the example of a phantom sample consisting of two materials. It was scanned using a speckle reference pattern and reconstructed with various different parameters. The quality of the resulting images is quantified in terms of angular sensitivity and spatial resolution.

6.4.1 Experimental Setup

Measurements were conducted at the I13-2 Manchester-Diamond imaging beamline at Diamond Light Source with an X-ray beam of energy 20 keV, extracted from the undulator spectrum with a double-crystal monochromator.

The experimental arrangement is illustrated in Fig. 6.13. The setup consisted of a diffuser, here two stacked pieces of sandpaper of FEPA granularities P800 and P5000, see Sect. 6.3.2.1, the sample mounted on a translation stage 4 cm downstream of the diffuser and a camera system located another $d = 40$ cm downstream. The camera system was a pco.4000 CCD camera (pixel size: 9 μm) coupled to a scintillation screen (250 μm cadmium tungstate), a relay optics system and a 4× magnifying objective lens. This gave an effective detector pixel size of $p_{\text{eff}} = 1.12$ μm.

The sample was a silicon sphere (diameter: 480 μm) mounted on a wooden toothpick using superglue. For data acquisition with the UMPA scheme, the diffuser was moved to several different transverse positions, at which images were recorded with and without the sample in the beam.

The reconstruction of the differential phase and dark-field signals was performed with the UMPA processing algorithm described in the previous section (Sect. 6.3).

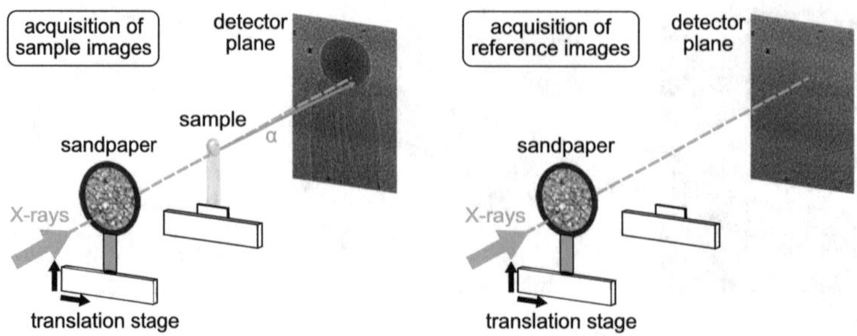

Fig. 6.13 Experimental setup for the UMPA measurements. The diffuser was moved to N different transverse positions and speckle images were recorded with (left) and without (right) the sample in the beam. Figure reprinted with permission from Zdora et al. (2018), licensed under CC BY 3.0

6.4.2 *Phase-Contrast and Dark-Field Imaging of a Demonstration Sample*

6.4.2.1 Acquired Speckle Images

Speckle images with and without the sample in the beam were acquired as described in Sect. 6.4.1. Figure 6.14a, b show one of the acquired sample and reference speckle images, respectively, corrected for the dark current of the detector. The visibility of the interference pattern was calculated as the ratio of standard deviation and mean intensity of a 500×500 pixels area in the centre of the reference pattern and is on average 29%. The line profile plots in Fig. 6.14c, d through the sample (red) and reference (blue) speckle images illustrate the modulations of the interference pattern induced by the silicon sphere, see Fig. 6.14c, and the wooden toothpick, see Fig. 6.14d. On the left side of the plots, reference and sample patterns agree almost perfectly, as no sample is present. Inside the sphere a reduction in intensity as well as a displacement of the speckles can be observed in Fig. 6.14c, while the visibility of the speckle pattern is essentially not affected. On the contrary, in Fig. 6.14d the speckle peaks are broadened and their visibility is strongly reduced due to the scattering nature of the wooden fibres. Furthermore, a slight displacement can be seen, while the change in mean intensity is only subtle as wood is a low-absorbing material. It can be observed that a few smaller additional peaks appear in the sample plot, which are due to edge-enhancement fringes around the wooden fibres.

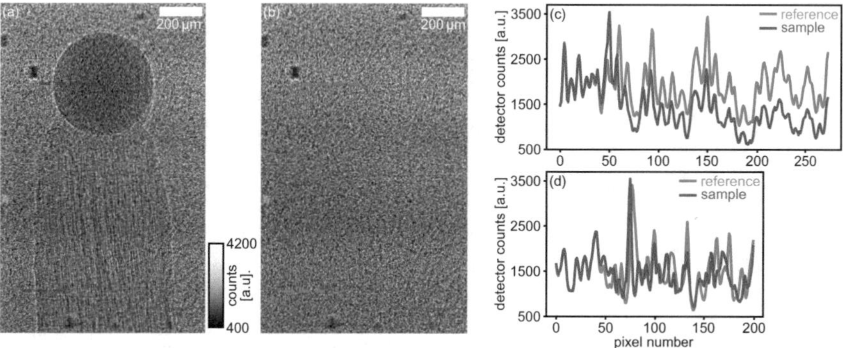

Fig. 6.14 Near-field speckle pattern used as a wavefront marker. Speckle interference patterns **a** with and **b** without the sample in the beam. **c**, **d** Line plots through the patterns at the positions of the sphere and the toothpick, as indicated in (**a**) and (**b**). Figure reprinted with permission from Zdora et al. (2018), licensed under CC BY 3.0

6.4.2.2 Reconstructed Multimodal Image Signals

A displacement and reduction in visibility of the speckle pattern could be observed in areas of the sample. From this information, the refraction angle signals and dark-field small-angle scattering information can be retrieved, as described in Sect. 6.3.2.2.

The refraction angle signals α_y in the vertical and α_x in the horizontal direction as well as the phase shift Φ after integration of the two differential signals are shown in Fig. 6.15a, b, r, respectively, obtained with $N = 20$ steps of the diffuser and an analysis window of size $w \times w = 5 \times 5$ pixels. It can clearly be seen that the refraction of the sphere is symmetric in the two directions, as expected from the geometry of the sample. Some uneven features originate from glue on the surface of the sphere.

To demonstrate the tunable character of the UMPA approach, a region of interest of the horizontal refraction α_x centred around the sphere is shown for different reconstruction parameters in Fig. 6.15c–q. The size w of the analysis window and the number N of diffuser steps were varied. The window size sets a limit for the achievable spatial resolution of the reconstructed images (Zdora et al. 2017b) and hence a loss in resolution can be observed for increasing w. However, a larger window w also improves the angular sensitivity σ, i.e. the sensitivity of the refraction angle measurement (Zdora et al. 2017b). When increasing N at constant w, it can be observed that the spatial resolution is unchanged while the noise level is reduced and there are less artefacts at sharp phase changes such as the edge of the sphere.

The sensitivity values for the different parameter combinations are summarised in Table 6.2. They were determined as the standard deviation in a region of 150 × 150 pixels in the air background of α_x and α_y. The vertical sensitivity σ_y is reduced

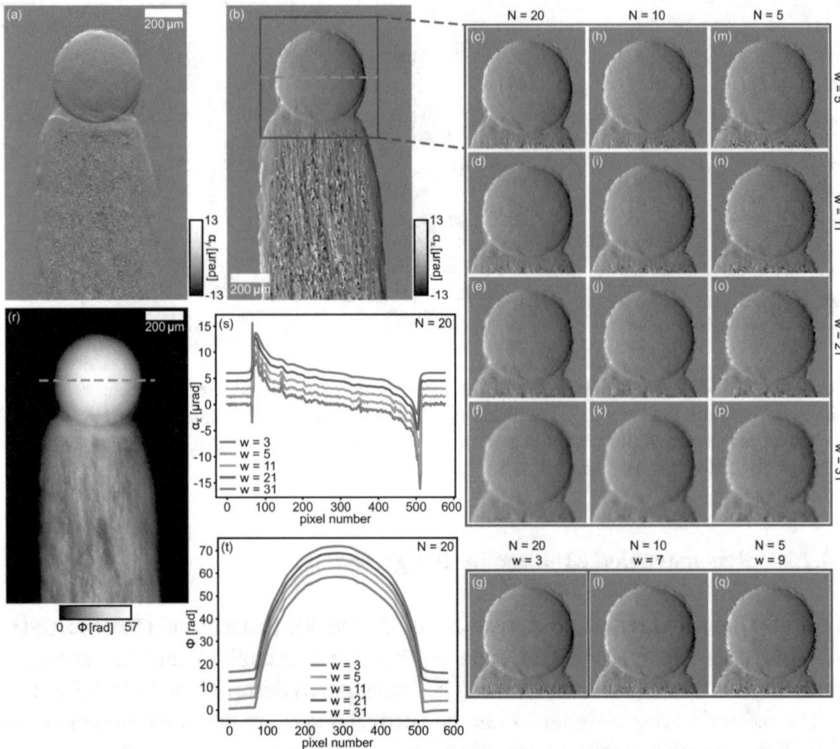

Fig. 6.15 Evaluation of the effects of different scan and reconstruction parameters on the differential phase signal. Refraction angles in **a** the vertical and **b** the horizontal direction for $N = 20$, $w = 5$. **c**–**q** Horizontal refraction of the sphere reconstructed for different window sizes w and numbers of steps N. **r** Phase shift integrated from (**a**) and (**b**). **s**, **t** Line plots through the horizontal refraction and integrated phase in the sphere for $N = 20$ and different w. Lines are offset by 1.5 μrad in (**s**) and 4 rad in (**t**) for clarity. Figure reprinted with permission from Zdora et al. (2018), licensed under CC BY 3.0

compared to σ_x because of the lower vertical coherence of the X-ray beam. The first row of Table 6.2 also gives an indication of the limit Δ in spatial resolution (in pixels), given by twice the FWHM of the Hamming-type analysis window. If it is desired to decrease N, e.g. for reducing scan time or dose, one should be aware of the simultaneous decrease in σ for constant w. The angular sensitivity σ is indirectly proportional to the factor $w\sqrt{N}$ (Zdora et al. 2017b), see Eq. (6.4). Using this relation for a given number of steps N, it is possible to calculate the window size w that is required to achieve the same sensitivities as $\sigma_x^{N=20,w=5}$, $\sigma_y^{N=20,w=5}$. This gives a required window size of $w \approx 7$ for $N = 10$ and a size of $w = 10$ for $N = 5$. As windows of an uneven number of pixels across are required by the reconstruction algorithm, the parameter sets $N = 10, w = 7$ and $N = 5, w = 9$ were used here. The horizontal refraction signals α_x for these settings are shown in Fig. 6.15l, q and the corresponding σ values are highlighted in blue in Table 6.2.

Table 6.2 Angular sensitivity values σ_x in the horizontal and σ_y in the vertical (in units of nrad) for different reconstruction parameters used in Fig. 6.15. The parameter combinations that are expected to have similar sensitivity values are highlighted in blue. The first row gives the limit of the spatial resolution in pixels, imposed by the window size. Table reprinted with permission from Zdora et al. (2018), licensed under CC BY 3.0.

Δ [px]	2	4	6	8	10	20	30
N \ W	3	5	7	9	11	21	31
5	–	σ_x 134.6	–	σ_x 50.1	σ_x 34.7	σ_x 12.7	σ_x 7.5
	–	σ_y 185.0	–	σ_y 66.4	σ_y 45.4	σ_y 16.3	σ_y 9.6
10	–	σ_x 93.6	σ_x 55.1	–	σ_x 24.6	σ_x 9.0	σ_x 5.3
	–	σ_y 122.0	σ_y 72.6	–	σ_y 32.6	σ_y 11.7	σ_y 7.1
20	σ_x 112.6	σ_x 64.5	–	–	σ_x 17.6	σ_x 7.0	σ_x 4.4
	σ_y 143.0	σ_y 84.3	–	–	σ_y 25.0	σ_y 11.8	σ_y 9.0

Figure 6.15g shows the image with highest resolution at a window size of $w = 3$ with $N = 20$. The higher spatial resolution comes at the cost of a reduced angular sensitivity, see Table 6.2.

In Fig. 6.15s, t the horizontal line profiles through the centre of the sphere for the horizontal refraction angle α_x and the integrated phase Φ are plotted for $N = 20$ and the different window sizes w used in Fig. 6.15c–g. Note that the curves are shifted vertically against each other by 1.5 µrad in Fig. 6.15s and 4 rad in Fig. 6.15t for clarity. The plots demonstrate clearly the gain in angular sensitivity and loss in spatial resolution with increasing window size at fixed N. From the maximum of the curve in Fig. 6.15t, the phase shift in the centre of the sphere can be estimated, which was measured to be $\Phi = 58.4$ rad for $N = 20$, $w = 5$. This compares well with the theoretically expected phase shift of 20 keV X-rays after passing through $t = 480$ µm of silicon (density: 2.33 g/cm^3): $\Phi = (2\pi/\lambda)\,\delta t \approx 58.9$ rad, where $\lambda_{20\,\text{keV}} = 6.1992 \times 10^{-11}$ m and $\delta_{\text{Si},20\,\text{keV}} = 1.2162 \times 10^{-6}$ (Henke et al. 1993).

In addition to the refraction signal, the UMPA approach also allows for the reconstruction of the dark-field signal, shown in Fig. 6.16. As the sphere is made from a single material with homogeneous electron density distribution, only the edges of the sphere and the glue show up in the dark-field signal. The wooden fibres of the toothpick, however, lead to strong scattering. The signal intensity increases from the top to the bottom of the toothpick due to the increase in thickness and hence scattering material. As visible in Fig. 6.16b–m, the scatter images show reduced spatial resolution with increasing window size w, as observed for the refraction signal. Furthermore, the decreased signal sensitivity for a smaller number N of diffuser steps leads to a more grainy appearance of the dark-field signal.

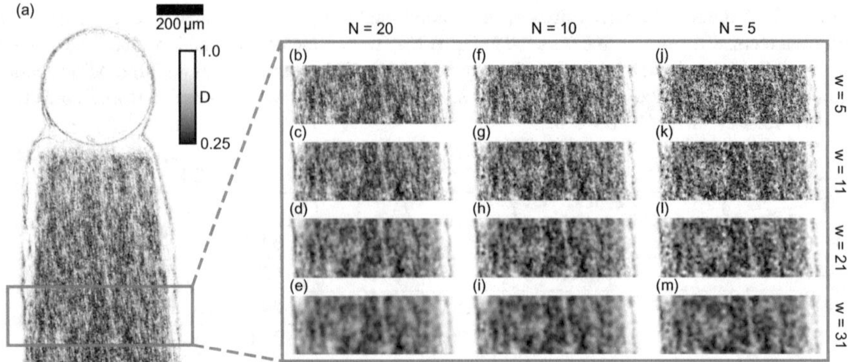

Fig. 6.16 Evaluation of the effects of different scan and reconstruction parameters on the dark-field signal. **a** Dark-field signal of the entire sample obtained with $N = 20$ steps and a window size $w = 5$. **b–m** Dark-field signal in a region of interest in the wooden toothpick reconstructed with different parameters. Figure reprinted with permission from Zdora et al. (2018), licensed under CC BY 3.0

6.4.3 Conclusions and Outlook

In this study the tunable character and potential of the recently proposed UMPA approach was demonstrated for versatile, accurate multimodal X-ray imaging. Using a phantom sample consisting of a silicon sphere and a wooden toothpick, the refraction angle signals in both directions, the integrated phase shift, and the dark-field images were retrieved for different reconstruction parameters. In agreement with the theoretical considerations of the algorithm, an improvement in angular sensitivity and reduction in noise with increasing number of steps and increasing window size were observed. Moreover, the influence of the window size on the spatial resolution of the images was discussed. The flexibility of the UMPA approach allows for adapting the scan and reconstruction parameters to specific experimental requirements, either pushing the spatial resolution with a smaller window size, or keeping a shorter scan time by reducing the number of steps while trading off in resolution. Hence, UMPA bridges the gap between the single-shot and scanning methods, not only for speckle-based, but also for grating-based imaging, opening new avenues for quantitative, high-contrast multimodal X-ray imaging.

6.5 Concluding Remarks

In this chapter, a new data acquisition and reconstruction approach for X-ray speckle-based imaging was presented, which was developed and later optimised as one of the main objectives of this Ph.D. project. It was demonstrated that the UMPA approach is significantly more flexible than previously proposed operational modes and the scan and reconstruction parameters can be tuned to individually match the given experimental conditions and requirements. Furthermore, as UMPA is compatible with any kind of reference pattern, including periodic, it bridges the gap between grating—and speckle-based imaging for a unified treatment of both techniques.

In the following two chapters, the first applications of UMPA with random and periodic phase modulators are demonstrated. It is illustrated that the method has high potential for X-ray optics characterisation (see Chap. 7) and 3D virtual histology of biomedical specimens (see Chap. 8). Other fields of applications that have been identified recently, such as geology and materials science, will be discussed in Chap. 9.

References

Acton DS, Atcheson PD, Cermak M, Kingsbury LK, Shi F, Redding DC (2004) James Webb Space Telescope wavefront sensing and control algorithms. Proc SPIE 5487:887–896

Almoro PF, Hanson SG (2008) Wavefront sensing using speckles with fringe compensation. Opt Express 16(11):7608–7618

Bérujon S, Wang H, Sawhney K (2012a) X-ray multimodal imaging using a random-phase object. Phys Rev A 86(6):063813

Bérujon S, Ziegler E (2015) Near-field speckle-scanning-based x-ray imaging. Phys Rev A 92(1):013837

Bérujon S, Ziegler E (2016) X-ray multimodal tomography using speckle-vector tracking. Phys. Rev. Appl 5(4):044014

Bérujon S, Ziegler E (2017) X-ray multimodal tomography using speckle-vector tracking. Phys. Rev. A 95(6):063822

Bérujon S, Ziegler E, Cerbino R, Peverini L (2012b) Two-Dimensional X-Ray Beam Phase Sensing. Phys. Rev. Lett. 108(15):158102

Bevins N, Zambelli J, Li K, Qi Z, Chen G-H (2012) 11Multicontrast x-ray computed tomography imaging using Talbot-Lau interferometry without phase stepping". Phys Rev Lett 39(1):424–428

Cloetens P, Guigay JP, De Martino C, Baruchel J, Schlenker M (1997) Fractional Talbot imaging of phase gratings with hard x rays. Opt Lett 22(14):1059–1061

Cloetens P, Ludwig W, Baruchel J, Dyck DV, Landuyt JV, Guigay JP, Schlenker M (1999) Holotomography: quantitative phase tomography with micrometer resolution using hard synchrotron radiation x rays. Appl Phys Lett 75(19):2912–2914

David C, Nöhammer B, Solak HH, Ziegler E (2002) Differential x-ray phase contrast imaging using a shearing interferometer. Appl Phys Lett 81(17):3287–3289

Federation of European Producers of Abrasives (2019) FEPA P-grit sizes coated abrasives. https://www.fepa-abrasives.com/abrasive-products/grains. Accessed 21 June 2019

Henke BL, Gullikson EM, Davis JC (1993) X-Ray interactions: photoabsorption, scattering, transmission, and reflection at E = 50–30,000 eV, Z = 1–92. Atomic Data Nucl Data Tables 54(2):181–342

Hipp A, Herzen J, Hammel JU, Lytaev P, Schreyer A, Beckmann F (2016) Single-grating interferometer for high-resolution phase-contrast imaging at synchrotron radiation sources. Proc SPIE 9967:996718

Koch FJ, Detlefs C, Schröter TJ, Kunka D, Last A, Mohr J (2016) Quantitative characterization of X-ray lenses from two fabrication techniques with grating interferometry. Opt Express 24(9):9168–9177

Kottler C, David C, Pfeiffer F, Bunk O (2007) A two-directional approach for grating-based differential phase contrast-imaging using hard x-rays. Opt Express 15(3):1175–1181

Marschner M, Willner M, Potdevin G, Fehringer A, Noël PB, Pfeiffer F, Herzen J (2016) Helical X-ray phase-contrast computed tomography without phase stepping. Sci Rep 6:23953

Miao H, Chen L, Bennett EE, Adamo NM, Gomella AA, DeLuca AM, Patel A, Morgan NY, Wen H (2013) Motionless phase stepping in X-ray phase contrast imaging with a compact source. Proc Natl Acad Sci 110(48):19268–19272

Misell DL (1973) A method for the solution of the phase problem in electron microscopy. J Phys D Appl Phys 6(1):L6–L9

Momose A, Kawamoto S, Koyama I, Hamaishi Y, Takai K, Suzuki Y (2003) Demonstration of X-Ray Talbot interferometry. Jpn J Appl Phys 42:L866–L868

Morgan KS, Paganin DM, Siu KKW (2012) X-ray phase imaging with a paper analyzer. Appl Phys Lett 100(12):124102

Nazmov V, Reznikova E, Mohr J, Snigirev A, Snigireva I, Achenbach S, Saile V (2004) Fabrication and preliminary testing of X-ray lenses in thick SU-8 resist layers. Microsys Technol 10(10):716–721

Rau C, Wagner U, Pešić Z, De Fanis A (2011) Coherent imaging at the Diamond beamline I13. Phys Status Solidi A 208(11):2522–2525

Revol V, Kottler C, Kaufmann R, Straumann U, Urban C (2010) Noise analysis of grating-based x-ray differential phase contrast imaging. Rev Sci Instrum 81(7):073709

Rodenburg JM, Hurst AC, Cullis AG (2007a) Transmission microscopy without lenses for objects of unlimited size. Ultramicroscopy 107(2–3):227–231

Rodenburg JM, Hurst AC, Cullis AG, Dobson BR, Pfeiffer F, Bunk O, David C, Jefimovs K, Johnson I (2007b) Hard-x-ray lensless imaging of extended objects. Phys Rev Lett 98(3):034801

Selwyn EWH (1935) A theory of graininess. Photogr J 75:571–580

Stockmar M, Cloetens P, Zanette I, Enders B, Dierolf M, Pfeiffer F, Thibault P (2013) Near-field ptychography: phase retrieval for inline holography using a structured illumination. Sci Rep 3:1927

Tanaka J, Nagashima M, Kido K, Hoshino Y, Kiyohara J, Makifuchi C, Nishino S, Nagatsuka S, Momose A (2013) Cadaveric and in vivo human joint imaging based on differential phase contrast by X-ray Talbot-Lau interferometry. Z Med Phys 23(3):222–227

Thibault P, Dierolf M, Menzel A, Bunk O, David C, Pfeiffer F (2008) High-resolution scanning x-ray diffraction microscopy. Science 321(5887):379–382

Wang H, Kashyap Y, Sawhney K (2016) From synchrotron radiation to lab source: advanced speckle-based X-ray imaging using abrasive paper. Sci Rep 6:20476

Weitkamp T, Diaz A, David C, Pfeiffer F, Stampanoni M, Cloetens P, Ziegler E (2005) X-ray phase imaging with a grating interferometer. Opt Express 13(16):6296–6304

Wilkins SW, Gureyev TE, Gao D, Pogany A, Stevenson AW (1996) Phase-contrast imaging using polychromatic hard X-rays. Nature 384:335–337

Yashiro W, Takeda Y, Momose A (2008) Efficiency of capturing a phase image using cone-beam x-ray Talbot interferometry. J Opt Soc Am A 25(8):2025–2039

Zanette I, Weitkamp T, Donath T, Rutishauser S, David C (2010) Two-dimensional x-ray grating interferometer. Phys Rev Lett 105(24):248102

Zanette I, Zhou T, Burvall A, Lundström U, Larsson DH, Zdora M-C, Thibault P, Pfeiffer F, Hertz HM (2014) Speckle-based x-ray phase-contrast and dark-field imaging with a laboratory source. Phys Rev Lett 112(25):253903

Zdora M-C, Thibault P, Deyhle H, Vila-Comamala J, Rau C, Zanette I (2018) Tunable X-ray speckle-based phase-contrast and dark-field imaging using the unified modulated pattern analysis approach. J Instrum 13(05):C05005

Zdora M-C, Thibault P, Pfeiffer F, Zanette I (2015) Simulations of x-ray speckle-based dark-field and phase-contrast imaging with a polychromatic beam. J Appl Phys 118(11):113105

Zdora M-C, Thibault P, Rau C, Zanette I (2017a) Characterisation of speckle-based X-ray phase-contrast imaging. J Phys Conf Ser 849(1):012024

Zdora M-C, Thibault P, Zhou T, Koch FJ, Romell J, Sala S, Last A, Rau C, Zanette I (2017b) X-ray phase-contrast imaging and metrology through unified modulated pattern analysis. Phys Rev Lett 118(20):203903

Zhang Y, Pedrini G, Osten W, Tiziani HJ (2003) Whole optical wave field reconstruction from double or multi in-line holograms by phase retrieval algorithm. Opt Express 11(24):3234–3241

Chapter 7
At-Wavelength Optics Characterisation via X-ray Speckle- and Grating-Based Unified Modulated Pattern Analysis

7.1 Introductory Remarks

Following the first demonstration and working principle of the UMPA approach in the previous chapter, this chapter presents the first application of UMPA for X-ray optics characterisation, which was already shortly addressed in Chap. 6, Sect. 6.3.5. The work in this chapter was published in a peer-reviewed paper as (Zdora et al. 2018b): M.-C. Zdora, I. Zanette, T. Zhou, F. J. Koch, J. Romell, S. Sala, A. Last, Y. Ohishi, N. Hirao, C. Rau, and P. Thibault, "At-wavelength optics characterisation via X-ray speckle—and grating-based unified modulated pattern analysis," Opt. Express **26**(4), 4989–5004 (2018), licensed under CC BY 4.0. The following sections of this chapter are based on a modified reprint of this publication.

Two polymer X-ray refractive lenses were studied using UMPA and the effects of beam damage and shape errors on their refractive properties are evaluated. Already mentioned in the previous chapter, it is, furthermore, shown here that UMPA can be applied not only to random speckle, but also periodic reference patterns. Measurements were performed with speckle—and grating-based UMPA setups and validated with conventional X-ray grating interferometry. The results obtained on the lenses demonstrate that UMPA is a promising candidate for high-throughput quantitative optics characterisation and wavefront sensing that can be performed at-wavelength and with an easily implemented, cost-effective and adaptable setup.

7.2 Background and Motivation

The current trend towards extremely brilliant and coherent X-ray beams has initiated a new era for X-ray science, in particular in the fields of X-ray imaging and diffraction, allowing for unprecedented spatial and temporal resolution (Eriksson and van der Veen 2014; Ueda 2017). To fully exploit the coherence properties of the X-ray beam,

M.-C. Zdora, *X-ray Phase-Contrast Imaging Using Near-Field Speckles*,
Springer Theses, https://doi.org/10.1007/978-3-030-66329-2_7

precise and aberration-free optical elements, such as refractive lenses, mirrors and diffractive optics, are desired. The need for testing and continuously improving these components has led to an increasing demand for accurate, precise and sensitive methods for optics characterisation and wavefront measurement.

Many established ex-situ metrology techniques, mostly in the visible light regime, such as the Fizeau interferometer (Malacara 1992; Hariharan 1997) and slope-measuring profilers (Takacs et al. 1987; Siewert et al. 2004, 2014), can provide detailed and accurate information about some types of optics, e.g. mirrors, while they are less suitable for others, such as X-ray refractive lenses. Furthermore, it is important to analyse the performance of the optical elements in-situ and at-wavelength, under the same or similar conditions as they are typically used. In particular, mechanical stress, beam-induced heat load, vibrations and drifts can have an influence on the performance of the optics and should hence be comparable to the typical experimental conditions when performing the quality assessment.

Methods like the (Hartmann 1900) or Shack-Hartmann (Shack and Platt 1971) sensor have been translated from the visible to the soft (Flöter et al. 2010, 2011) and hard X-ray regimes (Mayo and Sexton 2004; Idir et al. 2010). However, due to their limited spatial resolution and elaborate calibration procedure, other techniques have proven more suitable for at-wavelength metrology and wavefront measurements in the hard X-ray regime.

Ptychographic coherent diffractive imaging (Kewish et al. 2010; Vila-Comamala et al. 2011) and X-ray grating interferometry (XGI) (Weitkamp et al. 2005b; Engelhardt et al. 2007; Diaz et al. 2010; Wang et al. 2011; Bérujon and Ziegler 2012; Kayser et al. 2017) are successfully being used for metrology and provide high sensitivity. Drawbacks of the techniques include the need for elaborate alignment procedures of the setup components, as well as the computational expense and vast number of frames for the former and the limited sensitivity in only one direction for the latter. Despite these limitations, XGI has become a popular method to analyse the quality of X-ray refractive lenses that are challenging to measure using ex-situ visible light methods (Engelhardt et al. 2007; Rutishauser et al. 2011; Wang et al. 2012, 2013; Koch et al. 2016). Two-dimensional information can be achieved with grating interferometry by taking two data sets with the grating oriented in orthogonal directions (Engelhardt et al. 2007)—at the cost of doubling the acquisition time. Alternatively, a 2D grating can be used (Zanette et al. 2010; Rutishauser et al. 2011), which is, however, elaborate and expensive to produce.

Other approaches for 2D phase-sensitive imaging with a periodic pattern have been proposed, e.g. using an absorption grid with an analysis in Fourier space (Wen et al. 2010) or a phase grid with a correlation analysis in real space (Morgan et al. 2013). These methods allow for fast single-shot acquisition, although at the cost of a reduced spatial resolution.

In the last five years, X-ray speckle-based phase-contrast imaging (SBI) (Morgan et al. 2012; Bérujon et al. 2012) has found increasing use as a cost-effective, easily implemented alternative to XGI. Compared to the other methods, SBI is more relaxed regarding alignment requirements and does not rely on a high degree of longitudinal nor lateral coherence of the X-ray beam. It, furthermore, simultaneously delivers the differential phase in two orthogonal directions. SBI has successfully been applied to wavefront measurements and metrology of X-ray refractive lenses, mirrors and Fresnel zone plates (Bérujon et al. 2012, 2013, 2014, 2015; Wang et al. 2015a, b, c; Kashyap et al. 2016a, b), see also Chap. 5, Sect. 5.8.1.

Recently, the unified modulated pattern analysis (UMPA) has been proposed (see previous chapter), which overcomes some of the main drawbacks of the common implementations of XGI and SBI and allows flexible tuning of signal sensitivity and spatial resolution (Zdora et al. 2017, 2018a). These properties make UMPA a suitable candidate for wavefront measurements and metrology applications with a simple setup that can be adapted to most experimental conditions. UMPA, furthermore, does not impose strong restrictions on the properties of the investigated optics as it is often the case for XGI (Koch et al. 2016).

A first demonstration of optics characterisation using the UMPA method has been presented recently on a polymer X-ray refractive lens (Zdora et al. 2017), see previous chapter. Here, the results of beam damage on the performance of this lens are studied in more detail and shape errors of another undamaged polymer lens are investigated using the UMPA technique with a random as well as a periodic reference pattern.

7.3 Experimental Setup and Data Acquisition

Measurements were conducted at the beamline I13-1 at Diamond Light Source, UK (Rau et al. 2011), which is described in Chap. 3. A horizontally deflecting monochromator was used to select a beam with an X-ray energy of 18.4 keV from the undulator spectrum.

The results presented here were acquired with three different setups, as sketched in Fig. 7.1: (a) A setup using the UMPA method with a 1D beam-splitter grating (material: SU-8 photoresist polymer, period: 5.4 µm, line height: 26.7 µm, fabricated by KIT/IMT, Germany) as a phase modulator, (b) a setup using the UMPA method with a piece of sandpaper (combination of granularities P800 and P5000 (Federation of European Producers of Abrasives, 2019)) as a phase modulator, and (c) for comparison a common X-ray grating interferometer consisting of a 1D phase grating (material: nickel, period: 2.4 µm, line height: 3 µm, fabricated by KIT/IMT, Germany) and a 1D absorption grating (material: gold, period: 2.4 µm, line height: 80 µm, fabricated by microworks, Germany).

The sample, here a refractive lens, was located about 220 m downstream of the X-ray source. For the UMPA setups, the detector was placed at a distance of $d_{S1} = 0.34$ m downstream of the sample. For the first setup in Fig. 7.1a the 1D grating was placed 0.16 m downstream of the lens, i.e. $d_G = 0.18$ m upstream of the detector, while for the second setup in Fig. 7.1b the sandpaper was situated 0.17 m upstream of the lens. For practical reasons the lens-detector distance was kept constant for the two setups, while the order and distance between lens and phase modulator was changed. This does not affect the absolute refraction angle values, but can have an influence on the sensitivity. For the grating interferometer in Fig. 7.1c an inter-grating distance

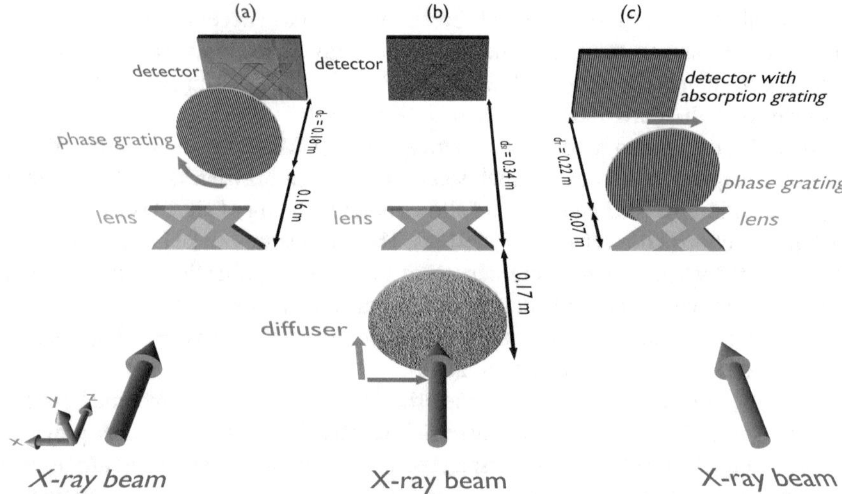

Fig. 7.1 Experimental arrangements for the characterisation of the refractive lenses. Setups using the UMPA method with **a** a 1D phase grating rotated around the optical axis in 46 steps and **b** a piece of sandpaper, which was translated perpendicular to the beam direction on a spiral pattern in 24 steps. **c** X-ray grating interferometry setup for validation, consisting of a 1D phase grating and a 1D absorption grating placed in front of the detector. Phase-stepping was performed with five steps over one grating period. Figure reprinted with permission from Zdora et al. (2018b), licensed under CC BY 4.0

of $d_T = 0.22$ m was chosen, corresponding to the 5th fractional Talbot distance (Cloetens et al. 1997). The refractive lens was located at a distance of 0.07 m upstream of the phase grating G1 (giving a lens-detector distance of $d_{S2} = 0.29$ m) and the absorption grating G2 was placed as close as possible to the detector.

The detector system consisted of a scintillation screen, a 10× objective lens, a magnifying relay optics system and a pco4000 CCD camera (pixel size: 9 µm), leading to an effective pixel size of $p_{\text{eff}} = 0.40$ µm.

Data acquisition with the UMPA method was performed for the setup in Fig. 7.1a by rotating the grating around the beam axis from 0° to 90° in increments of 2° between each of the 46 acquired projections. For the setup in Fig. 7.1b the sandpaper was moved to 24 diffuser positions on a spiral pattern in the plane perpendicular to the beam direction with step sizes of about 28 µm. The X-ray grating interferometer in Fig. 7.1c was operated in phase-stepping mode, scanning the phase grating in the horizontal direction, perpendicular to the grating lines, in five steps over one grating period. For each of the scans, a set of reference projections without the lens in the beam and a set of sample projections with the lens in the beam were acquired. The differential phase shift signal was obtained by analysing the displacement of the interference pattern created by the phase modulator in the detector plane as described in the next section.

Two parabolic refractive lenses made from SU-8-based photo-resist polymer (Nazmov et al. 2004b) (refractive index decrement: $\delta_{\text{SU8}} = \left(2.7142 \times 10^{-4}\right)/E^2$

with E being the X-ray energy in keV[1]) were tested with the setups in Fig. 7.1: A line-focus lens (in the following referred to as *lens A*) with a radius of curvature $R_A = 6.000\,\mu\text{m}$ and a point-focus compound refractive lens (in the following referred to as *lens B*) with a radius of curvature $R_B = 19.625\,\mu\text{m}$, consisting of two crossed line-focussing elements each mounted with an angle of 45° to the substrate surface. Lens A was intact and functional, while lens B had experienced beam damage due to intense X-ray exposure (see information in Sect. 7.7.2).

7.4 Signal Reconstruction

As discussed in previous chapters of this thesis, both imaging methods, UMPA and classic XGI, allow for the reconstruction of multimodal signals: The differential phase shift or the refraction angle in the sample, the transmission through the specimen and the dark-field signal, which is a measure of small-angle scattering in the sample (Pfeiffer et al. 2008; Zdora et al. 2017). For this study, only the differential phase signal is analysed as it is the relevant modality for characterising optical elements such as X-ray refractive lenses.

UMPA—as operated here—can deliver the refraction information in the horizontal as well as the vertical direction, while XGI in its most common implementation with a 1D grating in phase stepping mode only delivers the refraction in the direction perpendicular to the grating lines.

7.4.1 Unified Modulated Pattern Analysis

For the measurements using the UMPA setups in Fig. 7.1a, b with the rotating grating or the sandpaper, respectively, data analysis was performed following the formalism described in detail by Zdora et al. (2017) and in Chap. 6. A mathematical derivation can be found in Appendix B. The refraction angle (α_x, α_y) and differential phase signal $\left(\frac{\partial\Phi}{\partial x}, \frac{\partial\Phi}{\partial y}\right)$ can be obtained from the speckle displacement vector $\mathbf{u}$ in small-angle approximation, see Eq. (6.3) in Chap. 6, Sect. 6.3.2.2.

7.4.2 X-ray Grating Interferometry

The grating interferometry measurements were performed with a two-grating setup in phase-stepping mode and the data was analysed with the commonly used approach based on a Fourier series representation (Weitkamp et al. 2005a), as explained in

[1]The refractive index decrement of SU-8 was measured at KIT/IMT in the X-ray range of 10–40 keV and a fit to the data was performed.

Chap. 4, Sect. 4.2.2. The differential phase shift $\partial\Phi/\partial x$ and refraction angle α_x in the direction perpendicular to the grating lines are given by Eq. (4.8).

7.4.3 Phase Integration

The absolute phase shift of the X-rays induced by the lens was obtained by Fourier integration of the horizontal and vertical differential phase signals (Kottler et al. 2007) for the UMPA reconstructions. For the XGI case, the phase shift was determined by simple 1D integration of the differential phase signal in the horizontal direction.

7.5 Lens Characterisation

State-of-the-art X-ray refractive lenses typically have a parabolic shape to achieve line—or point-focussing of the beam without spherical aberrations (Lengeler et al. 1999, 2002). The refraction angle α in the lens is then a linear function along the lens profile in the focussing direction and the focal length f is directly related to the refraction angle:

$$\begin{aligned} 1/f_x &= \frac{\partial \alpha_x}{\partial x} \\ 1/f_y &= \frac{\partial \alpha_y}{\partial y} \end{aligned} \tag{7.1}$$

for the focal lengths f_x, f_y in the horizontal direction x and vertical direction y, respectively. For a perfect point-focussing parabolic lens without astigmatism, f_x and f_y should have the same value. Furthermore, the focal length f is given by the radius of curvature R at the apex of the lens parabola and the refraction index decrement δ of the lens material (Snigirev et al. 1996):

$$f = \frac{R}{2\delta}. \tag{7.2}$$

In the last decade, SU-8 polymer lenses fabricated via deep X-ray lithography (Backer et al. 1982; Saile et al. 2009) have found increasing applications (Snigirev et al. 2003; Nazmov et al. 2004b, c, a; Reznikova et al. 2007, 2009). The fabrication step allows for high aspect ratios and precise alignment of the lens elements. Its transparency and stability to X-ray radiation make SU-8 a suitable material for compound refractive lenses (CRLs).

Here, two X-ray refractive lenses fabricated from SU-8 polymer are investigated: a line-focussing lens (lens A) that had not been subject to any external damage and a point-focussing lens (lens B) that had experienced prolonged exposure to an intense X-ray beam at SPring-8 synchrotron, see Sect. 7.7.2. Using the UMPA technique, the changes in focussing behaviour of the two lenses are quantified, which were caused by extreme X-ray beam exposure and shape errors, and their implications on the focussing properties are investigated.

7.5.1 Refraction Angle and Wavefront

The refraction angle signals in the horizontal (focussing) direction of the line-focus lens A are presented in Fig. 7.2a, b for the UMPA sandpaper setup and UMPA grating setup, respectively. Figure 7.2c shows the phase shift integrated from the differential phase images obtained with the UMPA grating setup. The horizontal line profiles through the middle of the lens in Fig. 7.2d, e show a globally linear behaviour of the refraction angle along the lens aperture and a parabolic shape of the phase shift. Excellent agreement between the results of the two setups can be observed.

The structures visible on the left and right side of the focussing aperture are support elements for stabilisation of the lens.

Figure 7.3 shows the refraction angle signals and integrated phase shift of the inner focussing aperture of the point-focus CRL (lens B) for the different imaging setups.

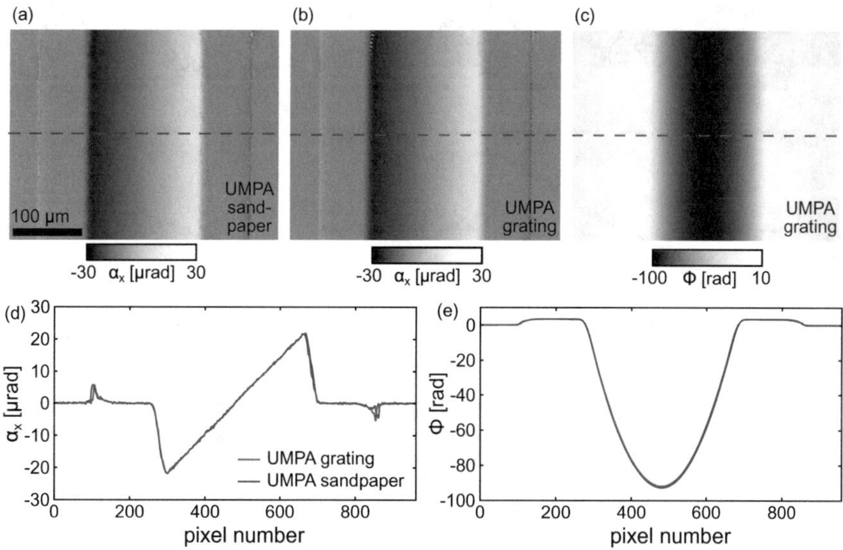

Fig. 7.2 Refraction angle signal α and integrated phase shift Φ of the line-focus lens A. Refraction angle α_x in the horizontal direction measured with UMPA using **a** a piece of sandpaper, **b** a periodic phase grating as a phase modulator and **c** total phase shift integrated from (**b**). Line profiles along the mid-line of the lens for **d** the horizontal refraction angles and **e** the integrated phase shift signals. Support elements can be observed on both sides of the lens aperture. Figure reprinted with permission from Zdora et al. (2018b), licensed under CC BY 4.0

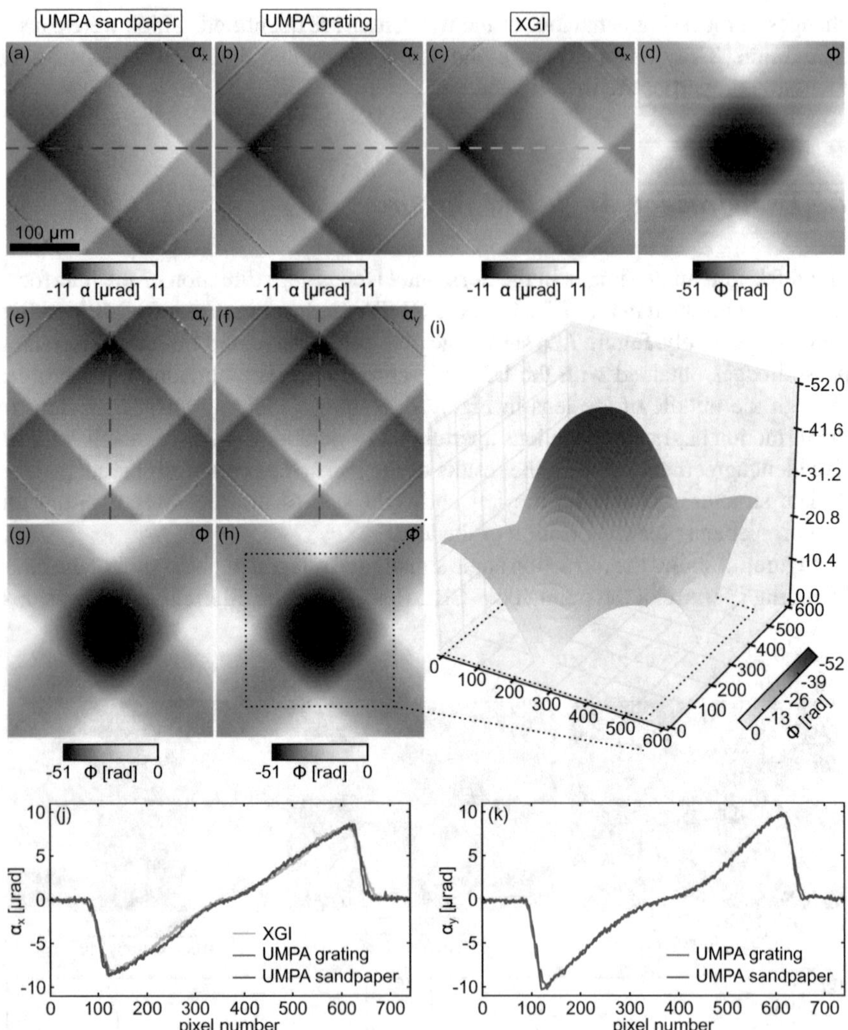

Fig. 7.3 Refraction angle signal α and integrated phase shift Φ of the point-focus lens B. Refraction angle α_x in the horizontal direction measured with UMPA using **a** sandpaper, **b** a periodic phase grating and **c** conventional two-grating XGI. **d** Absolute phase shift determined by 1D integration of (**c**). Refraction angle α_y in the vertical direction measured with UMPA using **e** sandpaper, **f** a periodic phase grating and **g**, **h** absolute phase shift signals integrated from (**a**) and (**e**) or (**b**) and (**f**), respectively. **i** 3D surface plot of the beam phase profile after lens B in panel (**h**) measured in the detector plane with the UMPA grating setup. **j**, **k** Line profiles through the horizontal and vertical refraction angle signals in (**a–c**) and (**e–f**), respectively. Figure reprinted with permission from Zdora et al. (2018b), licensed under CC BY 4.0

The refraction angles in the horizontal direction measured with the UMPA sandpaper setup, the UMPA grating setup and the XGI setup, respectively, in Fig. 7.3a–c show good agreement between the different data sets, as confirmed by the line plots through the centre of the lens in Fig. 7.3j. Slight deviations are due to the fact that the lens was removed from the sample stage between the scans causing small differences in its position. Furthermore, it should be noted that the propagation distance from the lens to the detector was different for the UMPA setups ($d_{S1} = 0.34\,\text{m}$) and the XGI setup ($d_{S2} = 0.29\,\text{m}$) due to practical limitations. This leads to a small difference in the demagnification of the lens aperture caused by the focussing effect of the lens, which hence appears slightly wider for the XGI case. This, however, does not affect the quantitative analysis of the refraction properties. Furthermore, it can be observed that the images obtained from the setups using gratings (Fig. 7.3b–d, f and h) appear more blurred than for the UMPA sandpaper setup. This is likely to be caused by the beam-splitting effect of the grating, which leads to a reduced resolution when the resolving power of the detector system is better than the separation of the first order diffracted beams.

The line profiles in Fig. 7.3j, k through the refraction angle signals in the horizontal and vertical directions, respectively, clearly demonstrate a deviation from the expected linear focussing behaviour, in particular for the vertical case. Due to the limitation of the phase sensitivity in the horizontal direction, the strong defects in the vertical focussing direction cannot be measured using XGI in this configuration, while UMPA can deliver the information in both directions. This also leads to an improved integrated phase signal with significantly reduced artefacts for the UMPA setups, see Fig. 7.3g, h, compared to the phase signal obtained from XGI in Fig. 7.3d. The parabolic shape of the wavefront after the lens is visualised in Fig. 7.3i as a surface plot of the integrated phase signal from the UMPA grating setup in Fig. 7.3h.

7.5.2 *Lens Defects and Aberrations*

In the previous section, the refraction behaviour of the X-ray lenses was analysed qualitatively. While by visual inspection of the line plots through the differential phase images, lens A showed an overall linear refraction angle along the lens profile, as expected, deviations from the design refraction behaviour were observed for the damaged lens B. In this section, the effects of defects and shape errors of the lenses on the wavefront gradient are investigated in a quantitative manner.

7.5.2.1 Deviations from the Expected Refraction Behaviour

The lenses are designed to produce a known parabolic phase shift of the X-ray beam, i.e. a linear refraction angle. A straightforward way to analyse defects and errors is to look at the deviation from this expected refraction signal, i.e. the wavefront gradient error. Here, the residuals from a linear fit to the reconstructed refraction angle were calculated, as proposed by Koch et al. (2016). Figure 7.4a, d and g show the measured deviation in the aperture of the damaged point-focus lens B from the expected linear refraction in the horizontal direction for the UMPA sandpaper, the UMPA grating and the XGI setups, respectively. The results agree well for the different setups, confirming the validity and reproducibility of the UMPA method. The same analysis was performed for the refraction in the vertical direction for the two UMPA setups in Fig. 7.4b, e, revealing even stronger aberrations from a linear behaviour. Figure 7.4c, f and h display the absolute deviation, i.e. the square root of the summed squares of the contributions from the horizontal and vertical signals. For the XGI case, the absolute of the deviation in the horizontal is shown, as the setup is only sensitive in one direction.

The small grain-like artefacts visible in Fig. 7.4a–f are a result of remaining features of the reference pattern in the reconstructed refraction signals. This is caused by the demagnification effect of the lens on the interference pattern, which is here neglected, but could in the future be included by considering higher-order subset distortions in the UMPA analysis for further improvement of the technique.

Large discrepancies from the expected focussing behaviour of lens B can be observed in both directions. Thanks to the two-dimensional sensitivity of the UMPA approach, the change in the lens properties can be fully characterised from a single data set. The deviations, which reach up to 3.0 µrad, are more pronounced in the vertical direction and largest in the areas around the centre of the aperture, indicating an alteration of the refractive index and/or thickness of the lens material in this region. This significant divergence within the lens aperture would result in strong changes of the desired focussing behaviour making the lens useless for further applications as an optical element.

The absolute deviation is also shown for lens A in Fig. 7.4i, j for the UMPA sandpaper and grating setups, respectively. The linear fit to the refraction angle was performed in the upper part of the lens. It can clearly be seen that the divergence from the expected refraction is significantly smaller than for the damaged lens. However, in the lower part deviations can be noted, which strongly increase when moving closer to the substrate. This has been observed previously for refractive lenses made from SU-8 material and has been attributed to shape errors of the lens due to tilted side walls and thermal stress during the manufacturing process (Koch et al. 2016; Koch 2017). These effects are more pronounced close to the substrate, where as a consequence the refraction behaviour cannot be modelled as a linear curve anymore, which leads to significant deviations of the measured refraction angle data from a linear fit. It has been shown that the shape errors result in a decrease of the measured focal length from the top to the bottom of the lens (Koch et al. 2016; Last et al. 2015; Koch 2017). This effect can clearly be observed for lens A. Figure 7.5 illustrates the variation of the focal length over the height of the lens. The focal length f was determined by performing a linear fit of the refraction angle for each row of the image and applying Eq. (7.1) to obtain f. As seen from Fig. 7.5, the focal length clearly decreases from the top to the bottom of the lens.

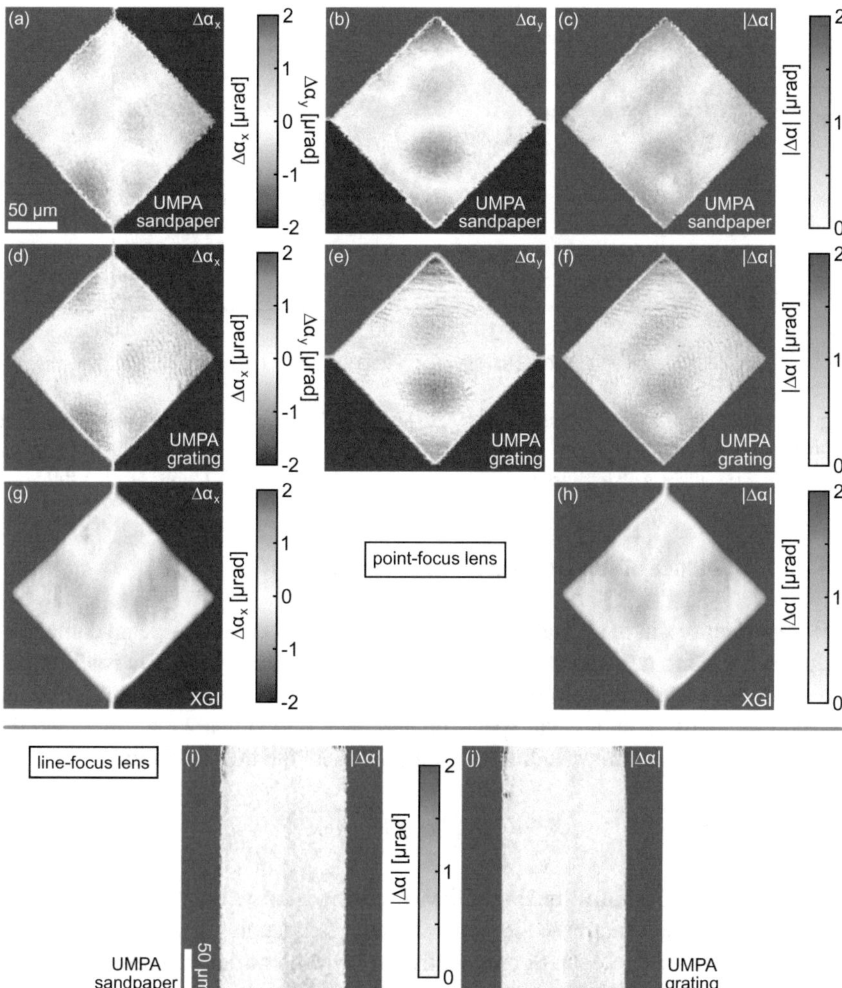

Fig. 7.4 Evaluation of the refraction aberrations of the lenses. Residuals from a linear fit to the refraction angle of lens B (wavefront gradient error) in **a**, **d** the horizontal, **b**, **e** the vertical direction, and **c**, **f** absolute deviation, measured with the UMPA sandpaper and grating setups, respectively. Residuals from **g** the horizontal refraction angle and **h** absolute deviation in the horizontal, obtained from the two-grating interferometer. **i**, **j** Absolute deviation from the horizontal refraction angle signal of lens A for the UMPA sandpaper and grating setups, respectively. Figure reprinted with permission from Zdora et al. (2018b), licensed under CC BY 4.0

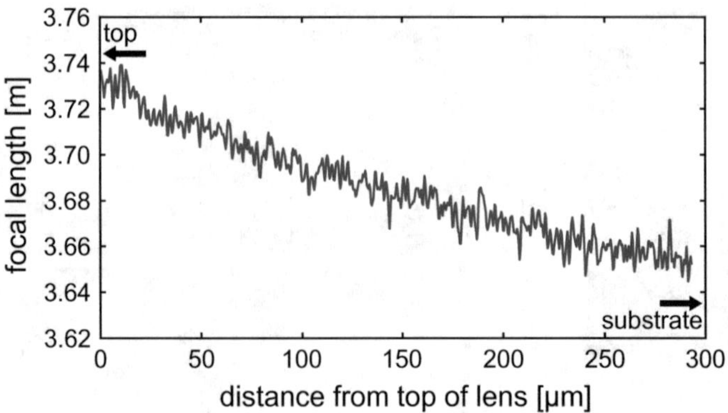

Fig. 7.5 Variation of the determined focal length along the height of lens A (line-focus lens). The decrease of the focal length closer to the substrate is due to shape errors more pronounced in this area. Figure reprinted with permission from Zdora et al. (2018b), licensed under CC BY 4.0

7.5.2.2 Influence of the Deviations on the Focal Spot

The imperfections in the wavefront created by the lenses caused by beam damage or shape errors led to deviations from the desired point or line focus. To visualise this, the measured wavefront was propagated along the beam direction z starting from the detector plane using an angular spectrum approach (Goodman 2004), in which the Fourier transform of the wave field is multiplied with the transfer function:

$$\mathscr{H} = \exp\left(ikz\sqrt{1 - q^2\lambda^2}\right), \tag{7.3}$$

where z is the propagation direction of the wavefront, λ the wavelength, $k = 2\pi/\lambda$ the wave vector and q^2 the sum of the squared spatial frequencies in x and y, transverse to the propagation direction. Subsequently, an inverse Fourier transform is applied to obtain the propagated wavefront in real space.

Figure 7.6 shows cuts through the propagated wavefronts, obtained from the UMPA setup using the rotating grating, through the centre of the lens position. For lens B, an asymmetry of the longitudinal cuts in the horizontal and vertical directions in Fig. 7.6a, c, respectively, can be observed. Theoretically, the same focal length $f_B = 12.24$ m would be expected in both focussing directions for the point-focussing lens B according to Eq. (7.2). The beam damage, however, gave rise to astigmatism of the lens, i.e. the focal planes in the two directions are located at different distances. The minimum beam size was found at 11.91 m (orange line) downstream of the lens, i.e. 11.57 m from the detector, for the horizontal and at 12.40 m (cyan line) from the lens, i.e. 12.06 m from the detector, for the vertical focussing direction. The transverse beam profiles at these distances are shown in Fig. 7.6b, d, which, furthermore, reveal that the diameter of the focal spot differs for the horizontal and vertical directions. Some spurious features can be observed in these images around the area of

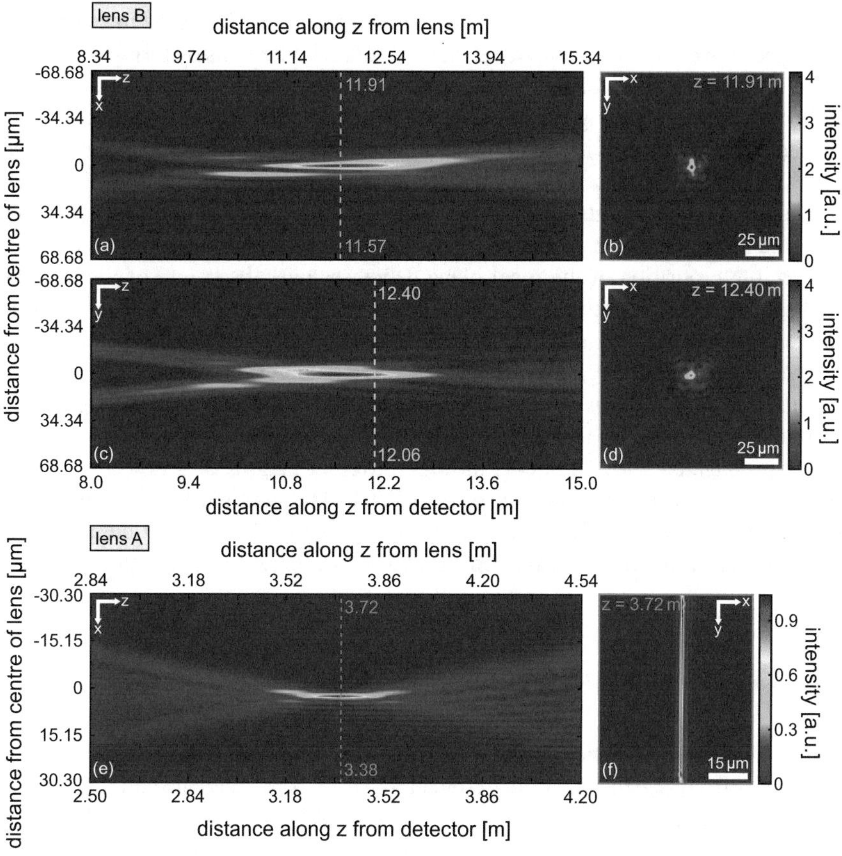

Fig. 7.6 Analysis of the focal spot and focal line qualities via wavefront propagation. Cuts through the propagated wave fields after the point-focus lens B (beam damage) and line-focus lens A (shape errors), calculated from the phase signal measured with the UMPA grating setup in the detector plane by using an angular spectrum propagator. **a**, **c** Longitudinal cuts through the centre of lens B in the xz—and yz-planes, respectively. The dashed lines indicate the distance from the detector where the focal length was found for the horizontal (orange line) and the vertical (cyan line) focussing direction (numbers give the distance from the detector (bottom) and from the lens (top)). **b**, **d** Transverse slices (xy-plane) through the wave field of lens B at the focal distances for the horizontal and vertical directions, respectively. **e** Longitudinal cut through the propagated wave field of lens A in the top part of the lens (pixel row 39 from the top) and distance of the focal line from the detector (magenta line) and **f** corresponding transverse xy-slice at this z-position. Figure reprinted with permission from Zdora et al. (2018b), licensed under CC BY 4.0

highest intensity, which are caused by the astigmatism. The size of the focal spot was estimated separately in the horizontal and the vertical directions by fitting a Gaussian curve to the line profile of the propagated wave field through the centre of the lens in the respective focussing direction, while neglecting the spurious features around the spot. The FWHM of the fitted Gaussian can be seen as an estimate for the spot size. The FWHM is 4.7 µm for the horizontal and 5.4 µm for the vertical direction.

The propagation of the wave field of lens A is shown as a longitudinal cut in the xz-plane through the top part of the lens in Fig. 7.6e. As discussed in the previous section, the z-position of the focal plane decreases over the height of the lens due to shape errors. It is located at 3.72 m downstream of the lens (3.38 m from the detector) for the upper part of the lens as indicated by the magenta line in Fig. 7.6e. The transverse cut through the propagated wave field at this distance can be found in Fig. 7.6f, which shows a sharp focus at the top of the lens. The change of the position of the focal line for different heights of the lens is illustrated in more detail by the propagated wave field cuts through different heights of the lens in Fig. 7.7 in the supplementary information in Sect. 7.7.1. The nominal focal length of the lens calculated with Eq. (7.2) is given by $f_A = 3.74$ m. The width of the focal line, estimated by fitting of a Gaussian line as above, is approximately 1.3 µm.

7.6 Conclusions and Outlook

It was demonstrated that the recently proposed UMPA method is suitable for X-ray wavefront sensing and optics characterisation. Beam damage and shape errors of two polymer X-ray refractive lenses were successfully detected by analysing the wavefront downstream of the lens qualitatively as well as quantitatively. Furthermore, the excellent agreement of the results obtained from an UMPA setup with a periodic and a random reference interference pattern was shown and they were validated with conventional X-ray grating interferometry measurements. Combined with the flexibility and ease of implementation, this illustrates that UMPA is readily adaptable to many existing X-ray phase-sensitive setups and it is expect to find widespread applications for wavefront analysis and optics characterisation. As UMPA is a robust technique that does not require a high degree of spatial and lateral coherence and does not impose stringent restrictions on the setup length or alignment, it is suitable for the investigation of various kinds of optics and could in the future be transferred to laboratory sources without major efforts, see Chap. 9, Sect. 9.4. This makes UMPA a perfect candidate for cost-effective routine inspection and quality testing of optics such as refractive lenses, allowing near at-wavefront characterisation without the need for costly and limited synchrotron access. This will be an important step for the improvement and development of optics fabrication.

7.7 Supplementary Information

7.7.1 *Focussing Properties of the Line Focus Lens A*

As observed above, the line focus lens A shows some shape errors caused by the manufacturing process that lead to a decrease of the focal length from the top to the bottom part of the lens. The effect on the focussing behaviour of the lens can be visualised by propagating the wavefront measured in the detector plane in a simulation using an angular spectrum propagator (see Eq. (7.3)). Figure 7.7 shows longitudinal and transverse cuts through the propagated wave field of lens A for different heights of the lens and different propagation distances. It can be seen from the longitudinal sections through the top, middle and bottom (close to the substrate) of the lens in Fig. 7.7a, c and e, respectively, that the position of the focal line

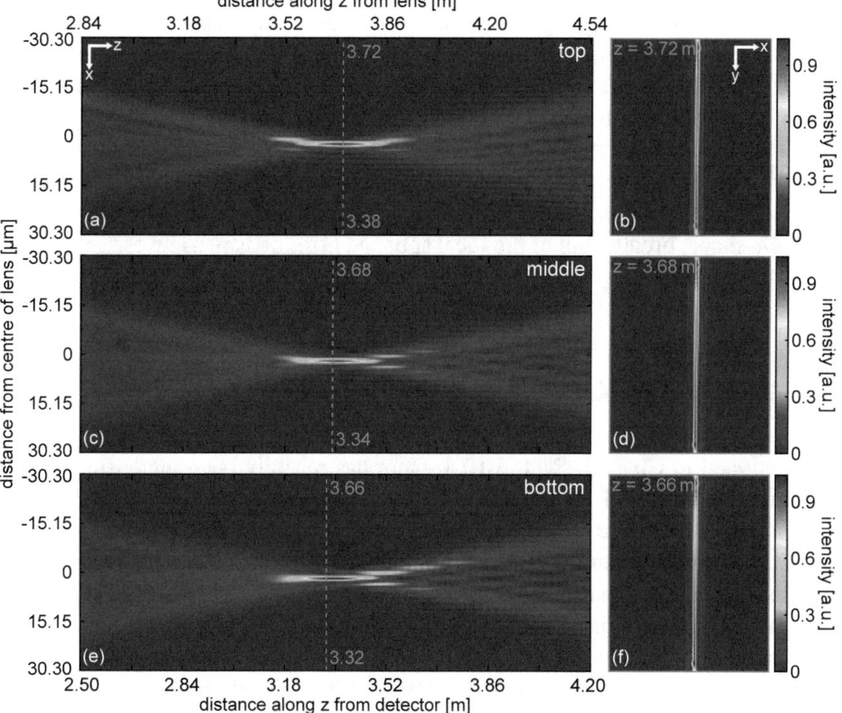

Fig. 7.7 Analysis of the influence of the shape error of lens A on the focal line. **a, c, e** Longitudinal cuts (xz-plane) through the propagated wave fields of lens A in the top, middle and bottom part of the lens. Magenta lines indicating the position of the calculated focal length measured from the detector plane (bottom value) or lens position (top value). **b, d, f** Transverse cuts at the focal distances at $z = 3.72$ m, $z = 3.68$ m, and $z = 3.66$ m from the lens. Figure reprinted with permission from Supplementary Material of Zdora et al. (2018b), licensed under CC BY 4.0

changes over the height of the lens. The magenta dashed lines indicate the distance of the focal line from the detector for each height, which decreases closer to the substrate. In Fig. 7.7b, d and f the corresponding transverse cuts of the wave field at the corresponding focal lengths are shown.

7.7.2 Information about the X-ray Exposure of the Damaged Lens B

Beam damage occurred in the point-focussing SU-8 refractive lens (lens B) due to intense X-ray radiation over a prolonged period of time prior to the characterisation measurements presented here. It was used at beamline BL10XU at SPring-8 synchrotron as part of a CRL array for beam focussing at an energy of 30 keV. Pre-focussing of the X-ray beam upstream of the lens under study was performed with a 2 mm-diameter beryllium CRL resulting in a beam diameter of 150 µm. The beam was then cut down using a pair of slits, so that an X-ray beam of $100 \times 100\,\mu m^2$ effectively impinged onto the SU-8 lens. The photon flux on the investigated lens was 10^{13} photons/(s·mm^2). The CRL remained in the X-ray beam for 523 days with a daily X-ray irradiation of approximately 4 h, summing up to about 2100 h of exposure. The total dose absorbed by the lens during this time was approximately 27.3 kJ/mm^3. This accumulated dose led to a change in the properties of the lens material and hence the refraction behaviour, resulting in deviations of the wavefront from a parabolic shape, broadening of the focal spot and astigmatism, as investigated in the previous sections.

7.8 Concluding Remarks

As mentioned in Chap. 5, Sect. 5.10.4, there has recently been increasing interest in X-ray speckle-based imaging for metrology and beam phase sensing (Bérujon et al. 2019), which also led to the routine implementation of the technique at the synchrotron optics beamline (B16) at Diamond Light Source and the development of a user-oriented software for automated mirror alignment (Zhou et al. 2018, 2019). Thanks to the speed of data collection and its simplicity, the speckle-based technique is, furthermore, now routinely used at the ESRF for testing newly acquired or fabricated refractive lenses (Bérujon et al. 2019) and has recently been applied to beam and optics characterisation at the European XFEL (Seaberg et al. 2019).

In this chapter, the promising potential of the UMPA approach for optics characterisation was demonstrated. With its flexible setup and adaptability it is expected to soon find its way into routine use for CRL characterisation as well as other metrology and beam phase sensing applications at many synchrotrons. However, as UMPA can also be implemented at the laboratory, it is a promising candidate for routine optics testing and quality control at laboratory setups that could be performed during and after the fabrication process at the production facility itself.

References

Backer EW, Ehrfeld W, ünchmeyer DM, Betz H, Heuberger A, Pongratz S, Glashauser W, Michel HJ, Siemens RV (1982) Production of separation-nozzle systems for uranium enrichment by a combination of X-ray lithography and galvanoplastics. Naturwissenschaften 69(11):520–523

Bérujon S, Ziegler E (2012) Grating-based at-wavelength metrology of hard x-ray reflective optics. Opt Lett 37(21):4464–4466

Bérujon S, Wang H, Sawhney K (2012) X-ray multimodal imaging using a random-phase object. Phys Rev A 86(6):063813

Bérujon S, Wang H, Sawhney KJS (2013) At-wavelength metrology using the X-ray speckle tracking technique: case study of a X-ray compound refractive lens. J Phys Conf Ser 425(5):052020

Bérujon S, Wang H, Alcock S, Sawhney K (2014) At-wavelength metrology of hard X-ray mirror using near field speckle. Opt Express 22(6):6438–6446

Bérujon S, Ziegler E, Cloetens P (2015) X-ray pulse wavefront metrology using speckle tracking. J Synchrotron Radiat 22(4):886–894

Cloetens P, Guigay JP, De Martino C, Baruchel J, Schlenker M (1997) Fractional Talbot imaging of phase gratings with hard x rays. Opt Lett 22(14):1059–1061

Diaz A, Mocuta C, Stangl J, Keplinger M, Weitkamp T, Pfeiffer F, David C, Metzger TH, Bauer G (2010) Coherence and wavefront characterization of Si-111 monochromators using double-grating interferometry. J Synchrotron Rad 17(3):299–307

Engelhardt M, Baumann J, Schuster M, Kottler C, Pfeiffer F, Bunk O, David C (2007) Inspection of refractive x-ray lenses using high-resolution differential phase contrast imaging with a microfocus x-ray source. Rev Sci Instrum 78(9):093707

Eriksson M, van der Veen JF eds (2014) Special issue on diffraction-limited storage rings and new science opportunities. J. Synchrotron Rad 21(5)

érujon SB, Cojocaru R, Piault P, Celestre R, Roth T, Barrett R, Ziegler E (2019) X-ray optics and beam characterization using random modulation. arXiv 1902.09418

Federation of European Producers of Abrasives (2019) FEPA P-grit sizes coated abrasives. https://www.fepa-abrasives.com/abrasive-products/grains. Accessed 21 June 2019

Flöter B, Juranić P, Kapitzki S, Keitel B, Mann K, Plönjes E, Schäfer B, Tiedtke K (2010) EUV Hartmann sensor for wavefront measurements at the Free-electron LASer in Hamburg. New J Phys 12(8):083015

Goodman JW (2004) Introduction to fourier optics 3rd edn. Roberts & Company Publishers, Englewood, CO, United States

Hariharan P (1997) Interferometric testing of optical surfaces: absolute measurements of flatness. Opt Eng 36(9):2478–2481

Hartmann J (1900) Bemerkungen über den Bau und die Justierung von Spektrographen. Z Instrumentenkd 20:47

Idir M, Mercere P, Modi MH, Dovillaire G, Levecq X, Bucourt S, Escolano L, Sauvageot P (2010) X-ray active mirror coupled with a Hartmann wavefront sensor. Nucl Instrum Methods Phys Res A 616(2):162–171

Kashyap Y, Wang H, Sawhney K (2016a) Development of a speckle-based portable device for in situ metrology of synchrotron X-ray mirrors. J Synchrotron Radiat 23(5):1131–1136

Kashyap Y, Wang H, Sawhney K (2016b) Speckle-based at-wavelength metrology of X-ray mirrors with super accuracy. Rev Sci Instrum 87(5):052001

Kayser Y, David C, Flechsig U, Krempasky J, Schlott V, Abela R (2017) X-ray grating interferometer for in situ and at-wavelength wavefront metrology. J Synchrotron Rad 24(1):150–162

Kewish CM, Guizar-Sicairos M, Liu C, Qian J, Shi B, Benson C, Khounsary AM, Vila-Comamala J, Bunk O, Fienup JR, Macrander AT, Assoufid L (2010) Reconstruction of an astigmatic hard X-ray beam and alignment of K-B mirrors from ptychographic coherent diffraction data. Opt Express 18(22):23420–23427

Koch FJ (2017) X-ray optics made by X-ray lithography: process optimization and quality control. Ph.D. thesis, Karlsruher Institut für Technologie (KIT), Karlsruhe, Germany

Koch FJ, Detlefs C, Schröter TJ, Kunka D, Last A, Mohr J (2016) Quantitative characterization of X-ray lenses from two fabrication techniques with grating interferometry. Opt Express 24(9):9168–9177

Kottler C, David C, Pfeiffer F, Bunk O (2007) A two-directional approach for grating-based differential phase contrast-imaging using hard x-rays. Opt Express 15(3):1175–1181

Last A, árkus OM, Georgi S, Mohr J (2015) Röntgenoptische Messung des Seitenwandwinkels direktlithografischer refraktiver Röntgenlinsen. In: Proceedings of MEMS, Mikroelektronik, Systeme, Mikrosystemtechnik-Kongress 6, VDE, Frankfurt am Main, Germany, pp 508–510

Lengeler B, Schroer C, Tümmler J, Benner B, Richwin M, Snigirev A, Snigireva I, Drakopoulos M (1999) Imaging by parabolic refractive lenses in the hard X-ray range. J Synchrotron Rad 6(6):1153–1167

Lengeler B, Schroer C, Benner B, Gerhardus A, Günzler TF, Kuhlmann M, Meyer J, Zimprich C (2002) Parabolic refractive X-ray lenses. J Synchrotron Rad 9(3):119–124

Malacara D (ed) (1992) Optical shop testing, chapter 1. Roberts & Company, Englewood, CO, United States

Mayo SC, Sexton B (2004) Refractive microlens array for wave-front analysis in the medium to hard x-ray range. Opt Lett 29(8):866–868

Morgan KS, Paganin DM, Siu KKW (2012) X-ray phase imaging with a paper analyzer. Appl Phys Lett 100(12):124102

Morgan KS, Modregger P, Irvine SC, Rutishauser S, Guzenko VA, Stampanoni M, David C (2013) A sensitive x-ray phase contrast technique for rapid imaging using a single phase grid analyzer. Opt Lett 38(22):4605–4608

Nazmov V, Reznikova E, Boerner M, Mohr J, Saile V, Snigirev A, Snigireva I, Di Michiel M, Drakopoulos M, Simon R, Grigoriev M (2004a) Refractive lenses fabricated by deep SR lithography and LIGA technology for X-ray energies from 1 keV to 1 MeV. AIP Conf Proc 705(1):752–755

Nazmov V, Reznikova E, Mohr J, Snigirev A, Snigireva I, Achenbach S, Saile V (2004b) Fabrication and preliminary testing of X-ray lenses in thick SU-8 resist layers. Microsys Technol 10(10):716–721

Nazmov V, Reznikova E, Somogyi A, Mohr J, Saile V (2004c) Planar sets of cross x-ray refractive lenses from SU-8 polymer. Proc SPIE 5539:235–243

öter BFl, Juranić P, Großmann P, Kapitzki S, Keitel B, Mann K, önjes EPl, Schäfer B, Tiedkte K (2011) Beam parameters of FLASH beamline BL1 from Hartmann wavefront measurements. Nucl Instrum Methods Phys Res A 635(1, Supplement):108–S112

Pfeiffer F, Bech M, Bunk O, Kraft P, Eikenberry EF, Brönnimann C, Grünzweig C, David C (2008) Hard-X-ray dark-field imaging using a grating interferometer. Nat Mater 7:134–137

Rau C, Wagner U, Pešić Z, De Fanis A (2011) Coherent imaging at the Diamond beamline I13. Phys Status Solidi A 208(11):2522–2525

Reznikova E, Weitkamp T, Nazmov V, Last A, Simon M, Saile V (2007) Investigation of phase contrast hard X-ray microscopy using planar sets of refractive crossed linear parabolic lenses made from SU-8 polymer. Phys Status Solidi A 204(8):2811–2816

Reznikova E, Weitkamp T, Nazmov V, Simon M, Last A, Saile V (2009) Transmission hard X-ray microscope with increased view field using planar refractive objectives and condensers made of SU-8 polymer. J Phys Conf Ser 186(1):012070

Rutishauser S, Zanette I, Weitkamp T, Donath T, David C (2011) At-wavelength characterization of refractive x-ray lenses using a two-dimensional grating interferometer. Appl Phys Lett 99(22):221104

Saile V, Wallradbe U, Tabata O, Korvink JG (2009) LIGA and its applications, volume 7 of advanced micro & nanosystems. Wiley-VCH, Weinheim, Germany

Seaberg M, Cojocaru R, érujon SB, Ziegler E, Jaggi A, Krempasky J, Seiboth F, Aquila A, Liu Y, Sakdinawat A, Lee HJ, Flechsig U, Patthey L, Koch F, Seniutinas G, David C, Zhu D, Mikeš L, Makita M, Koyama T et al (2019) Wavefront sensing at X-ray free-electron lasers. J Synchrotron Rad 26(4)

Shack RV, Platt BC (1971) Production and use of a lenticular Hartmann screen. J Opt Soc Am 61:656

Siewert F, Noll T, Schlegel T, Zeschke T, Lammert H (2004) The nanometer optical component measuring machine: a new Sub-nm topography measuring device for X-ray optics at BESSY. 3 AIP Conf Proc 705(1):847–850

Siewert F, Buchheim J, Zeschke T, Störmer M, Falkenberg G, Sankari R (2014) On the characterization of ultra-precise X-ray optical components: advances and challenges in ex situ metrology. J Synchrotron Rad 21(5):968–975

Snigirev A, Kohn V, Snigireva I, Lengeler B (1996) A compound refractive lens for focusing high-energy X-rays. Nature 384:49–51

Snigirev A, Snigireva I, Drakopoulos M, Nazmov V, Reznikova E, Kuznetsov S, Grigoriev M, Mohr J, Saile V (2003) Focusing properties of x-ray polymer refractive lenses from SU-8 resist layer. Proc SPIE 5195:21–31

Takacs PZ, Qian S-N, Colbert J (1987) Design of a long trace surface profiler. Proc SPIE 0749:59–64

Ueda K ed (2017) Special issue on X-Ray free-electron laser. Appl Sci 7(11)

Vila-Comamala J, Diaz A, Guizar-Sicairos M, Mantion A, Kewish CM, Menzel A, Bunk O, David C (2011) Characterization of high-resolution diffractive X-ray optics by ptychographic coherent diffractive imaging. Opt Express 19(22):21333–21344

Wang H, Sawhney K, Bérujon S, Ziegler E, Rutishauser S, David C (2011) X-ray wavefront characterization using a rotating shearing interferometer technique. Opt Express 19(17):16550–16559

Wang H, Bérujon S, Sawhney K (2012) Characterization of a one dimensional focusing compound refractive lens using the rotating shearing interferometer technique. AIP Conf Proc 1466(1):223–228

Wang H, Bérujon S, Sawhney K (2013) Development of at-wavelength metrology using grating-based shearing interferometry at Diamond Light Source. J Phys Conf Ser 425(5):052021

Wang H, Kashyap Y, Laundy D, Sawhney K (2015a) Two-dimensional in situ metrology of X-ray mirrors using the speckle scanning technique. J Synchrotron Radiat 22(4):925–929

Wang H, Kashyap Y, Sawhney K (2015b) Speckle based X-ray wavefront sensing with nanoradian angular sensitivity. Opt Express 23(18):23310–23317

Wang H, Sutter J, Sawhney K (2015c) Advanced in situ metrology for x-ray beam shaping with super precision. Opt Express 23(2):1605–1614

Weitkamp T, Diaz A, David C, Pfeiffer F, Stampanoni M, Cloetens P, Ziegler E (2005a) X-ray phase imaging with a grating interferometer. Opt Express 13(16):6296–6304

Weitkamp T, Nöhammer B, Diaz A, David C, Ziegler E (2005b) X-ray wavefront analysis and optics characterization with a grating interferometer. Appl Phys Lett 86(5):054101

Wen HH, Bennett EE, Kopace R, Stein AF, Pai V (2010) Single-shot x-ray differential phase-contrast and diffraction imaging using two-dimensional transmission gratings. Opt Lett 35(12):1932–1934

Zanette I, Weitkamp T, Donath T, Rutishauser S, David C (2010) Two-dimensional X-Ray grating interferometer. Phys Rev Lett 105(24):248102

Zdora M-C, Thibault P, Zhou T, Koch FJ, Romell J, Sala S, Last A, Rau C, Zanette I (2017) X-ray phase-contrast imaging and metrology through unified modulated pattern analysis. Phys Rev Lett 118(20):203903

Zdora M-C, Thibault P, Deyhle H, Vila-Comamala J, Rau C, Zanette I (2018a) Tunable X-ray speckle-based phase-contrast and dark-field imaging using the unified modulated pattern analysis approach. J Instrum 13(05):C05005

Zdora M-C, Zanette I, Zhou T, Koch FJ, Romell J, Sala S, Last A, Ohishi Y, Hirao N, Rau C, Thibault P (2018b) At-wavelength optics characterisation via X-ray speckle- and grating-based unified modulated pattern analysis. Opt Express 26(4):4989–5004

Zhou T, Wang H, Fox O, Sawhney K (2018) Auto-alignment of X-ray focusing mirrors with speckle-based at-wavelength metrology. Opt Express 26(21):26961–26970

Zhou T, Wang H, Fox OJL, Sawhney KJS (2019) Optimized alignment of X-ray mirrors with an automated speckle-based metrology tool. Rev Sci Instrum 90(2):021706

Chapter 8
3D Virtual Histology Using X-ray Speckle with the Unified Modulated Pattern Analysis

8.1 Introductory Remarks

Most of the implementations of X-ray speckle-based phase tomography reported in the literature to date are proof-of-principle demonstrations and mostly performed on samples of limited scientific interest. A major achievement of this Ph.D. project was the translation of speckle phase tomography to real-world applications, in particular for 3D virtual histology of unstained biomedical soft-tissue specimens. This was enabled by the benefits of the UMPA approach for speckle-based imaging, which was introduced in Chap. 6. UMPA has proven suitable for biomedical imaging applications thanks to its high sensitivity to small density differences, its relaxed requirements on the experimental conditions and its flexible character, amongst others.

In this chapter, the first demonstrations of X-ray speckle-based virtual histology are presented and its potential as a complementary or alternative method to conventional histology is illustrated. It is shown that UMPA speckle-based imaging can deliver comprehensive 3D structural and quantitative density maps of unstained soft-tissue specimens, even in hydrated state. It allows for revealing the tissue microstructure at high detail and at the same time enables quantitative discrimination of minute density differences at density resolutions down to $2\,\mathrm{mg/cm^3}$.

A first test of UMPA virtual histology was performed on an unstained and fully hydrated mouse testicle scanned at Diamond I13-2 beamline, which is presented in the first section of this chapter.

Following this successful demonstration, an optimised scan and in-depth analysis was conducted on a hydrated, unstained mouse kidney. The high contrast of the 3D density maps enabled semi-automatic segmentation and quantitative density analysis. The results on the mouse kidney were prepared in a manuscript, which has recently been published in the journal Optica.

To demonstrate that the method can be applied to a wide range of biomedical samples prepared in various ways, further measurements on human cerebellar brain

M.-C. Zdora, *X-ray Phase-Contrast Imaging Using Near-Field Speckles*,
Springer Theses, https://doi.org/10.1007/978-3-030-66329-2_8

tissue embedded in paraffin wax were performed and first results are shown in the last section of this chapter. UMPA virtual histology allowed for visualising details of the tissue down to single Purkinje nerve cells.

8.2 Background and Motivation

The visualisation of human and animal tissue is fundamental for expanding the knowledge about organ function and dysfunction as well as for routine clinical histopathology. Revealing the internal organ structure requires imaging methods with high spatial resolution and high contrast for soft tissue, ideally capable of providing undistorted 3D information from hydrated specimens.

To date, conventional histology is the biomedical standard. It consists in using a light microscope to image thin slices of a specimen, which were beforehand embedded in a hard matrix and stained. Combining 2D serial sections into a 3D volume is possible but time-consuming and often hindered by artefacts occurring during tissue preparation and sectioning (Ourselin et al. 2001; Pichat et al. 2018). Moreover, conventional histology is unable to provide unbiased isotropic sampling, essential for accurate and precise volumetric analysis, and is associated with sample shrinkage due to the required dehydration process.

A suitable alternative to histology for non-destructive volumetric imaging at isotropic spatial resolution is the use of X-ray micro computed tomography (microCT) (Stock 2008). While X-ray absorption-based microCT is broadly used in a variety of fields, contrast is often too low for biomedical applications and soft-tissue discrimination (Metscher 2009a, b; Pauwels et al. 2013; Martins et al. 2015; Shearer et al. 2016; Missbach-Guentner et al. 2018; Busse et al. 2018; Müller et al. 2018) or vessel visualisation (Zagorchev et al. 2010; Vasquez et al. 2011; Lundström et al. 2012; Ehling et al. 2016; Hlushchuk et al. 2018) require contrast agents that bear the risk of altering the tissue structure, are susceptible to incomplete tissue perfusion and vessel rupture, and are, furthermore, often incompatible with other imaging techniques.

Alternative X-ray methods that are based on extracting phase-contrast signals (Wilkins et al. 2014) can provide much better contrast and have been identified as a promising approach to achieve virtual histology (Töpperwien et al. 2018; Khimchenko et al. 2018; Wu et al. 2009; Shirai et al. 2014; Zanette et al. 2013; Velroyen et al. 2014). However, existing methods are typically limited by elaborate experimental setups and other constraints on the nature and size of samples.

In this chapter, it is demonstrated that X-ray phase tomography based on speckle-based imaging is a powerful, versatile method for 3D virtual histology of unstained and hydrated specimens, providing a robust pathway to a wealth of information from a single data set, which has not been feasible with existing methods.

8.3 First Demonstration: X-ray Speckle-Based Phase Tomography of a Mouse Testicle

As a first demonstration a mouse testicle was scanned and a preliminary analysis of the phase volume was performed.

The experiments presented in this section were performed with the support of Dr. Joan Vila-Comamala, ETH Zurich and Paul Scherrer Institute (Switzerland), and Dr. Hans Deyhle, University of Southampton (UK), now Diamond Light Source (UK).

8.3.1 Materials and Methods

8.3.1.1 Experimental Setup and Data Acquisition

Data was taken at the Diamond I13-2 imaging branchline at Diamond Light Source, see Chap. 3. The X-ray beam was delivered by an undulator insertion device with a gap of 9.5 mm and was filtered with several primary filters (280 µm pyrolitic graphite, 705 µm pyrolitic graphite, 3200 µm aluminium, 35 µm silver and 35 µm cadmium). A rhodium mirror was used to suppress higher harmonics. The resulting X-ray spectrum contained contributions mainly in the range of 15–25 keV with a weighted mean observed energy of 22.7 keV. For the calculation of the weighted mean observed energy, the effect of the filters and mirror as well as the scintillator response were taken into account, but beam hardening in the sample was not considered here. This effect is, however, expected to be small for biological samples in the given energy range, which is far from the absorption edges.[1]

The experimental setup was similar to the one shown in Fig. 8.5 in the next section. It consisted of a diffuser, here two stacked sheets of P800 sandpaper, on translation stages for stepping, the specimen mounted on a tomography stage 4 cm further downstream and a detector system located at a distance of 40 cm downstream of the sample. The detector system comprised a pco.4000 CCD camera (chip size: 4008×2672 pixels; pixel size: 9 µm) coupled to an infinity-corrected optical microscope (magnification: $2\times$) with a $2\times$ objective lens, leading to a total magnification of approximately $4\times$. A 250 µm-thick cadmium tungstate ($CdWO_4$) scintillation screen was mounted in front of the objectives for the conversion of X-rays to visible light. The 10% MTF value of the detector system, determined from a horizontal knife edge, was approximately 1.79 pixels. The effective pixel size was measured by vertical translation of the knife edge and is $p_{\text{eff}} = 2.26$ µm.

A tomographic scan of 1201 projections, equally spaced over 180° of sample rotation, was recorded at each of the $N = 20$ diffuser positions. The diffuser steps

[1]The effect of the specimen in the beam can be roughly estimated by assuming a sample composition of 5 mm of carbon (for the testicle), 2 mm of water (for the agarose gel) and 600 µm of polypropylene (for the sample container). The weighted mean observed energy would in this case be hardened to 23.0 keV.

were defined beforehand and followed a spiral pattern with step sizes larger than 100 μm.[2] Before each tomographic scan, 10 reference images without the sample in the beam were taken at the respective diffuser position. 20 dark images without X-ray beam were acquired after the entire scan. The exposure time for each frame of the scan was 1.5 s. The entire measurements took almost 16 h, which includes more than 6.5 h of overhead time due to motor movement.[3]

8.3.1.2 Sample Preparation

Testicles were removed from a C57BL/6 mouse that had been anaesthetised with isofluorane. Immediately after extraction, the testicles were immersion-fixed in 4% formaldehyde/PBS over night, then embedded in 3% agar/PBS in a 0.5 mL micro-centrifuge tube.

All animal experiments were approved by the cantonal veterinary office of Zurich (Switzerland) in accordance with the Swiss federal animal welfare regulations. All animals had unlimited access to water and standard diet at all times.

The mouse was kindly provided by Dr. Evelyne Kuster, Institute of Physiology, University of Zurich, Switzerland, and the testicle was extracted and prepared by Willy Kuo from the Interface Group, Institute of Physiology, University of Zurich, Switzerland.

8.3.1.3 Image Reconstruction

The reconstruction of the image signals was performed the same way as described in more detail in Sect. 8.4.1.3 below. In a pre-processing step, drifts and instabilities of the diffuser were corrected for by aligning the reference speckle images to the corresponding sample images before the UMPA reconstruction procedure. This was done by performing a cross-correlation in an air background area without sample, where the two images are expected to show perfect agreement. The measured displacements in this correction procedure were small and on the order of a fraction of a pixel, but this step still contributed to an improved image quality.

The differential phase and transmission signals for each projection were retrieved with the standard UMPA reconstruction approach (Zdora et al. 2017). Subsequently, Fourier integration (Kottler et al. 2007; Morgan et al. 2012) was applied to the differential phase signals to obtain the absolute phase shift of the X-rays in the sample.

---

[2]The diffuser scan pattern is not crucial for UMPA, as long as the steps are larger than the size of the analysis window used in the reconstruction, which ensures maximum diversity between the different reference speckle patterns. However, a spiral pattern has proven to be the best solution in terms of image quality and scan efficiency, in particular under conditions of high noise level, e.g. at laboratory sources.

[3]The long overhead time is due to the fact that the sample had to be rotated to 90° before moving it out of the beam using a 1D translation stage due to practical limitations of the setup.

The phase tomogram was reconstructed from the projections using filtered back-projection (Kak and Slaney 2001) with the Python-based *pyCT* processing package developed by the Chair of Biomedical Physics (E17) at Technical University Munich, Germany. A low-pass Butterworth filter was applied to the sinograms before the reconstruction to reduce ring artefacts.

8.3.2 Results and Discussion

8.3.2.1 Projections

The differential phase and transmission images for one of 1201 projections of the tomography scan are shown in Fig. 8.1.

The refraction angle signals in Fig. 8.1a, b show the outlines of the testicle in the plastic tube. Its inner structure comprised of tubules can faintly be observed in the projections, while the fatty tissue on top of the sample is visualised more clearly due to the large difference in density compared to the surrounding agar gel. The adipose tissue also gives the strongest signal in the transmission projection in Fig. 8.1c because of the significantly lower X-ray absorption compared to the testicular tissue and agar. Other features are harder to identify in the transmission than in the differential phase projections. As observed in previous measurements, it should be noted that the transmission signal also contains contributions from propagation effects (edge enhancement) in addition to pure absorption, which is most prominently evident here at the air-tube interface.

As discussed in Sect. 5.4.4 in Chap. 5, an important characteristic to assess the quality of the reconstructed differential phase projections is the angular sensitivity, which is typically determined as the standard deviation of the refraction angle in a homogeneous region of interest in the image. Here, it was measured in a 150×150 pixels region in the air background in Fig. 8.1a, b and is about $\sigma_x = 140$ nrad in the horizontal and $\sigma_y = 90$ nrad in the vertical, respectively. At synchrotron sources, including I13-2 beamline, the spatial coherence is better in the vertical direction than in the horizontal, leading to a sharper speckle pattern and better sensitivity in this direction.

8.3.2.2 Phase Tomogram

While some first qualitative observations could be made already from the differential phase projections in Fig. 8.1, 3D and quantitative density information can only be obtained from the tomographic volume. Longitudinal and transverse slices through the phase tomogram of the mouse testicle are shown in Fig. 8.2, visualising the distribution of the refractive index decrement δ. The shape and inner structure of the testicle, which is comprised of seminiferous tubules, the site of sperm production, are depicted with high contrast. Other internal features of the specimen are visible

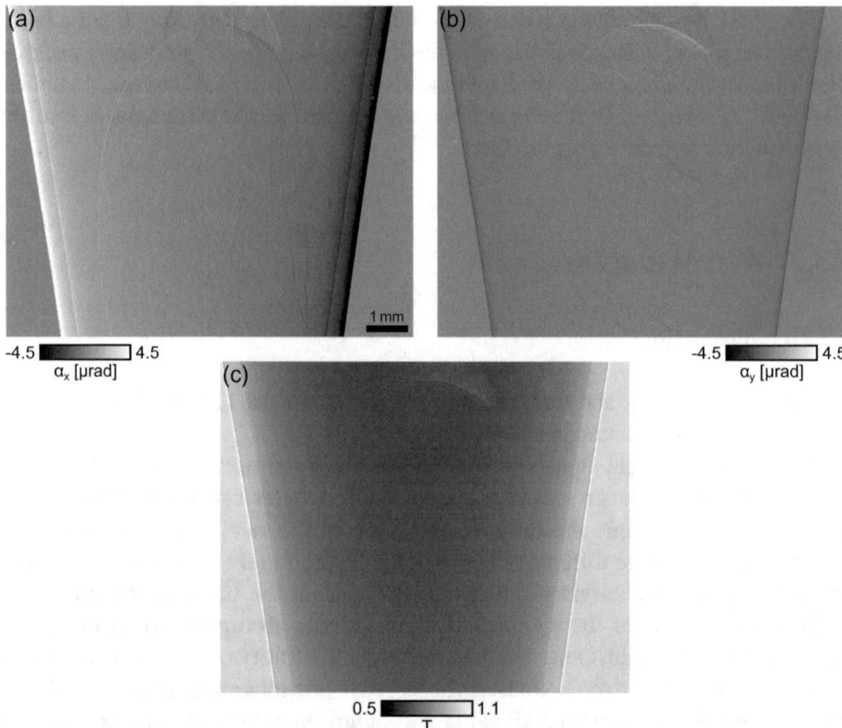

Fig. 8.1 One of the 1201 projections of the tomography scan of the mouse testis in agar gel obtained with multimodal UMPA speckle imaging. **a** Differential phase (refraction angle signal) in the horizontal and **b** the vertical direction. **c** Transmission projection reconstructed from the same data set. edge-enhancement effects are observed in addition to pure X-ray absorption

in the slices in Fig. 8.2 as labelled. Most prominent in contrast is the low-density adipose tissue surrounding the testicle and epididymis. The latter is an elongated structure attached to the testis that forms a coiled tubule in which the sperm matures. The rete testis pointed out at the bottom of the slice in Fig. 8.2a is a collecting pool for the sperm delivered from the seminiferous tubules. The tunica albuginea is the membrane that surrounds the whole testis and shows as a bright white line due to its high density. High-density features within the tissue are blood vessels that run through the organ, see Fig. 8.2b.

Additional slices through the phase tomogram of the testis, as indicated by the dashed lines in Fig. 8.2, are shown in Fig. 8.3 to demonstrate the high image quality of the volume. Previously observed features can be identified, such as the rete testis in Fig. 8.3d. The transverse slice reveals that some tubules close to the rete testis appear smaller than the seminiferous tubules (see black ellipse). These are known as straight tubules (tubuli recti). They are lined only by Sertoli cells, which aid in the progression of sperm cells to spermatozoa (Haschek et al. 2010). Hence, they have

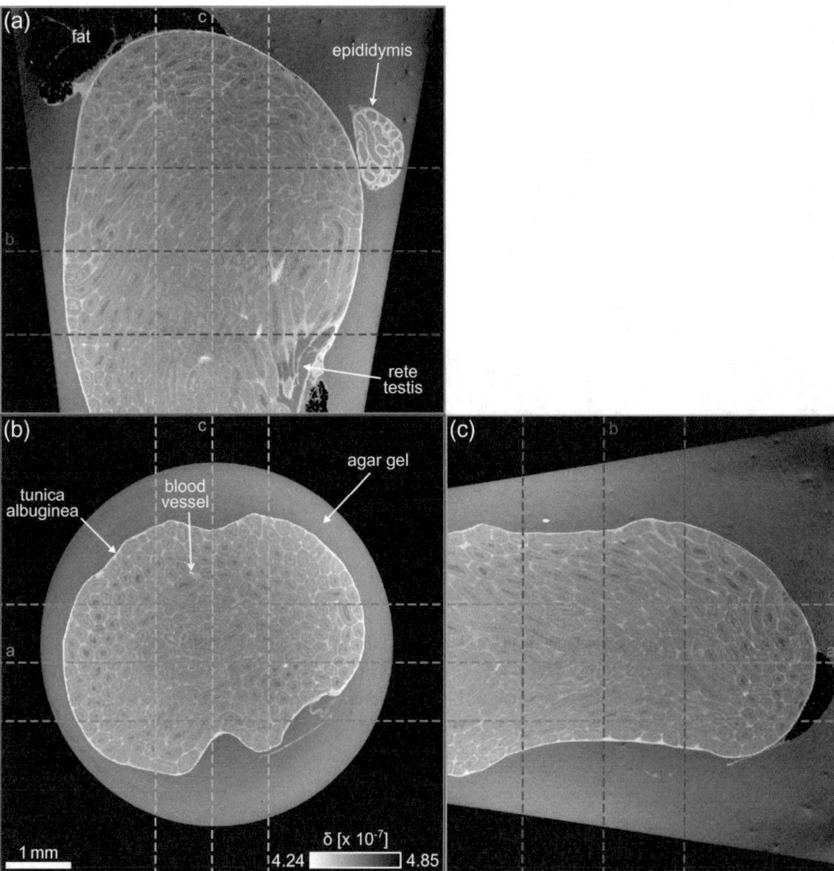

Fig. 8.2 Slices through the phase tomogram of the mouse testis obtained with UMPA. **a** Longitudinal slice close to the centre of the specimen. The epididymis, surrounding epididymal fat and the rete testis can be identified. **b** Transverse slice showing cross-sections of the seminiferous tubules revealing their structure comprised of an empty inner lumen and outer germ cells. Furthermore, blood vessels and the encapsulating tunica albuginea can be seen. The agar surrounding the testicle shows medium density values. **c** Central longitudinal slice in the direction orthogonal to panel (**a**). Additional dashed lines (not labelled) indicate the location of the slices shown in Fig. 8.3

a smaller diameter than the rest of the seminiferous tubules, as can be seen in the magnified view of the black circled ROI shown as an inset in Fig. 8.3d.

A closer look into the tissue structure of the seminiferous tubules is given in Fig. 8.4b, which shows a ROI of the transverse slice in Fig. 8.4a.

It can be seen that the seminiferous tubules are comprised of an inner empty space, the lumen, here darker, surrounded by sperm cells at various developmental stages. The sperm tails typically extend into the lumen, which can be observed as the bright areas directly around the lumen, while the sperm cells at earlier developmental stages have a lower density and hence appear darker.

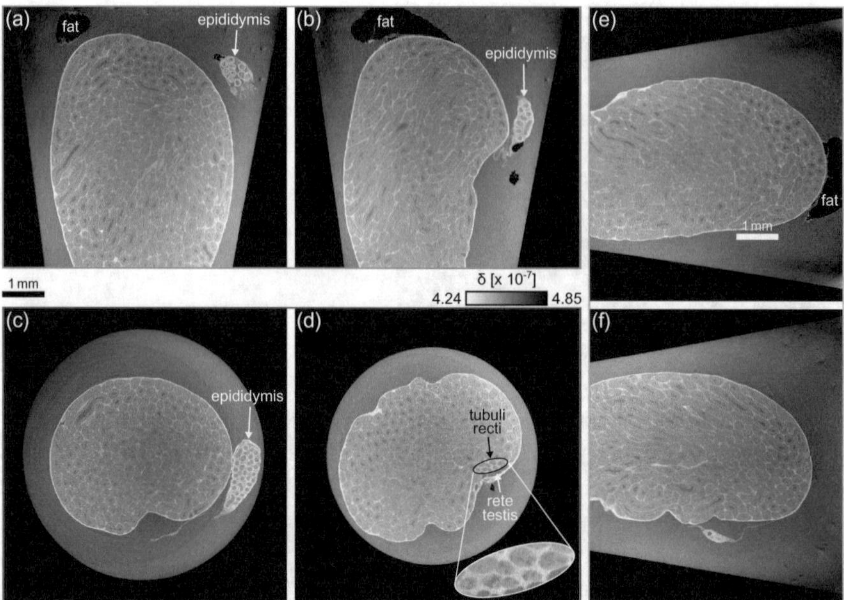

Fig. 8.3 Additional slices through the phase volume of the mouse testis as indicated by the dashed lines in Fig. 8.2. **a**, **b** Longitudinal slices, **c**, **d** transverse slices and **e**, **f** orthogonal longitudinal slices. Panel (**d**) shows a transverse view of the rete testis and the surrounding tubuli recti (straight tubules), black ellipse, which have a smaller diameter than the seminiferous tubules as they are only lined by Sertoli cells, see inset

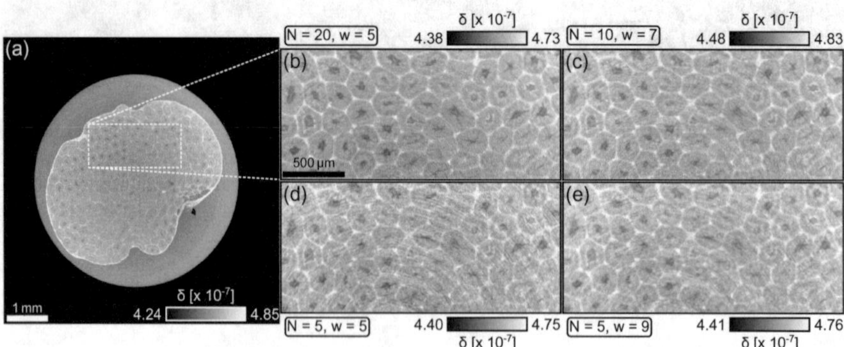

Fig. 8.4 Region of interest of a transverse slice in the phase tomogram of the testis reconstructed with different parameters. **a** Transverse slice of the phase tomogram. The location of the ROI showing the detailed structure of the seminiferous tubule network is indicated by the white dashed box. **b** ROI from panel (**a**) reconstructed with $N = 20$ steps of the diffuser and an analysis window of 5×5 pixels. **c** Same ROI reconstructed with $N = 10$ and a window of 7×7 pixels. Contrast and resolution are still sufficient to visualise small details of the tissue structure. **d** Same ROI reconstructed with $N = 5$ and a window of 5×5 pixels. While the spatial resolution is the same as in panel (**b**), tissue details are hidden by the high noise level. **e** Same ROI reconstructed with $N = 5$ and a window of 9×9 pixels. While contrast is high and the noise level reduced compared to panel (**d**), the finest tissue details cannot be resolved due to the low spatial resolution

8.3.2.3 Evaluation of Different Reconstruction Parameters

Figure 8.4, furthermore, illustrates the effect of varying the parameters (window size and number of steps) in the UMPA reconstruction on the image quality of the phase tomogram slices. The influence of the scan and reconstruction parameters was explored previously for 2D projections of a test sample in Chap. 6, Sect. 6.4, but not yet for 3D tomographic imaging.

The data shown in the previous figures and the ROI in Fig. 8.4b were reconstructed with $N = 20$ diffuser steps and an analysis window size of 5×5 pixels. This leads to good contrast and low noise at a spatial resolution sufficient to visualise details of the seminiferous tubules. Although $N = 20$ is already a significantly lower number of steps than required for most other operational modes such as 1D or 2D stepping (see Chap. 5), reducing N further would allow for shorter scan times and lower dose to the specimen. Hence, it was assessed here by how much the number of steps can be decreased to still obtain a good image quality. For this, the same data set was reconstructed only considering the images from the first 10 or first 5 diffuser positions.

An increase in noise and a decrease in signal sensitivity is expected when reducing the number of diffuser steps, see Chap. 6. This was compensated for by choosing a larger window size in the UMPA analysis. The results for $N = 10$ with a window size of 7×7 pixels and $N = 5$ with a window size of 9×9 pixels can be found in Fig. 8.4c, e, respectively, which both show high contrast. However, the spatial resolution in Fig. 8.4c, e is reduced compared to Fig. 8.4a due to the larger window sizes used in the reconstruction. While for the window of $w = 7 \times 7$ in Fig. 8.4c, the spatial resolution is still sufficient to visualise the detailed inner structure of the seminiferous tubules, the even larger window ($w = 9 \times 9$ pixels) in Fig. 8.4e leads to the loss of some of the smallest features. For comparison, the window size was kept constant at $w = 5 \times 5$ pixels in Fig. 8.4d and the spatial resolution is conserved. However, small details are here lost due to the increase in noise.

These results illustrate that the parameters in the UMPA reconstruction need to be chosen carefully and depending on the primary goal of the measurement. For the present data, the scan time could be significantly reduced at only minor loss of information by halving the number of steps while slightly increasing the window size. This flexibility in adapting the scan and reconstruction parameters is a major benefit of the UMPA approach.

While the spatial resolution was not explicitly measured here, a good indication is given by the features that are visible in the ROI in Fig. 8.4b. In some of the seminiferous tubules a few, but not all, of the germ cells seem to be resolved. The sperm cells at different development stages have diameters in the range of 6–13 μm (Ray et al. 2015). It can be assumed that the largest of these cells are resolved here and the spatial resolution of the phase tomogram can be estimated to be approximately 13 μm. It should, however, be considered that this is a very rough estimation. The spatial resolution of the differential phase images is in general determined by twice the FWHM of the analysis window, which is around 2 pixels ($\approx$4.52 μm) for the case of the 5 × 5 pixels Hamming window used here, by the point spread function (PSF) of the detector system and by the phase integration and tomographic reconstruction processes.

8.3.3 Conclusions and Outlook

The high-quality phase tomogram of a mouse testis presented in this section demonstrates the promising potential of the UMPA speckle-based technique for biomedical imaging and in particular 3D virtual histology of soft-tissue specimens. The contrast and spatial resolution of the phase volume obtained with UMPA were sufficient to visualise features of the testicle down to the fine structure of the seminiferous tubules.

Moreover, the effect of varying the reconstruction parameters was investigated as a way to reduce scan time and dose to the sample while maintaining a good image quality. Following this first qualitative analysis, in the future a more detailed and systematic quantitative analysis will be performed to evaluate the influence of the reconstruction parameters on the signal-to-noise ratio, contrast-to-noise ratio, density resolution and spatial resolution of the phase tomogram.

As can be seen in Figs. 8.2 and 8.3, some low-frequency artefacts remain in the reconstructed phase volume. Slight cupping artefacts in the transverse slices and a vertical phase ramp in the longitudinal slices originate mainly from the low-pass nature of the phase integration step, but possibly also slight beam hardening effects occurring under non-monochromatic illumination. Furthermore, the ring filter that was applied to the sinograms can introduce low-frequency artefacts. For the experiments presented in the next sections, artefacts were reduced by improving the setup and using a beam with a smaller X-ray energy bandwidth.

The promising results on the mouse testicle suggest that in the future speckle-based virtual histology could be used for investigating the 3D structure and arrangement of the seminiferous tubules. To date, the characteristics of the seminiferous tubules in rodents are still largely unknown, which is mainly due to the fact that 3D information about the structure of whole tubules and their relationships with each other are difficult to extract from the 2D images obtained with conventional histology. Early works on the study of the 3D structure of the seminiferous tubules only analyse a few tubules of a single animal (Curtis 1918; Clermont and Huckins 1961). In a later publication, all tubules in a single testis were reconstructed (Nakata et al. 2015), but this still required manual tracing, making it a laborious and time-consuming process. Semi-automatic segmentation was later achieved by a combination of fluorescence and bright-field imaging of histological slices treated with immunohistological and structural stains so that several specimens at different developmental stages could be studied (Nakata et al. 2017; Nakata 2019). However, the preparation, scanning and analysis processes of these techniques are still extremely time-consuming and the dehydration and staining of the specimens can have unknown effects on the tissue structure.

UMPA phase tomography, on the other hand, can provide a more robust, time-efficient and simpler pathway to obtain information on the structure and arrangement of seminiferous tubules without the need for staining. Analysis of the 3D phase volumes will allow for the quantification of their number, length and mutual relationships. Moreover, the quantitative information on the tissue density could aid in the identification of pathological conditions such as infertility, in particular in combination with the 3D structural information.

Although the scan time in the present example was relatively long, the overall time effort of UMPA speckle-based tomography is still significantly less than for conventional histology. Furthermore, the optimisation of the setup allowed for a major reduction in acquisition time. This is demonstrated in the following section, in which the total scan time was reduced by a factor of more than 16, whilst doubling the number of projections.

8.4 Comprehensive X-ray Speckle-Based Virtual Histology of a Murine Kidney

After the first demonstration on the mouse testicle, UMPA speckle-based phase tomography was applied to a murine kidney. A comprehensive analysis was performed illustrating the potential of X-ray speckle-based imaging for 3D virtual histology applications.

The results presented here were later (after submission of the thesis) prepared in a journal article that has recently been published as (Zdora et al. 2020): M.-C. Zdora, P. Thibault, W. Kuo, V. Fernandez, H. Deyhle, J. Vila-Comamala, M. P. Olbinado, A. Rack, P. M. Lackie, O. L. Katsamenis, M. J. Lawson, V. Kurtcuoglu, C. Rau, F. Pfeiffer, and I. Zanette, "X-ray phase tomography with near-field speckles for three-dimensional virtual histology," Optica **7**(9), 1221–1227 (2020), licensed under the OSA Open Access Publishing Agreement. The following sections are based on a previous version of this manuscript.

8.4.1 Materials and Methods

8.4.1.1 Experimental Arrangement and Data Acquisition for X-ray Speckle-Based Phase Tomography

Data was acquired at beamline ID19 at the European Synchrotron Radiation Facility (ESRF, Grenoble, France) using a U13 single-harmonic undulator (gap: 12.0 mm) with additional filtering by 5.6 mm aluminium, a 1.4 mm-thick diamond attenuator

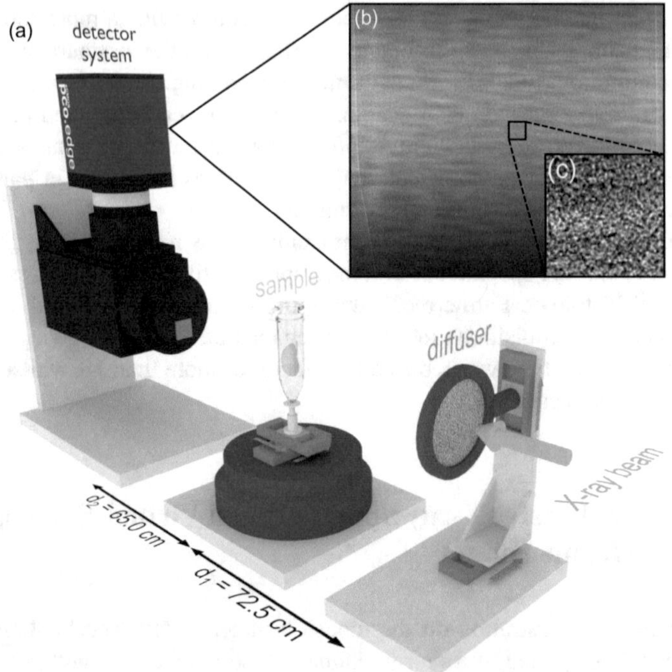

Fig. 8.5 Experimental setup for X-ray speckle-based phase tomography. **a** Setup used for X-ray speckle-based tomography of the mouse kidney specimen performed at beamline ID19, ESRF. **b** One of the raw speckle projections acquired with the sample in the beam. **c** Zoomed region of the speckle pattern in panel (**b**). The X-ray refraction angle can be recovered from the minute displacement of this pattern with respect to the reference pattern without sample. Reprinted with permission from Zdora et al. (2020) © The Optical Society

and a 0.5 mm-thick beryllium exit window. This delivers an X-ray beam with a narrow energy spectrum of approximately 1–2% with a peak energy of 26.3 keV.[4]

The experimental setup for phase tomography based on SBI is shown in Fig. 8.5a. It is simple and the only addition to the standard X-ray imaging apparatus was a piece of abrasive paper used as a phase modulating diffuser, which was fixed on translational stages for stepping in the plane transverse to the beam direction. The near-field speckle pattern created by scattering and subsequent interference of the X-rays from the diffuser particles can be seen in Fig. 8.5b and in more detail in Fig. 8.5c.

The sample was mounted on a tomographic stage located $d_1 = 72.5$ cm downstream of the diffuser. Reference images (without the specimen) and sample images (with the specimen) were acquired at $N = 20$ different transverse positions of the dif-

[4]The peak energy value is stated here rather than the detected weighted average as the exact X-ray spectrum coming from the undulator was not measured at the time of the experiment. However, the bandwidth of the spectrum is narrow and hence the peak energy can be used in good approximation.

fuser. The resulting interference patterns were recorded by a detection system placed $d_2 = 65.0\,\text{cm}$ downstream of the sample. It consisted of a pco.edge 5.5 sCMOS camera (PCO AG, Kelheim, Germany) coupled to a 2.1× magnification optics system, comprising two Hasselblad lenses (Hasselblad, Gothenburg, Sweden) in tandem configuration with a numerical aperture of 0.17 and a scintillation screen (250 µm-thick cerium-doped lutetium aluminium garnet (LuAG:Ce)). The effective pixel size of this system was $p_{\text{eff}} = 3.1\,\mu\text{m}$. The exposure time for each recorded frame was 50 ms.

Acquisition of the sample images was performed in tomographic mode by taking 2401 projections at equidistant viewing angles over 180 degrees of sample rotation. This was repeated for each of the 20 diffuser positions. Before each of the tomographic sample scans a set of 20 reference images was recorded, which were averaged to obtain a single reference image for each diffuser position. 20 dark images without the beam on the camera were taken before the entire scan and averaged.

Due to restrictions in the field of view, the scanning of the mouse kidney specimen was performed in two height steps with a vertical overlap of 143 pixels. Image reconstruction was performed separately for both height steps and subsequently the two reconstructed phase volumes were registered manually and concatenated. The total scan time for the two height steps was approximately 1.5 h.

8.4.1.2 Sample Preparation

One 8-week old male C57BL/6JRj mouse, purchased from Janvier Labs (Le Genest-Saint-Isle, France), was anaesthetised with a ketamine (120 mg/kg)/xylazine (25 mg/kg) combination. The kidneys were perfusion-fixed for optimal structural preservation, in particular prevention of the collapse of tubular lumina from post-mortem blood pressure loss. A blunted 21G butterfly needle was inserted retrogradely into the abdominal aorta and the kidneys were flushed with 10 mL, 37 °C phosphate-buffered saline (PBS) to remove the blood, then with 100 mL, 37 °C fixation solution (4% paraformaldehyde, 0.1% glutaraldehyde, PBS). The right kidney was excised and stored in the fixation solution at room temperature. Before X-ray tomography scanning, the kidney was embedded in 1.5% agar/PBS in a 0.5 mL microcentrifuge tube.

All animal experiments were approved by the cantonal veterinary office of Zurich (Switzerland) in accordance with the Swiss federal animal welfare regulations. All animals had unlimited access to water and standard diet at all times.

The sample was kindly provided and prepared by Willy Kuo from the Interface Group, Institute of Physiology, University of Zurich, Switzerland.

Conventional histological processing was performed on the same kidney specimen after the X-ray speckle-based phase tomography measurement. For this, the fixed kidney was first dehydrated using an ethanol series and embedded in paraffin wax. Subsequently, it was cut in half along the short axis and serial sections at different levels of the sample were taken along the transverse (short) axis for one half and along the sagittal (long) axis for the other half of the kidney (sections 4 µm thick). The sections on microscope slides were de-waxed, re-hydrated and stained using standard haematoxylin and eosin (H&E), periodic acid-Schiff (PAS) or Masson's trichrome (MTRI) protocols. Sections were then dehydrated and mounted in XTF mountant (CellPath, Newtown, Powys, UK). The slides were scanned with an Olympus VS120 Virtual Slide Scanner (Olympus-lifescience, Tokyo, Japan) using 10× (UPLSAPO10X2, numerical aperture: 0.4) and 20× (UPLSAPO20X, numerical aperture: 0.75) objectives. The pixel sizes in the sample plane were 692 nm for the 10× and 346 nm for the 20× objective.

8.4.1.3 Image Reconstruction for X-ray Speckle-Based Phase Tomography

The reconstruction of the 3D phase volume was performed in several steps, analogous to the data analysis of the mouse testis scan in the previous section. First, all acquired frames were corrected for the dark current by subtracting the average dark image. Then, the differential phase signals in the horizontal and the vertical directions were reconstructed from the dark-current corrected sample and reference images recorded at the different diffuser positions. This was done separately for each projection of the tomography scan by applying the UMPA approach, which is presented in detail by Zdora et al. (2017) and in Chap. 6. Although the UMPA formalism allows for multimodal signal retrieval, here the focus is on the phase-shifting properties of the specimen, which can be observed as a displacement of the speckle pattern, related to the refraction angle and differential phase signal via Eq. (6.3) in Chap. 6. From the differential phase signals, the phase shift images were retrieved using a Fourier phase-integration routine, combining the horizontal and vertical differential phase information (Morgan et al. 2012; Kottler et al. 2007).

The tomographic reconstruction was performed by applying a conventional filtered back-projection algorithm (Kak and Slaney 2001) using the Python-based *pyCT* processing package developed by the Chair of Biomedical Physics (E17) at Technical University Munich, Germany. To reduce ring artefacts in the tomographic slices originating mainly from scintillator defects and noise, a low-pass Butterworth filter was applied to the sinogram of each slice prior to the filtered back-projection step. After reconstruction of the tomogram, a 3D polynomial of second order was fitted to the agarose-filled background areas and subtracted from the phase volume to correct for low-frequency artefacts that can arise as a result of the phase integration routine.

All of the above procedures were performed for each of the two height steps separately and finally the two volumes were concatenated after manual correlation.

8.4.1.4 Conversion to Electron and Mass Density Values

The 3D δ-distribution obtained from the tomographic reconstruction can be directly converted to an electron density ρ_{el} map for X-ray energies far from the absorption edges of the elements composing the sample, see Eq. (2.31) in Chap. 2, Sect. 2.5.4. Furthermore, conversion to mass density values can be performed using Eq. (2.32).

8.4.1.5 Three-Dimensional Visualisation and Segmentation

The 3D visualisation and analysis of the results were performed using VG Studio Max 2.0, 2.1 and 3.0 (Volume Graphics, Heidelberg, Germany). For the segmentation of the blood vessel network and renal tubules, a semi-automatic approach using a combination of grey-value thresholding and the region-growing tool was employed.

8.4.2 Results and Discussion

8.4.2.1 Differential Phase Projections

As described in Sect. 8.4.1, the first step in the analysis is the reconstruction of the phase-contrast signal for all of the several hundreds of projections recorded at different viewing angles of the sample. An example is presented in Fig. 8.6a, b, showing the horizontal and vertical differential phase signal, respectively, at the same viewing angle for both of the separate tomography scans acquired for the two different height steps. The outlines of the kidney in the plastic tube[5] can be observed, but only little detail of the inner structure of the sample is visible.

8.4.2.2 Phase Tomogram and 3D Virtual Histology of the Murine Kidney

The 3D inner structure of the uncontrasted and fully hydrated kidney can be visualised in the phase tomogram enabling non-destructive virtual histology.

The complex functionally interconnected networks of renal tubules, blood vessels and collecting ducts at a micrometre length scale are shown in Fig. 8.7.

The micro-structural detail of the hydrated kidney tissue is clearly visualised in the phase tomogram slices in Fig. 8.7a, c–e, matching that seen by conventional histology (H&E stain), see Fig. 8.7b, f–h, performed on the same specimen after X-ray imaging. The phase volume slices are presented in histology-like false-colours for better visual comparison and their location within the volume is illustrated in Fig. 8.14 in Sect. 8.4.4. Note that the colours for the virtual phase volume slices were assigned based on tissue density, while conventional histology stains selectively attach to certain types of tissue structures.

[5]Note: The images were flipped upside down so that the kidney was in the correct anatomical orientation.

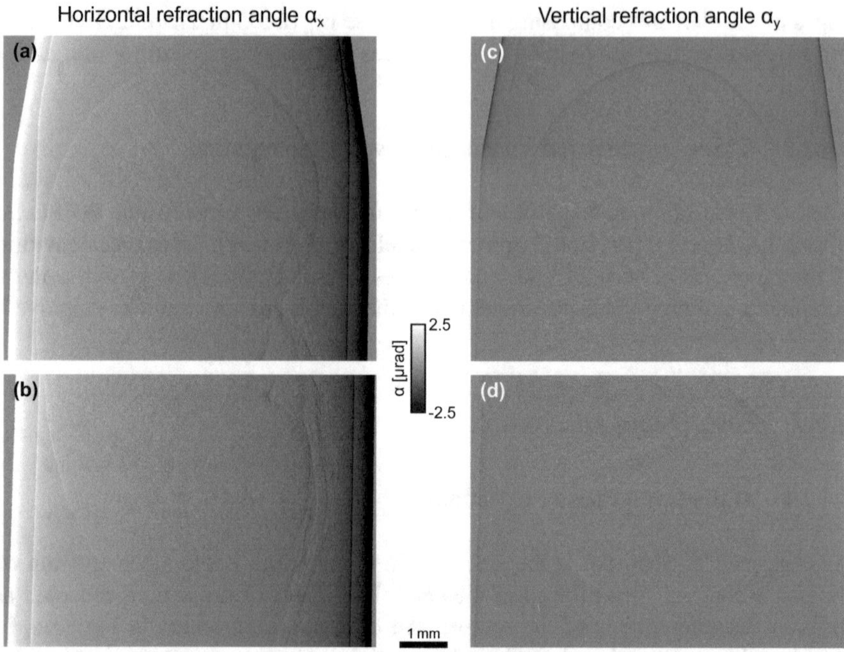

Fig. 8.6 Differential phase projections. One of the 2401 differential phase-contrast projections of the speckle-based phase tomography scan of the kidney. The specimen was scanned in two height steps. **a**, **b** Refraction angle signal in the horizontal direction and **c**, **d** the vertical direction. For each projection, phase integration was performed, combining the information from the horizontal and vertical differential phase projections. Filtered back-projection was then applied to obtain the phase volume. This was done separately for the two height steps, which were subsequently concatenated. Reprinted with permission from Zdora et al. (2020) © The Optical Society

The different regions of the kidney characterised by the presence of different segments of the renal tubules can be identified: inner medulla (IM), inner stripe of the outer medulla (ISOM), outer stripe of the outer medulla (OSOM) and cortex (COR).

Essential micro-structural features including small blood vessels, renal tubules and renal glomeruli can be identified in the regions of interest for both the virtual histology slice from the phase tomogram in Fig. 8.7e and the conventional histology slice in Fig. 8.7h.

The histological sections in Fig. 8.7f–h show some tissue damage, folding, distortion and overall shrinkage, here estimated to be around 20%, which can be encountered in conventional histology due to sample embedding and physical slicing (McInnes 2005).

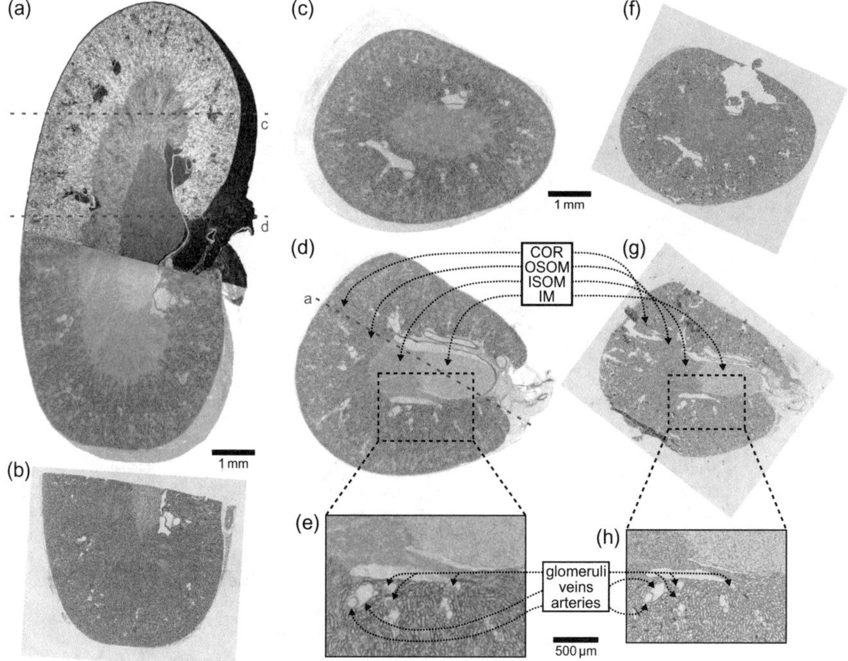

Fig. 8.7 Comparison of virtual phase volume slices and histological sections (H&E staining, 10× magnification). **a** Long-axis (sagittal) cut through the phase volume (partly in false colours) and **b** corresponding histological slice. **c**, **d** Short-axis (transverse) cuts through the phase volume (location indicated by the blue dashed lines in panel (**a**)) and **f**, **g** corresponding histological slices. The different kidney regions, cortex (COR), outer stripe of the outer medulla (OSOM), inner stripe of the outer medulla (ISOM) and inner medulla (IM), can be identified in both the phase volume and the histological cuts. **e**, **h** Zoomed view of regions of interest (ROIs) in panels (**d**) and (**g**) (histological image obtained with 20× magnification). Features such as the renal tubule network, glomeruli and blood vessels are visualised with both imaging modalities. The sample is smaller in the histology images due to the shrinkage during dehydration and processing. While standard H&E staining for general microstructural overview was used here, histological slices with other common stains are shown in Fig. 8.15 in Sect. 8.4.4. Reprinted with permission from Zdora et al. (2020) © The Optical Society

8.4.2.3 Quantitative Density Information from the Phase Volume

For X-ray energies far from the absorption edges of the sample materials, the refractive index decrement δ is directly proportional to the electron density ρ_{el} via Eq. (2.31) in Chap. 2, Sect. 2.5.4. Moreover, for specimens with moderate hydrogen content, a linear relationship between ρ_{el} and the mass density ρ_m holds in good approximation (see Eq. (2.32) in Chap. 2, Sect. 2.5.4). These relations allow for performing a quantitative analysis of the electron and mass density distribution within the specimen.

A section through the volume after conversion to electron density is shown in Fig. 8.8a and in more detail in Fig. 8.8b. The high density resolution enables the

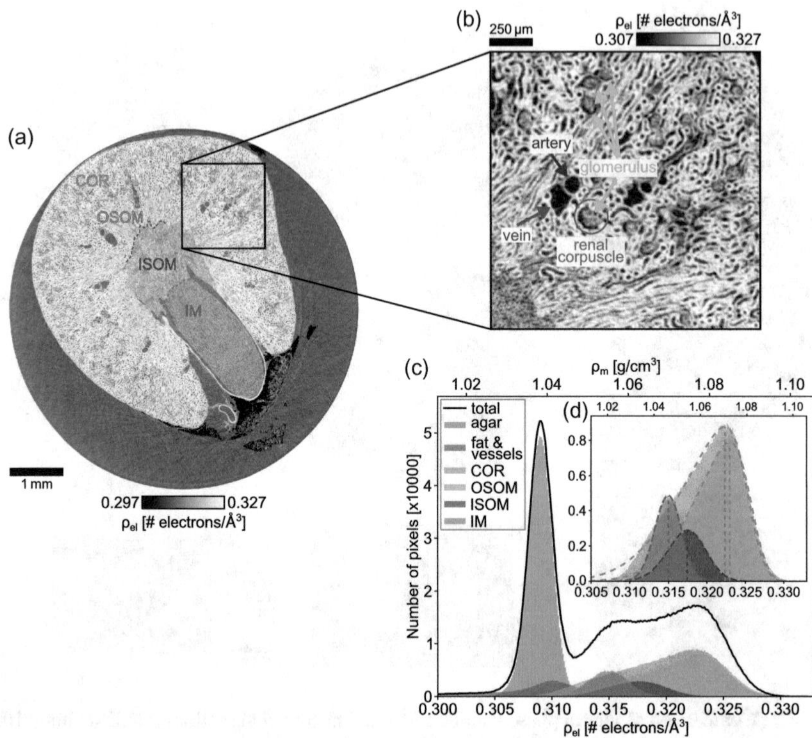

Fig. 8.8 Quantitative density analysis of the phase volume. **a** Slice through the phase volume of the murine kidney showing the electron density distribution. The regions (COR, OSOM, ISOM, IM) of the kidney can be distinguished by their density differences. Black areas are peri-renal fat. **b** Enlarged ROI in the COR/OSOM region showing the network of the renal tubules, arteries (red arrow), veins (blue arrow) and glomeruli (green arrows). A renal corpuscle (magenta circle) consists of a glomerulus, its surrounding fluid space (Bowman's space) and structure (Bowman's capsule). **c** Histogram of the electron and mass density distribution in the whole slice (black curve) and the different kidney areas separately (coloured bar plots). **d** Inset showing the histograms of the IM, ISOM, OSOM and COR regions only. A skewed Gaussian was fitted to each of the histograms (dashed curves) and the peak positions are indicated by vertical dashed lines. Peak positions and distribution widths can be found in Table 8.1. The long left shoulders of the OSOM and COR histograms are caused by contributions of tubular lumina and by partial volume effects at the tubule-lumen interfaces (see Sect. 8.4.4.4). Adapted with permission from Zdora et al. (2020) © The Optical Society

discrimination of not only the fat, agar gel and vessels in the sample, but also of the minute density differences between the kidney regions. This is illustrated in the histograms of the whole slice (Fig. 8.8c) and of the kidney tissue only (Fig. 8.8d). The peak positions (vertical dashed lines in Fig. 8.8d) were determined by fitting skewed Gaussians to the histograms.

The negative skewness of the distributions, most pronounced for the COR and OSOM, is caused by contributions from the tubule inner lumen as well as partial

Table 8.1 Mass density analysis of the kidney regions. Mass density values of the different regions, determined as the peak position of the skewed Gaussian curves fitted to the histograms in Fig. 8.8d, and peak widths given by the one-standard-deviation ranges of normal Gaussian fits to the right slope of the histograms, see Fig. 8.13. The uncertainty of the latter was calculated from the covariance matrix of the fit parameters via error propagation. Reprinted with permission from Zdora et al. (2020) © The Optical Society. Reprinted with permission from Zdora et al. (2020) © The Optical Society

Kidney region	Density [g/cm^3]	Range [mg/cm^3]
IM	1.046	5.651 ± 0.011
ISOM	1.054	7.792 ± 0.028
OSOM	1.073	8.001 ± 0.020
COR	1.071	8.524 ± 0.043

volume effects at the tubule wall—lumen border (see Sect. 8.4.4.4). The peak density values and peak ranges, determined from a normal Gaussian fit to the right side of the distribution, see Fig. 8.13b, are listed in Table 8.1. The mass density difference between the COR and OSOM was found to be 2.0 mg/cm^3, which is consistent with other measurements that report a density difference of 1.9 mg/cm^3 for a hydrated formalin-fixed rat kidney (Shirai et al. 2014).

In this experiment, the high density resolution of better than 1.9 mg/cm^3 (see Sect. 8.4.4.3) was obtained for an estimated spatial resolution of approximately 8 µm (see Sect. 8.4.4.2). The latter is essentially limited by the choice of reconstruction parameters in the UMPA analysis routine (Zdora et al. 2017). While this does not match light microscopy capable of sub-cellular resolution, features down to the looped structure of the glomerular capillaries can be observed, as can be seen in Fig. 8.8b. The diameter of the latter is estimated to be comparable to the diameter of red blood cells, i.e. approximately 6–7 µm (Smithies 2003; Engström and Taljedäl 1987), just at the spatial resolution limit of our scan.

8.4.2.4 3D Analysis and Segmentation of Complex Networks

As demonstrated above, UMPA speckle-based virtual histology allows for a quantitative analysis of the density distribution in the specimen. This can be combined with information on the tissue structure accessible thanks to the 3D character of the phase tomogram. The latter also provides a reliable way to perform further structural analysis on the shape, size and distribution of some functional elements of the organ such as the glomeruli or blood vessels, as illustrated in Fig. 8.9.

Retrieving this information reliably from 2D slices, as provided by conventional histology, can be extremely challenging. As an example for 3D size measurement, here the average glomerular diameter was determined and found to be (97 ± 19) µm,

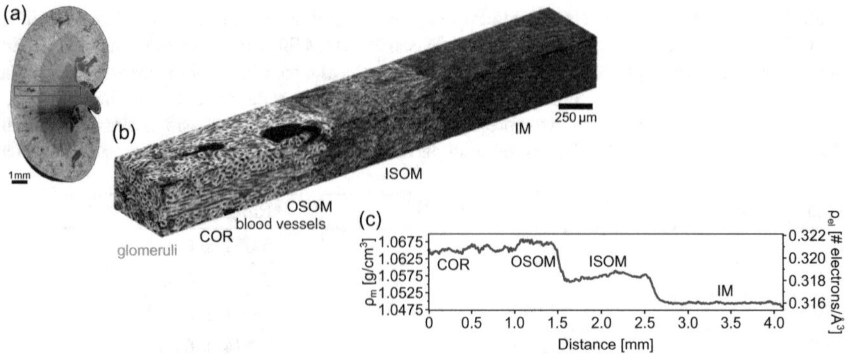

Fig. 8.9 Combination of 3D quantitative density and structural information. **a** Location of a cuboid of interest in the phase volume. **b** 3D rendering of the cuboid. The arrangement of the tubules in the different kidney regions is visualised and blood vessels (red) and glomeruli (green) can be identified and segmented. **c** Line profile along the long axis of the cuboid, averaged over all slices along the short axes. The density differences between the regions, including the minute difference between COR and OSOM, are clearly observable. Glomeruli and vessels were excluded for this analysis. Adapted with permission from Zdora et al. (2020) © The Optical Society

based on 20 glomeruli, in agreement with the value (99 ± 13) μm reported by Zhai et al. (2003).

The full 3D volume also reveals the interrelation and connectivity of the sample's functional elements. The high sensitivity of UMPA speckle-based phase tomography allowed for extracting various structural elements of the kidney using semi-automatic segmentation of the phase volume, see Fig. 8.10.

The specimen can be visualised from the outer renal capsule (Fig. 8.10a) to the complex interrelated networks inside, such as the blood vessel system (Fig. 8.10b). It should be highlighted that, in contrast to other methods such as absorption-based microCT that have been proposed to study the renal vascular network, no contrast agent nor blood was present in the vessels. The segmentation results can be combined with information on the tissue density, see Fig. 8.10c.

As shown in Fig. 8.10d, e, even a single renal nephron, the functional unit of the kidney, was successfully segmented out including the blood vessel supply, glomerulus and tubule. This is an important step towards the 3D visualisation of the functional and structural renal unit, allowing for the connectivity and interrelation of the blood vessels, tubules and collecting ducts in the kidney to be characterised.

While the segmentation of the main blood vessel network and a single nephron was demonstrated up to the end of the proximal straight tubule using commercially available software, the extraction of the complete tubule and vascular networks down to the smallest arterioles will require the development of advanced segmentation algorithms as well as data acquisition at a higher spatial resolution in the range of a micrometre, for which the large size of the data set will also have to be considered. As a future step, the extracted vessel and nephron networks could then be used for advanced functional studies such as image-based fluid dynamics simulations to

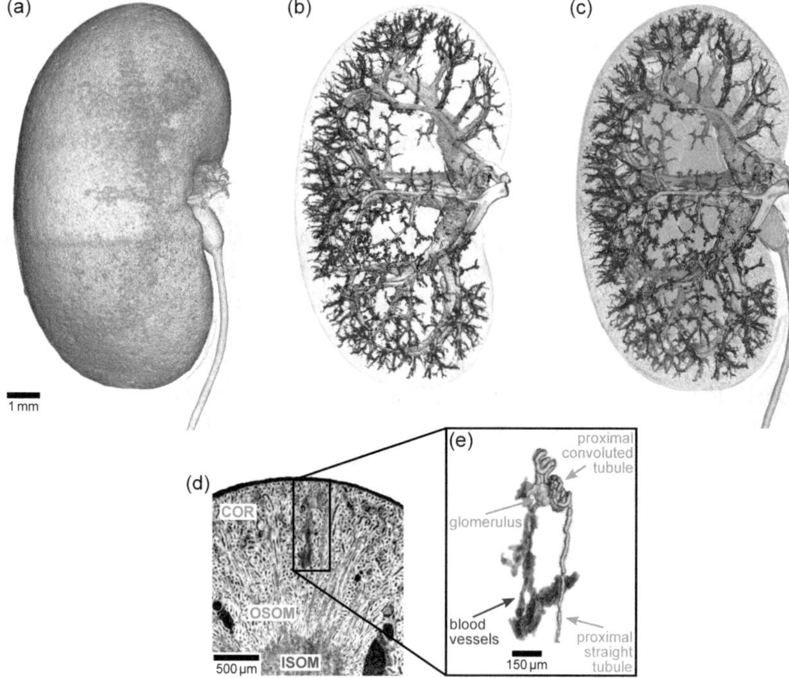

Fig. 8.10 3D visualisation and segmentation of the phase volume of the kidney. **a** Renal capsule with ureter (yellow) and surrounding fatty tissue (semi-transparent grey) obtained from the phase volume. **b** Vascular network (arteries and veins) of the kidney extracted from the same data set. **c** Combined visualisation of vascularisation and tissue structure. The tubular network of one slice through the phase volume is shown in semi-transparent colours (blue: IM, green/blue: ISOM, green: OSOM and COR). **d** Location of one renal nephron in the 3D phase volume (sagittal cut) and **e** segmented nephron consisting of blood vessel supply, glomerulus, and tubule. The segmentation of the tubule was terminated at the onset of the thin limb of the loop of Henle. Adapted with permission from Zdora et al. (2020) © The Optical Society

assess overall kidney function and investigate poorly known mechanisms like renal oxygenation (Layton 2012; Olgac and Kurtcuoglu 2016).

8.4.3 Conclusions and Outlook

It was demonstrated that X-ray speckle-based phase tomography provides high-sensitivity, high-resolution quantitative information on hydrated and unstained samples, making it a strong contender for wide adoption for future 3D virtual histology studies.

The high density resolution of better than 2 mg/cm^3 and spatial resolution of better than 8 μm obtained for the kidney tomogram allowed for the extensive segmentation of complex interconnected structures such as the renal vascular network and single renal nephrons. This will facilitate pathophysiological studies of the kidney and other organs, which can be validated and complemented by subsequent analysis with additional techniques including antibody-based and molecular biology-based approaches. In particular, SBI phase tomography could in the future become an invaluable tool for diagnostics and staging of diseases such as fibrosis and cancers that lead to changes in tissue density and structure.

As the spatial resolution and sensitivity of the measurement can be easily fine-tuned using the UMPA phase tomography approach, the method can be applied to samples containing structures with a wide range of sizes and compositions, extending to fossilised matter and materials science samples, see Chap. 9, Sect. 9.3.

Crucially, although the data presented here was acquired at a synchrotron source, SBI phase tomography is being implemented at laboratory X-ray sources without major efforts or costly equipment (Zanette et al. 2014, 2015), see also Chap. 9, Sect. 9.4. This will make the method widely accessible and suitable for advanced high-throughput clinical histopathology and research applications. X-ray speckle-based virtual histology hence has the potential to become part of routine investigations, and will enable studies on the interrelationship and connectivity of structural elements for a complete picture of organ function and pathology.

8.4.4 *Supplementary Information*

8.4.4.1 Speckle Visibility and Size

The properties of the reference speckle pattern that is used as a wavefront marker have an influence on the quality of the reconstructed projection images and the tomogram. The two important characteristics of the speckle pattern are the visibility and the average size of the speckles (Zdora 2018). An example of a reference speckle pattern used in the present experiment is shown in Fig. 8.11a. It is the average of 20 separate subsequently acquired frames and has been corrected for the detector dark current. The fine speckles can be observed in more detail in the enlarged ROI (150×150 pixels in the centre of the field of view) in Fig. 8.11b. Figure 8.11c shows the horizontal line profile across the centre of the ROI. Here, the orange broken line indicates the mean intensity value and the blue broke line one standard deviation of intensity in the region.

The visibility of the speckle pattern is typically quantified using the definitions given by Eqs. (5.1), (5.2) or (5.3) in Chap. 5, Sect. 5.3.1. The visibility was here determined with each of the three methods and for each of the $N = 20$ diffuser positions. Visibility maps were obtained by determining the visibility in a ROI of 150×150 pixels around each pixel of the field of view. The visibility maps for the first diffuser step, obtained with Eqs. (5.1), (5.2) and (5.3), are shown in Fig. 8.11d–f, respectively. The mean visibility values (with their standard deviations) over the whole map are shown in the second column of Table 8.2. Taking the average over

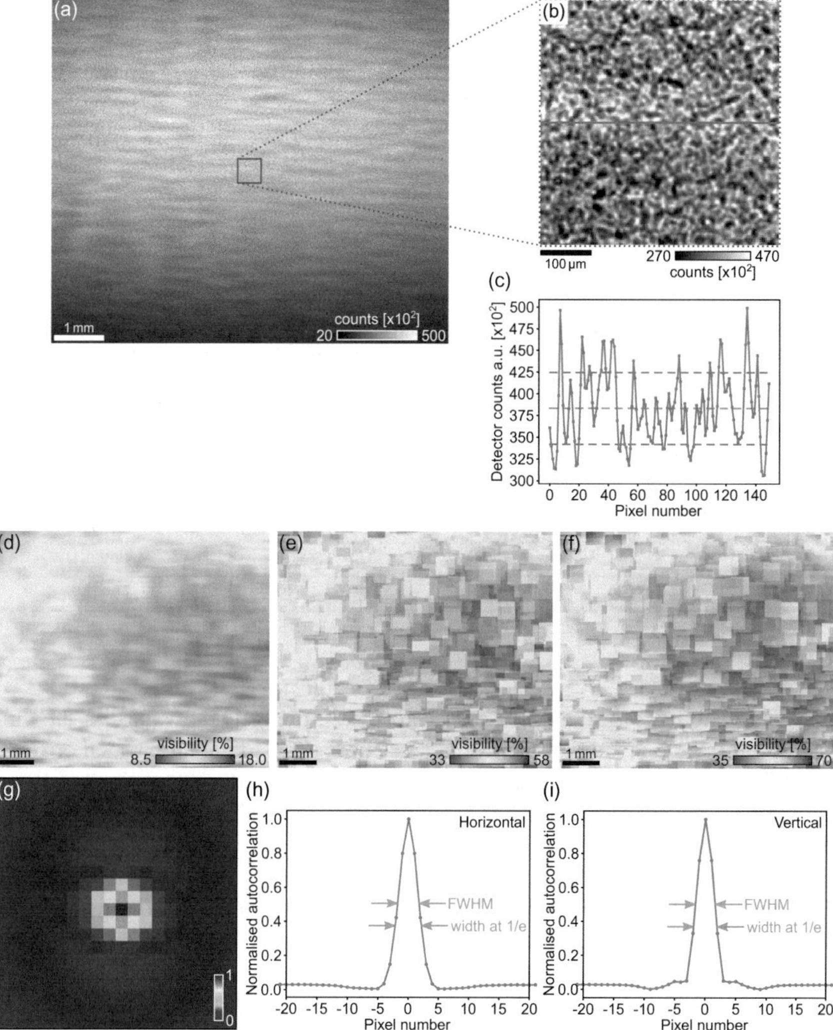

Fig. 8.11 Evaluation of speckle visibility and size. **a** Reference speckle pattern (without sample in the beam) at diffuser position $N = 20$ and **b** ROI in the centre of the image. **c** Horizontal line plot through the middle of the ROI showing the speckle intensity. The mean intensity of the pattern is indicated as an orange broken line and grey dashed lines illustrate the one-standard-deviation range. **d**–**f** Visibility maps calculated in regions of 150×150 pixels around each pixel using Eqs. (5.1)–(5.3), respectively. **g** Central region of the normalised 2D autocorrelation of the reference image in panel (**a**). **h** Horizontal and **i** vertical line profiles through the centre of panel (**g**). The FWHM or width at 1/e the peak value are commonly used as estimates for the speckle size, here: 3.6 pixels or 4.3 pixels in the horizontal and 3.2 pixels or 3.8 pixels in the vertical direction, respectively. Adapted with permission from Zdora et al. (2020) © The Optical Society

Table 8.2 Speckle visibility values determined using Eqs. (5.1)–(5.3). Mean visibility values at step number N = 20 averaged over the whole visibility map, see Fig. 8.11d–f, and average over all diffuser positions (with deviation errors). The low standard deviation σ_{avg} of the mean visibility values at the different diffuser positions indicates the consistency of the values

Visibility	N = 20	Average	σ_{avg}
v_1	(12.0 ± 1.2)%	(12.0 ± 0.5)%	0.1%
v_2	(41.8 ± 3.8)%	(41.9 ± 0.85)%	0.2%
v_3	(46.7 ± 5.5)%	(46.8 ± 1.2)%	0.0%

Table 8.3 Speckle sizes in the horizontal (x) and the vertical (y) direction, determined from a 2D autocorrelation analysis of the reference speckle pattern. Speckle sizes calculated as the FWHM or 1/e value of the autocorrelation peak for diffuser position N = 20, see Fig. 8.11g–i, and sizes averaged over all diffuser positions. The standard deviation of the sizes at the different diffuser positions was smaller than 0.05 pixels in all cases. Adapted with permission from Zdora et al. (2020) © The Optical Society

Speckle size	N = 20		Average	
	x	y	x	y
FWHM	3.6 px	3.2 px	3.5 px	3.1 px
1/e	4.3 px	3.8 px	4.2 px	3.7 px

the mean visibility values of all diffuser positions, one gets the values in the third column of Table 8.2. Here, the uncertainties correspond to the propagated errors of the visibilities at the different diffuser positions. The standard deviations of the N visibility values, see last column of Table 8.2, indicate that the results are consistent for the different diffuser positions.

To estimate the speckle size, a 2D autocorrelation analysis of the reference speckle pattern was performed for each diffuser position N. The width of the autocorrelation peak can be taken as a measure of the speckle size (Zdora 2018). Commonly, either the FWHM of the autocorrelation peak or the width at 1/e times the peak value are used for this purpose. Figure 8.11g shows the central part (41 × 41 pixels) of the 2D autocorrelation function of the reference speckle pattern for the first diffuser position (see speckle pattern in Fig. 8.11a). The horizontal and vertical line profiles through the centre of Fig. 8.11g are plotted in Fig. 8.11h, i, respectively. The peak widths at half maximum and at 1/e, indicated by green and orange arrows, are taken as a measure for the speckle size. The speckle sizes in units of pixels are summarised in Table 8.3 for diffuser position N = 20 and averaged over all positions. The standard deviation of the values at the different diffuser positions is less than 0.05 pixels for both directions, indicating a good agreement of the results. The speckle dimensions are larger in the horizontal direction than in the vertical due to the better transverse coherence of the X-ray beam in the vertical, which causes some blurring of the speckles in the horizontal.

8.4.4.2 Spatial Resolution Estimation

The spatial resolution in the 3D phase tomogram was estimated from the reconstructed slices using two different approaches.

One method that has been proposed to measure the spatial resolution in an image is based on analysing the Fourier power spectrum (FPS) (Modregger et al. 2007). The baseline flat tail of the FPS is considered to be the noise level of the image and the spatial frequency at which the FPS reaches twice the noise level is taken as a measure for the resolution. Here, a slice of the phase volume close to the centre of the kidney was chosen for the spatial resolution analysis, see Fig. 8.12a.

A ROI of 1000×1000 pixels covering all the different kidney regions (cortex, outer stripe and inner stripe of outer medulla and inner medulla) was selected from the slice, as indicated by the red dashed box in Fig. 8.12a. The azimuthally averaged FPS of this ROI is plotted in Fig. 8.12b. The noise level (green dashed line) was calculated as the average of the last 10 points of the FPS. A fourth-order polynomial was fitted to a 160 pixels region of the FPS plot, as shown in Fig. 8.12c, to determine the point of intersection with twice the noise level line (magenta dashed line). It was found at a spatial frequency of 127.61 line pairs per millimetre (lp/mm), corresponding to a feature size of approximately 7.8 µm (7.83 µm). To verify this, the same analysis was also performed on four adjacent slices of the tomogram and the average value over all five slices is 7.9 µm (7.86 ± 0.02) µm.

Another approach consists of looking at the profile of a sharp edge in the image. Typically, a Gaussian error function (ERFC) is fitted to the line profile of the edge. One can either directly consider half of the FWHM of the fitted ERFC as a measure for the spatial resolution, as done e.g. by Töpperwien et al. (2018), or determine the modulation transfer function (MTF) and take the spatial frequency value at 10% or 20% of the MTF maximum as an estimate of highest resolvable spatial frequency (Müller et al. 2017). The resolution analysis was performed on the same slice as above, see Fig. 8.12a. The inner edge of the plastic container (border agarose gel—container) was considered as a sharp edge and a line profile was taken through the centre of the container across the right side of the edge, as indicated by the red line in Fig. 8.12a. An ERFC was fitted to the edge profile, see Fig. 8.12d. From the fitted curve, the half-period resolution can be estimated directly as half of the FWHM, which gives values of 11.5 µm for the single slice and (11.4 ± 0.1) µm for the average of five slices as above.

The first derivative of the fitted edge delivers the line spread function (LSF) in Fig. 8.12e. The corresponding MTF can be calculated via Fourier transformation and is shown in Fig. 8.12f. A Gaussian fit of the MTF was performed to determine the spatial frequencies, at which the MTF reaches 10 and 20% of its maximum value, indicated by the magenta dash-dot and green dashed lines in Fig. 8.12f. The 10 and 20% MTF values lie at approximately 58 lp/mm and 48 lp/mm, respectively, which corresponds to half-period resolution values of 8.6 µm and 10.3 µm. The average half-period resolution for five adjacent slices (as above) is (8.5 ± 0.1) µm for the 10% and (10.2 ± 0.1) µm for the 20% criterion, indicating a good agreement over the slices.

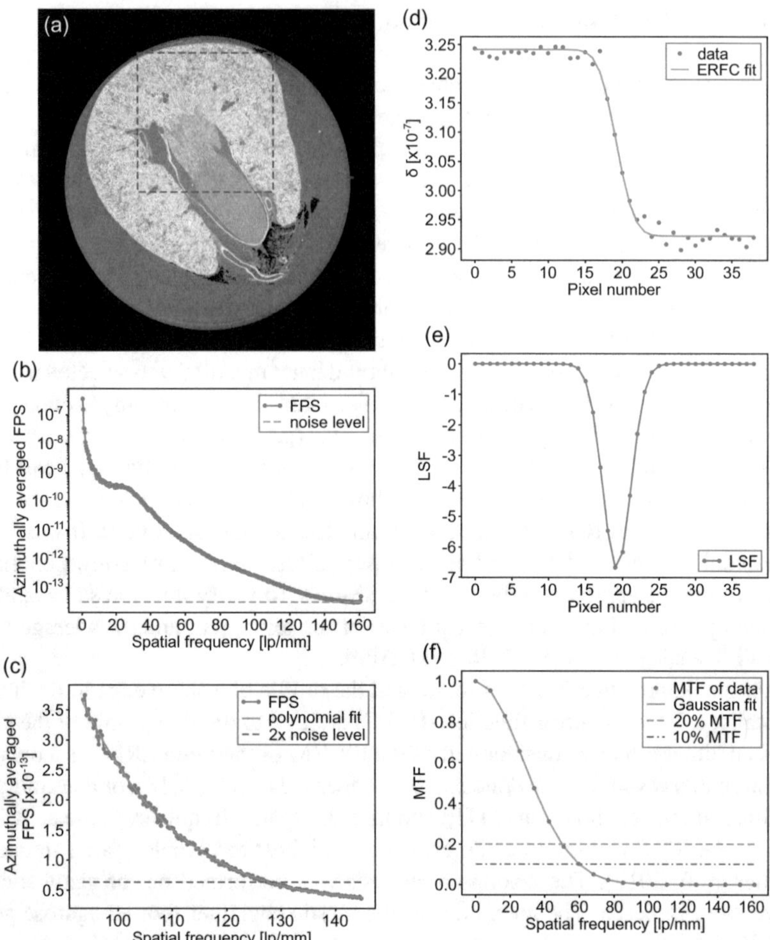

Fig. 8.12 Estimation of the spatial resolution in the phase tomogram. **a** Slice of the phase tomogram on which the spatial resolution estimation was performed. **b, c** First method based on the Fourier power spectrum (FPS). A ROI of 1000 × 1000 pixels was selected from the slice for the analysis (red dashed box). **b** The baseline of the azimuthally averaged FPS of the ROI is considered the noise level of the image (green dashed line). The spatial frequency at twice the noise level is taken as an estimate of the smallest resolvable features. **c** Magnified region at the high-frequency end of the FPS. The FPS in this region was fitted with a polynomial to determine the intersection with twice the noise level (magenta dashed line). It was found at 127.61 lp/mm corresponding to a smallest resolvable feature size of approximately 7.8 μm. **d–f** Second method based on the modulation transfer function (MTF) of a sharp edge. The inner edge of the plastic container was used for this analysis (see red line in panel (**a**)). **d** A Gaussian error function was fitted to the edge profile of the inner tube container. The FWHM of the edge is 11.5 μm. **e** The line spread function (LSF) was calculated from the edge via derivation. **f** The MTF is given as the Fourier transform of the LSF and was fitted with a Gaussian function (solid orange curve). The spatial frequencies at 10% or 20% MTF can be seen as a measure for the spatial resolution, here 8.6 μm or 10.3 μm, respectively. The spatial resolution values averaged over five slices are given in Table 8.4. Adapted with permission from Zdora et al. (2020) © The Optical Society

The values for the spatial resolution estimated with the different methods are summarised in Table 8.4.

8.4.4.3 Angular Sensitivity and Density Resolution

For X-ray speckle-based imaging, the angular sensitivity, i.e. the smallest refraction angle that can be measured with a setup under given scan and reconstruction conditions, is commonly estimated from the refraction angle projections by calculating the standard deviation in a small background region not containing any part of the sample. Here, the angular sensitivity was determined from a ROI of 600×18 pixels in air on the left side of the sample. The mean sensitivities averaged over all projections of the tomography scan are $\sigma_x \approx (95 \pm 1)$ nrad in the horizontal and $\sigma_x \approx (82 \pm 1)$ nrad in the vertical direction.

The smallest resolvable density difference in the specimen was estimated from the 3D phase volume by determining the standard deviation of the signal in a $50 \times 50 \times 50$ pixels region within the agarose gel above the kidney. This gives a δ-resolution of $\Delta\delta \approx 6.0 \times 10^{-10}$, which corresponds to an electron density resolution of approximately 0.59×10^{-3} el/Å^3 (electrons per cubic Angstroem) and a mass density resolution of approximately 1.9 mg/cm^3.

It should be noted that this estimation of density resolution is a very conservative approach, in particular as perfect homogeneity of the agar gel is assumed. It should be seen as a limit of the minimum density resolution. The actual density resolution of the data set is likely to be better.

The angular sensitivity and density resolution values are summarised in Table 8.4.

Table 8.4 Spatial resolution, angular sensitivity and density resolution values of the reconstructed data. The spatial resolution was evaluated from the phase volume using the azimuthally averaged Fourier power spectrum as well as half the FWHM or MTF of a sharp edge, as shown in Fig. 8.12. The given value is the average over five adjacent tomogram slices (with standard deviation). The angular sensitivity was determined from the 2D phase projections before tomographic reconstruction. The values are an average of the sensitivities of all projections (with standard deviation). The density resolution was calculated from the phase volume as the standard deviation in a $50 \times 50 \times 50$ pixels region within the agar gel above the kidney. Reprinted with permission from Zdora et al. (2020) © The Optical Society

Spatial resolution		Angular sensitivity		Density resolution	
Edge FWHM	$(11.4 \pm 0.1)\,\mu$m	x	(95 ± 1) nrad	$\Delta\delta$	6.0×10^{-10}
Edge MTF (10%)	$(8.5 \pm 0.1)\,\mu$m	y	(82 ± 1) nrad	$\Delta\rho_{el}$	0.59×10^{-3} el/Å^3
Edge MTF (20%)	$(10.2 \pm 0.1)\,\mu$m			$\Delta\rho_m$	1.9 mg/cm^3
FPS	$(7.9 \pm 0.0)\,\mu$m				

8.4.4.4 Notes on the Density Histograms of the Kidney

It was observed in Fig. 8.8c, d that the density distributions in the different kidney regions do not follow a normal Gaussian curve as it would be expected, but show long tails towards lower density values. These can be explained by the structure of the kidney tissue itself, which mainly consists of renal tubules with an empty lumen inside. The lumen areas have a lower density than the tubule walls and the histograms hence show a mixture of lumen and tubule grey values. The latter is, however, dominant as the walls cover a larger percentage of the analysed area. It can therefore be assumed that the peak position of the histograms represents the density value of the tubule walls in the respective tissue region.

In addition to the contribution of the lumen, partial volume effects play a role, which lead to an averaging of grey values at the interface of two sample materials, here the lumen and tubule walls, when parts of both components fall within one pixel, see e.g. Müller et al. (2006). The tails are most pronounced for the OSOM and COR regions as the proportion of lumina and interfaces in these regions are larger than in the ISOM and IM.

To find the peak positions corresponding to the density values in the regions, a skewed Gaussians was fitted to the histograms, see Figs. 8.8d and 8.13a. In order to estimate the distribution of density values associated with the tubule walls only, a normal Gaussian curve was fitted to the right slope of the histogram with the centre at the previously determined peak position, see Fig. 8.13b. The peak position, standard deviation range and skewness of each skewed Gaussian fit curve as well as the standard deviation range of the normal Gaussian fit curves can be found in Table 8.5. Furthermore, the ratio of the total number of pixels in the measured histograms and the area underneath the fitted normal Gaussian curves is given as an estimation for the influence of lumen density and partial volume effects.

8.4.4.5 Supplementary Figures

See Figs. 8.14 and 8.15.

8.5 X-ray Speckle-Based Virtual Histology of Brain Tissue

As a third type of biomedical sample, two pieces of human cerebellum were investigated with UMPA speckle-based imaging and the tissue structure is visualised at different length scales. It is illustrated in the following that speckle-based virtual histology using the UMPA technique is a promising candidate for studying brain tissue with high contrast and down to cellular level. While the testicle and kidney specimens were hydrated and scanned in agar gel, the cerebellum samples were embedded in paraffin wax, which is shown to also be compatible with speckle-based virtual histology.

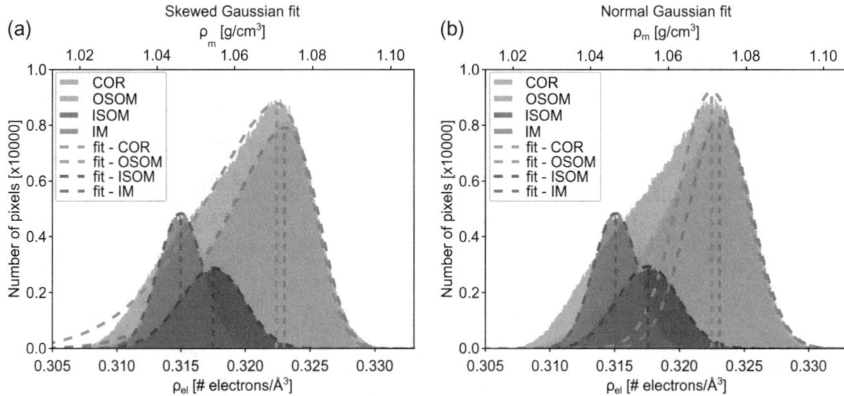

Fig. 8.13 Effects of tubule lumina and partial volume contributions on the density histograms. Histograms of the kidney regions (IM, ISOM, OSOM, COR) for the slice in Fig. 8.8a. **a** Histograms with fitted skewed Gaussians (dashed curves) as shown in Fig. 8.8d. The maximum position of the fit function is indicated by the vertical dashed line and is taken as the density value of the respective kidney region. **b** Same histograms but with normal Gaussians (also indicated by dashed curves) fitted to the right side of the histograms. The maximum positions and standard deviations of the Gaussians and the ratio of the number of pixels in the entire histogram of each region and the area underneath the normal Gaussian fit curves are summarised in Table 8.5. Reprinted with permission from Zdora et al. (2020) © The Optical Society

Table 8.5 Parameters of the skewed Gaussian and normal Gaussian functions fitted to the histograms of the kidney regions in Fig. 8.8. Second column: Mass density values determined as the peak position of the skewed Gaussian fit curves. Third column: Standard deviations of the skewed Gaussians with one-standard-deviation errors calculated from the propagated errors of the curve fit parameters. Fourth column: Skewnesses of the skewed Gaussian fit curves. Fifth column: Standard deviations of the normal Gaussians with one-standard-deviation errors calculated from the propagated errors of the curve fit parameters. Sixth column: Ratio of the total number of pixels in the measured histograms and the area underneath the normal Gaussian curves. The ratio gives an indication of the fraction of pixels that lie within the tubule walls, i.e. not inside the lumen or at the immediate interface of lumen and renal tubule (partial volume effect). Reprinted with permission from Zdora et al. (2020) © The Optical Society

Kidney region	Peak position [g/cm^3]	$\sigma_{\text{skewGauss}}$ [mg/cm^3]	Skewness	σ_{Gauss} [mg/cm^3]	Tubule ratio
IM	1.046	5.719 ± 0.007	-0.0337	5.651 ± 0.011	0.99
ISOM	1.054	8.333 ± 0.020	-0.143	7.792 ± 0.028	0.98
OSOM	1.073	12.990 ± 0.028	-0.476	8.001 ± 0.020	0.70
COR	1.071	15.805 ± 0.069	-0.511	8.524 ± 0.043	0.65

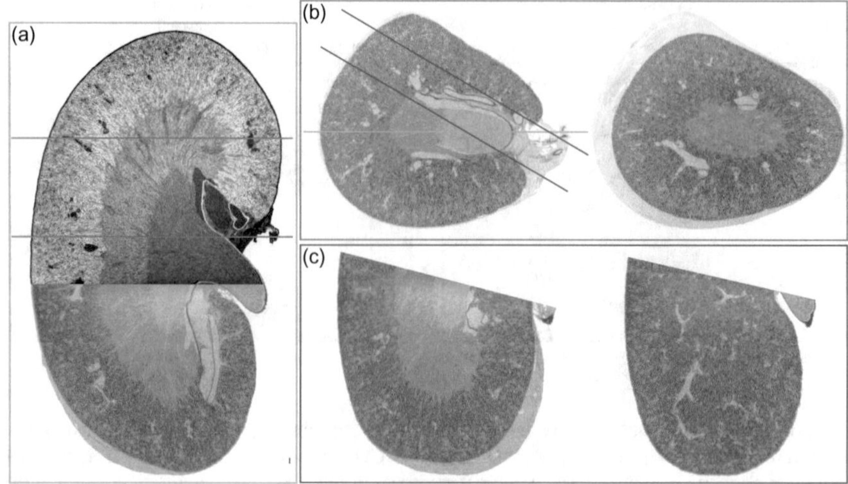

Fig. 8.14 Location of the virtual slices in the phase volume. **a** Virtual sagittal slice through the centre of the kidney. Bottom part shown in false-colours similar to H&E stain. The different kidney regions can clearly be distinguished due to their different densities. The locations of the transverse slices shown in Fig. 8.7c, d are indicated by orange lines. **b** Transverse slices through the phase volume as shown in Fig. 8.7c, d. The location of the sagittal slice in Fig. 8.7a as well as an additional sagittal slice shown in panel (**c**) are indicated by blue lines. The location of the central slice in panel (**a**) is shown as a green line. **c** Sagittal slices cutting through the COR and OSOM regions, as indicated by the blue lines in panel (**b**). Multiple bifurcating blood vessels are visible in the right slice. Reprinted with permission from Zdora et al. (2020) © The Optical Society

The experiments presented in the following were conducted with the support of Dr. Hans Deyhle, University of Southampton (UK), now Diamond Light Source (UK), Dr. Joan Vila-Comamala, ETH Zurich and Paul Scherrer Institute (Switzerland), Dr. Margie Olbinado, ESRF (France), now PSI (Switzerland), and Dr. Alexander Rack, ESRF (France).

8.5.1 Materials and Methods

8.5.1.1 Experimental Setups and Data Acquisition

Data was acquired at beamline ID19 at the ESRF, see Chap. 3, with two different setup configurations: one for medium spatial resolution with a larger field of view and a second one for higher spatial resolution but with a smaller field of view.

The first setup was the same as described in the previous section on the mouse kidney specimen, see Sect. 8.4.1.1, Fig. 8.5, with a peak energy of 26.3 keV of the X-ray spectrum from a single-harmonic undulator and an effective pixel size of 3.1 µm. The only difference in the setup was the sample-detector distance, which was slightly

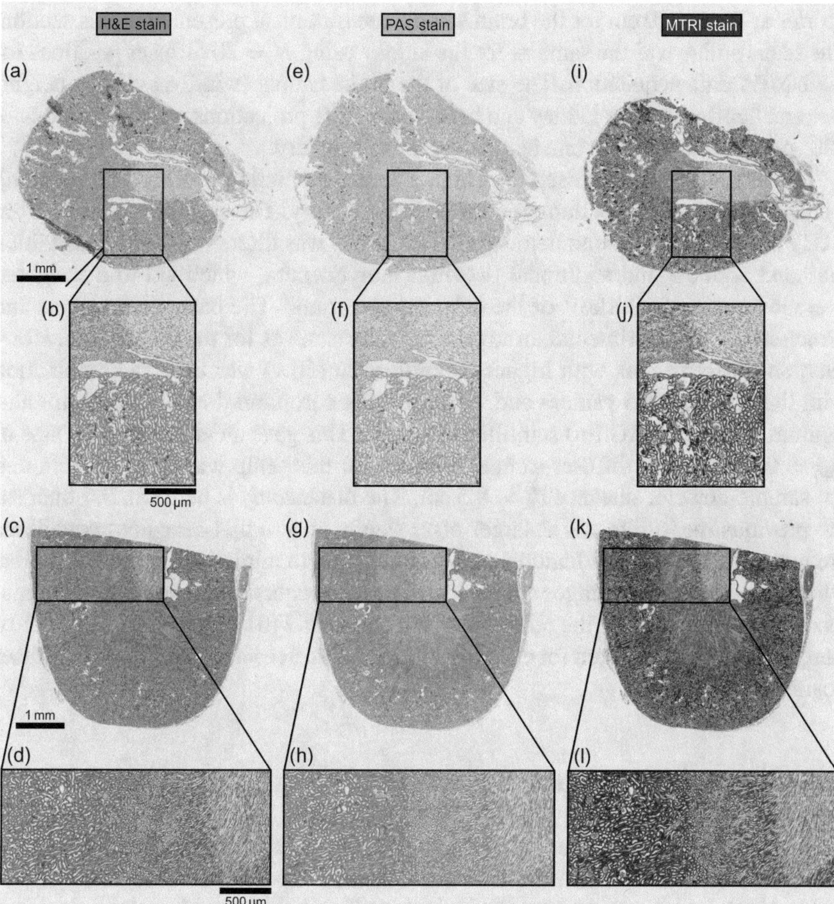

Fig. 8.15 Conventional histological sections with different stains. The sections correspond to the tomogram slices in Fig. 8.14(b, Left), (c, Left) and were treated with different stains commonly used for kidney tissue. **a–d** Haematoxylin and eosin (H&E) stain is used for general structural overview, which highlights cell nuclei in blue and the rest of the tissue in pink/red. **e–h** Periodic acid-Schiff (PAS) stain shows glycogen, mucins and muco-polysaccharids and highlights basement membranes. **i-l** Masson's trichrome (MTRI) stain colours nuclei blue/black, muscle, erythrocytes and cytoplasm red and connective tissue blue/green. The MTRI-stained slices appear to be most similar to the phase tomogram slices, which suggests that mostly connective tissue is visualised with X-ray phase-contrast imaging. Note that the PAS and MTRI slices are cuts at slightly different levels than the H&E slices and hence show marginally different structures. The ROIs in panels (**b**), (**f**) and (**j**) and panels (**d**), (**h**) and (**l**) illustrate in more detail how the different stains visualise various features in the kidney regions. Reprinted with permission from Zdora et al. (2020) © The Optical Society

shorter at $d_2 = 45.0\,\text{cm}$ for the brain sample measurement presented in this section. The scan routine was the same as for the kidney using $N = 20$ diffuser positions for the UMPA data acquisition. The size of the brain sample (width as well as height) was smaller than for the kidney and hence only 1801 projections were acquired over 180° of rotation and only one height step was necessary.

The second setup optimised for a higher spatial resolution was used for scanning a smaller piece of cerebellum at a lower beam energy. The spectrum coming from a U17 single-harmonic undulator (gap: 11.52 mm) was filtered by the 1.4 mm-thick diamond window and additional 0.14 mm molybdenum, which led to a weighted average energy of 22.8 keV of the detected spectrum.[6] The basic components and principle of the experimental arrangement is the same as for the other setup. However, an objective lens with higher magnification (10×) was used in combination with the pco.edge 5.5 camera and a 10 µm-thick europium-doped gadolinium aluminium garnet (GGAG:Eu) scintillation screen. This gave an effective pixel size of $p_{\text{eff}} = 0.65\,\mu\text{m}$. The diffuser-sample distance for this setup was $d_1 = 52.5\,\text{cm}$ and the sample-detector distance $d_2 = 8.5\,\text{cm}$. The distance d_2 is here smaller than for the previous measurements at larger pixel size to ensure that near-field conditions are given for the smallest features in the sample and to minimise artefacts that arise when multiple Fresnel fringes are present. The data acquisition scheme and parameters were the same as for the other setup, but a total of 2401 projections over 180° of sample rotation were taken for each tomography scan at each of the $N = 20$ diffuser positions.

8.5.1.2 Sample Preparation

Post-mortem specimens of a human cerebellum were excised from a donated brain. Informed consent for scientific use was obtained. Specimens were fixed in 4% histological-grade buffered formalin, dehydrated in ethanol, transferred to xylene and embedded in a paraffin/plastic polymer mixture (Surgipath Paraplast, Leica Biosystems, Switzerland). Cylinders of 4 mm diameter (for the medium-resolution scan) and 1.4 mm diameter (for the high-resolution scan) were excised from the paraffin blocks with a metal punch.

All procedures were conducted in accordance with the Declaration of Helsinki and were approved by the ethics committee of north-western Switzerland (Ethikkommission Nordwestschweiz).

The cerebella were kindly provided by the Department of Neuropathology, Universitätsspital Basel, Switzerland, and samples were prepared by Dr. Christos Bikis from the Department of Biomedicine at Universitätsspital Basel, Switzerland.

[6]The detected spectrum was determined from the undulator spectrum at the given conditions by taking into account the ID gap, the filters and the scintillator response. As for the scan of the testicle above, beam hardening in the specimen was not considered here. However, given the small bandwidth of the spectrum from the single-harmonic undulator, it can be neglected for this measurement.

8.5.1.3 Image Reconstruction

Data reconstruction with the UMPA reconstruction approach was performed analogous to the kidney data as described in Sect. 8.4.1.3.

8.5.2 Results and Discussion

The two pieces of human cerebellum were scanned at two different pixel sizes and energies in order to visualise the tissue structure at different length scales. The medium-resolution scan at an effective pixel size of 3.1 µm gives an overview of the different tissue components of the cerebellum, while the high-resolution scan at an effective pixel size of 0.65 µm allows for the visualisation down to individual neurons.

8.5.2.1 Phase Tomography of the Cerebellum at Medium Spatial Resolution

Transverse and longitudinal slices through the phase tomogram of the 4 mm-diameter human cerebellum sample scanned with the medium-resolution setup are shown in Fig. 8.16a, c, e–f.

The high contrast of the phase volume allows for an easy identification of grey and white matter and the separation of the molecular (ml) and granular layer (gl) of the grey matter. Furthermore, features such as the paraffin (p) inside the cerebellum folds, the pia mater (pm), which is the inner membrane surrounding the brain, and blood vessels (v) can be seen. Between the molecular layer and granular layer some bright spots can be observed. These are the Purkinje cells (Pc), which are the largest neurons in the body (20–40 µm in diameter). They are subject of a wide range of research studies as they play an important role in the neural circuit. Other cells are located in the granular layer, where some cell clusters can be observed, and in the molecular layer, where the cells are too small to be visualised at the given spatial resolution. The granular layer contains an abundance of granule cells, which have a cell body of 5–8 µm diameter and appear in clusters (Llinas and Negrello 2015). Due to the clustering they are visible in Fig. 8.16, even though individual cells are at the spatial resolution limit. The molecular layer contains mainly stellate and basket cells, which are similar in size. The smallest stellate cells have a diameter of approximately 5–9 µm (Llinas and Negrello 2015). As they do not occur in clusters and are much less abundant, they are harder to visualise at the given spatial resolution.

For comparison, a slice through the corresponding transmission volume from the same data set is presented in Fig. 8.16d. Due to the nature of the setup, it is here dominated by edge-enhancement effects and the contrast between the different

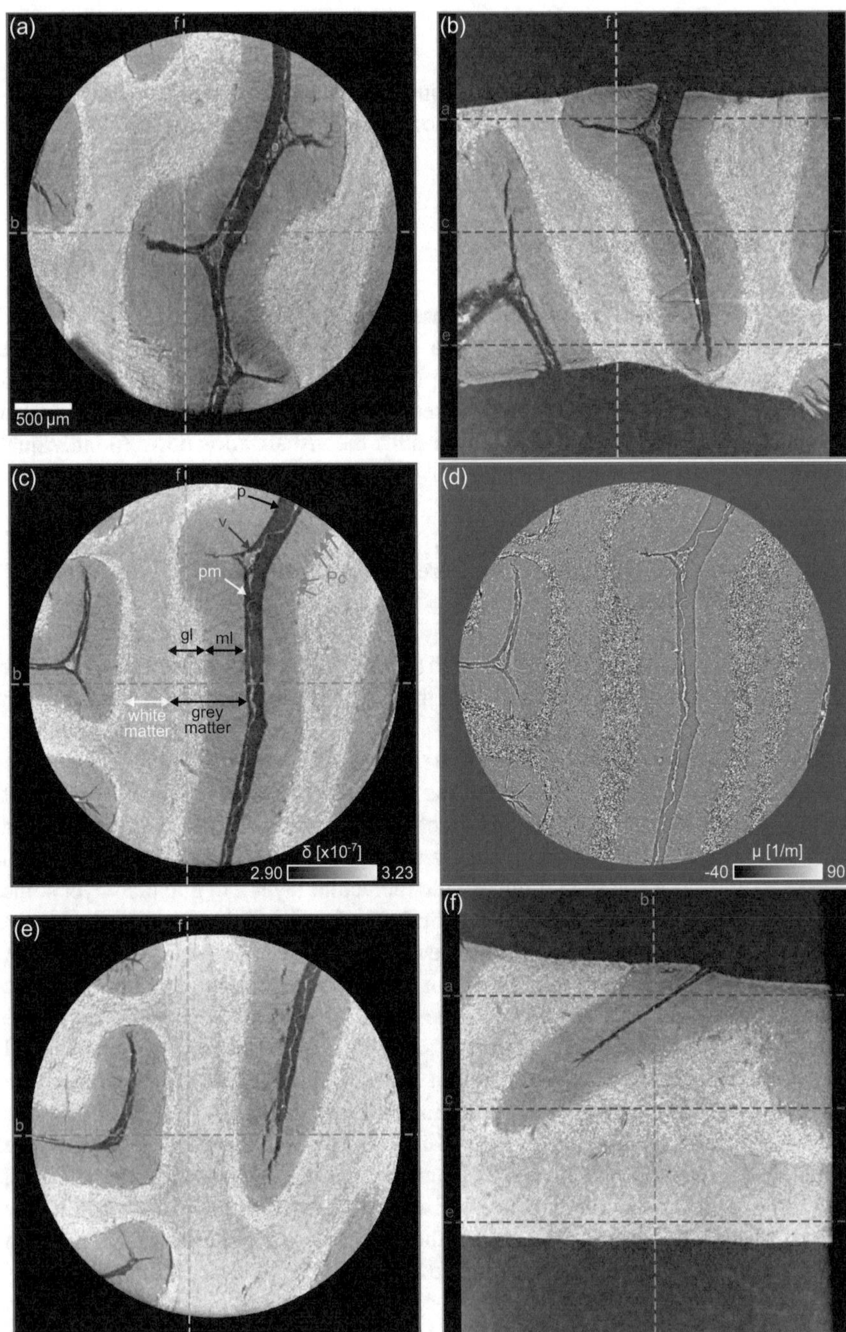

Fig. 8.16 (Continued)

◄ **Fig. 8.16** Slices through the volumes of a piece of human cerebellum embedded in paraffin. **a, c, e** Transverse and **b, f** longitudinal slices through the phase volume. White and grey matter can be distinguished and the molecular (ml) and granular layer (gl) can be identified. Purkinje cells (Pc) between the molecular and granular layer show as bright spots due to their higher density. The folds of the cerebellum are filled with paraffin (p). The pia mater (pm) shows as a white line and blood vessels (v) can be identified. **d** Transverse transmission slice. The contrast between the layers is significantly lower than for the corresponding phase slice in panel (**c**). The visibility of cells in the ml and gl is enhanced due to propagation fringes. UMPA was performed with $N = 20$ diffuser steps and an analysis window of 5×5 pixels

regions of the cerebellum is relatively low. Features with sharp density transitions such as the interface between paraffin and grey matter, the blood vessels and the cells are clearly distinguished from the rest of the tissue due to the Fresnel fringes around them.

8.5.2.2 Effect of Changing Scan and Reconstruction Parameters

Similar to the previous study for the scan of the hydrated mouse testis in Sect. 8.3.2.3, it is now evaluated how different parameters in the UMPA reconstruction affect the quality of the reconstructed phase tomogram slices. This can give a good indication of the required number of diffuser positions for a given experimental setup, which is essential for optimising scan time and dose.

The slice in Fig. 8.16c, which was analysed with the parameters $N = 20$, $w = 5 \times 5$ pixels, was reconstructed with a few different parameter combinations. First, it was investigated how changing the number of steps affects the image quality. For this, the window size was kept at $w = 5 \times 5$ pixels, but only $N = 10$ steps were considered for the results in Fig. 8.17b and $N = 5$ steps for the results in Fig. 8.17c. As expected, the noise in the images increases, while the spatial resolution, determined by the FWHM of the window, remains the same. Despite the higher noise level, the signal sensitivity in Fig. 8.17b for $N = 10$, $w = 5 \times 5$ pixels is still sufficient to identify all of the features visible in Fig. 8.17a, including the Purkinje cells pointed out by orange arrows. This means that, as observed for the testis scan, the number of steps, and hence the total dose to the sample, can be significantly reduced by a factor of two without losing valuable information. However, it can be observed in Fig. 8.17c that reducing the number of steps down to $N = 5$ leads to a noise level that occludes some of the smaller features such as the Purkinje cells. While the noise can be reduced by increasing the window size, see Fig. 8.17d, this comes at the cost of a reduced spatial resolution. As a consequence, small features such as the Purkinje cells cannot be observed anymore.

For the reconstruction in Fig. 8.17d, a window size of 10×10 pixels was deliberately chosen to achieve a similar angular sensitivity of the differential phase projections as for the reconstruction with $N = 20$, $w = 5 \times 5$ pixels. The window size was determined using the theoretical sensitivity model in Eq. (6.4) in Sect. 6.3 of Chap. 6. It should be noted, however, that the sensitivity of the differential phase projections

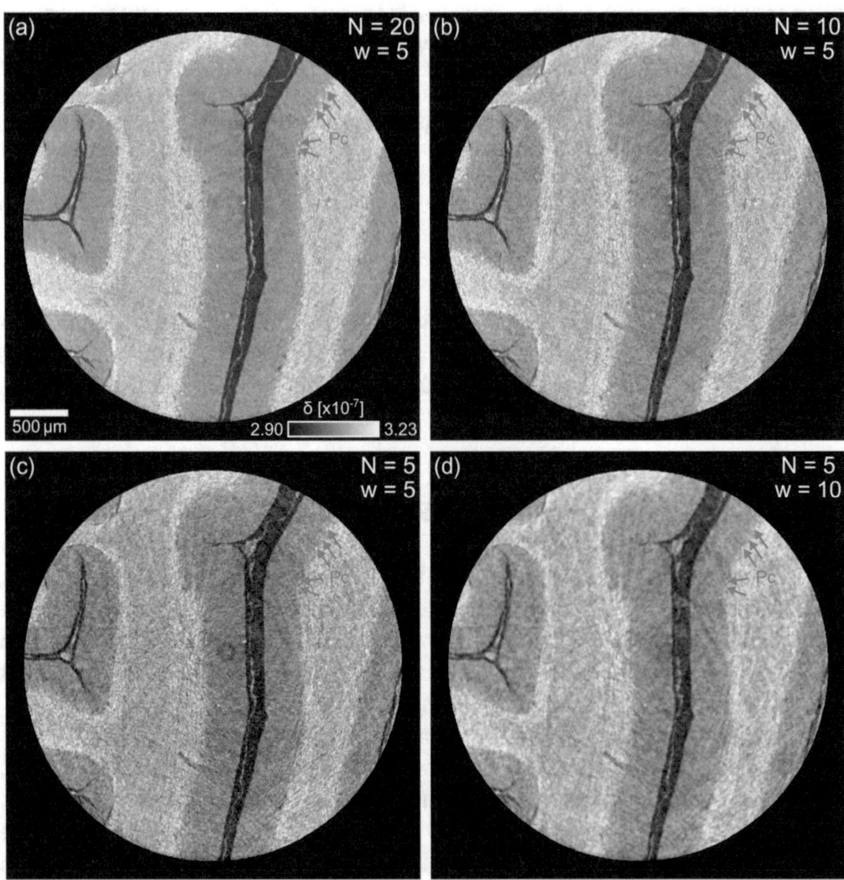

Fig. 8.17 Central transverse slice through the phase tomogram of the larger piece of human cerebellum reconstructed with different numbers of steps (N) and window sizes (w). **a** $N = 20$, $w = 5 \times 5$ pixels as presented in Fig. 8.16c. **b** $N = 10$, $w = 5 \times 5$ pixels and **c** $N = 5$, $w = 5 \times 5$ pixels. For smaller N at constant w, the noise increases. For $N = 5$ and $w = 5 \times 5$ pixels, Purkinje cells (Pc) are hard to distinguish from noise. **d** When increasing the window to $w = 10 \times 10$ pixels, noise is reduced but the Purkinje cells cannot be visualised due to the low spatial resolution

does not directly translate to the density resolution in the tomogram slices due to the additional steps for tomographic reconstruction that do not propagate the noise level linearly.

While the spatial resolution was not measured directly for the presented data set, a good indication of the size of the smallest resolvable features is given by the size of the smallest observable cells. For the scan with $N = 20$, $w = 5 \times 5$ pixels, Purkinje cells can clearly be observed and a few, but not many, of the granule cells can be resolved. This indicates that the spatial resolution is in the range of the size of the largest granule cells, i.e. approximately 10 µm. This is consistent with the spatial resolution estimation performed for the mouse kidney scan in Sect. 8.4.4.2, giving a smallest resolvable feature size between 7.9–11.4 μm, depending on the analysis method. As this data was taken under the same conditions with a similar experimental setup and the same scan procedure and was reconstructed with the same parameters, a comparable spatial resolution is expected.

8.5.2.3 Phase Tomography of the Cerebellum at Higher Spatial Resolution

For the scan of the cerebellum at medium spatial resolution, the gross structure and different regions of the tissue were visualised with high contrast. While the large Purkinje cells could be observed, individual smaller neurons were not clearly resolvable. For many studies on neurons and on cerebellar diseases and disorders, it is, however, essential to visualise individual cells smaller than the Purkinje cells. Furthermore, it is of interest to extract the size and shape of the Purkinje cells. Both requires a higher spatial resolution of the data set. To achieve this, a small piece of cerebellum (1.4 mm diameter) was scanned with UMPA speckle-based phase tomography at a smaller effective pixel size.

A slice through the reconstructed phase volume is shown in Fig. 8.18a.

Again, white and grey matter as well as the molecular and granular layers of the grey matter can clearly be distinguished. More clearly than in Fig. 8.16 one can now see the Purkinje cell layer (Pcl) between the molecular layer (ml) and the granular layer (gl). The ROIs in Fig. 8.18b, c at the transition zones of molecular layer, Purkinje cell layer and granular layer show some of the different types of neurons mentioned above, which can now be resolved clearly: The Purkinje cells (Pc), stellate cells (sc) in the molecular layer and granule cells (gc) in the granular layer. Figure 8.18e shows the same ROI as Fig. 8.18b, but with inverse contrast. This makes it easier for the human eye to visually identify the different cells, which are now dark on light background, a contrast the human eye is generally more accustomed to. In Fig. 8.18e, one can even see the nucleoli (central, most dense part of the nucleus) in the Purkinje cells, indicated by a white arrow. Khimchenko et al. (2018) report an average size of 3.5 μm for the nucleolus of Purkinje cells determined from holotomography measurements.[7] As the nucleoli can be resolved in our data set, the spatial resolution of the phase tomogram is expected to be better than 3.5 μm.

For comparison, a slice through the transmission volume, corresponding to the phase slice in Fig. 8.18a, is shown in Fig. 8.18d. It is once again mainly dominated by edge-enhancement fringes, which makes the stellate and granular cells show up clearly due to the sharp density transition to the surrounding tissue. On the other hand, the Purkinje cells are not easily identified in the transmission slice as their absorption properties are similar to the background. For visualising the Purkinje cells, the phase slices can reach much better contrast, allowing for an easier segmentation.

[7]Note that the size of the cell soma in this case was on average 54 μm—larger than reported by Llinas and Negrello (2015).

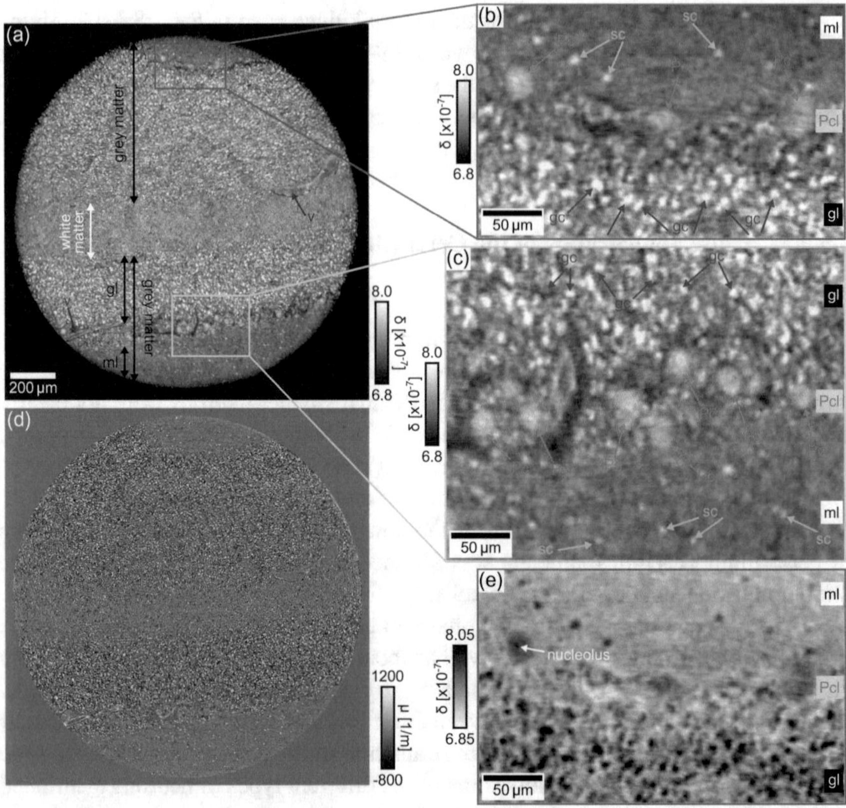

Fig. 8.18 Slice through the phase tomogram of a smaller piece of human cerebellum at higher spatial resolution. **a** Transverse phase slice. Grey and white matter as well as the molecular (ml), granular (gl) and Purkinje cell (Pcl) layer can clearly be distinguished. Furthermore, large blood vessels (v) can be seen. **b**, **c** ROIs of the slice in panel (**a**) at the transition zones of the granular layer, Purkinje cell layer and molecular layer. Purkinje cells (Pc) and stellate cells (sc) in the molecular layer and granule cells (gc) in the granular layer can be identified. **d** Corresponding slice through the transmission tomogram reconstructed from the same data set. It is mainly dominated by edge enhancement, which aids in showing the smaller cells. However, Purkinje cells are depicted with reduced contrast compared to panel (**a**). **e** ROI of panel (**b**), but with inverse colour coding for a more intuitive visual inspection. The nucleoli (black) of the Purkinje cells can be seen more clearly

8.5.3 Conclusions and Outlook

In this section, the phase tomograms of pieces of human cerebellum embedded in paraffin were studied at two different length scales. The results presented here are preliminary and segmentation and quantitative analysis have not been performed yet. However, the good contrast and the small details visible in the phase volume slices illustrate the promising potential of UMPA virtual histology for this type of sample and will allow for easy segmentation of the various features and cells in the cerebellum.

The phase volumes of the cerebellum pieces showed better contrast than the corresponding transmission volumes reconstructed from the same data sets. The latter mainly highlighted edges due to Fresnel fringes in the images. However, it should be noted that the scans were not adapted to absorption imaging. Excellent results have been obtained for imaging of paraffin-embedded brain tissue with absorption-based microCT using optimised setups at synchrotron and also laboratory X-ray sources (Khimchenko et al. 2016a, b; Bikis et al. 2019). On the other hand, it has been shown that for brain specimens scanned in formalin the phase-contrast signal is nonetheless superior to the absorption image (Schulz et al. 2010).

In the phase tomograms of the medium-resolution scan, the white matter and the molecular and granular layer of the grey matter were easily distinguished. Therefore, simple grey-value-based thresholding of UMPA phase tomograms will enable a fast segmentation of the different brain regions. From the higher-resolution scan, single neurons such as the Purkinje cells, stellate cells and granule cells were extracted. This information can in the future be used to study the number and distribution of neurons in the cerebellum. A similar analysis was for example performed recently by Töpperwien et al. (2018) using single-distance propagation-based phase-contrast imaging. UMPA speckle phase-contrast imaging could allow for a broader analysis as it delivers real quantitative information on the tissue density in addition to high-contrast 3D structural information. This will enable the correlation of the tissue density with information on its structure, function and cell distribution, which will be especially interesting for the study and diagnosis of neurodegenerative diseases.

In particular, it has been suggested that the structure, size or cell density of Purkinje cells changes in diseases like Friedreich's ataxia (Kemp et al. 2016), multiple sclerosis (Redondo et al. 2015), Huntington (Jeste et al. 1984), schizophrenia (Tran et al. 1998) and autism (Fatemi et al. 2002). Furthermore, neurodegenerative genetic diseases like Niemann-Pick disease have been reported to lead to cell-autonomous death of Purkinje cells (Ko et al. 2005). Many other diseases and genetic defects are correlated with changes to the Purkinje cells, which are very susceptible to genetic and environmental factors. The proposed virtual histology method based on UMPA speckle phase tomography could in the future greatly aid in the reliable and accurate study of such changes of Purkinje cells and other neurons.

The next step in validating the method for this purpose will be to study a larger number of samples, perform extensive segmentation and density analysis and later compare measurements of diseased and healthy specimens.

8.6 Concluding Remarks

In this chapter, it was demonstrated that X-ray speckle-based imaging, in particular using the UMPA approach that was developed during this Ph.D. project, is a promising method to perform 3D virtual histology of biomedical specimens. The UMPA phase tomograms show excellent contrast, which allows for a straightforward and simple segmentation of different components in the specimens, visualising their 3D structure and interrelationship. This is complemented by quantitative information on the tissue density that is given directly by the phase volume and quantifies the distribution of electron density values in the specimen.

It was demonstrated on three different specimens that X-ray speckle-based virtual histology can be applied to various kinds of biomedical soft-tissue samples prepared in different ways, including fully hydrated in almost native state. This is a major advantage over other techniques, which often require a specific form of tissue preparation. Hence, speckle-based phase tomography can be applied to a wide range of samples and can be easily combined with other methods for complementary analysis.

Further detailed studies on the data presented in this chapter will be conducted, in particular for the brain tissue specimens. This will include 3D segmentation based on density values and subsequent analysis of the distribution, average size, number density and connectivity of features such as the different types of cells in the cerebellum.

The results presented here were conducted on healthy murine and human tissue specimens. Following this first demonstration of its promising potential, X-ray speckle-based virtual histology will also be applied to diseased tissue, in particular for studying pathologies that result in 3D structural and tissue density changes. This has great potential for aiding the diagnosis and staging of diseases and monitoring treatment effectiveness. Furthermore, a larger number of samples will be scanned to obtain statistically relevant information, which is important for fundamental biological research on the mechanisms and origins of diseases.

As UMPA speckle-based imaging can be performed at laboratory X-ray sources, as will be demonstrated in Chap. 9, Sect. 9.4 of this thesis, it will in the future be possible to perform speckle-based virtual histology in the laboratory, which makes it promising for routine use for medical and biological studies.

With its excellent image quality and 3D quantitative character, combined with the simple, robust experimental setup that is compatible with widely accessible laboratory sources, X-ray speckle-based virtual histology is expected to receive great interest in the medical community and is well positioned to open up new possibilities for biomedical research and clinical histopathology.

References

Bikis C, Rodgers G, Deyhle H, Thalmann P, Hipp A, Beckmann F, Weitkamp T, Theocharis S, Rau C, Schulz G, Müller B (2019) Sensitivity comparison of absorption and grating-based phase tomography of paraffin-embedded human brain tissue. Appl Phys Lett 114(8):083702

Busse M, Müller M, Kimm MA, Ferstl S, Allner S, Achterhold K, Herzen J, Pfeiffer F (2018) Three-dimensional virtual histology enabled through cytoplasm-specific X-ray stain for microscopic and nanoscopic computed tomography. Proc Natl Acad Sci 115(10):2293–2298

Clermont Y, Huckins C (1961) Microscopic anatomy of the sex cords and seminiferous tubules in growing and adult male albino rats. Am J Anat 108(1):79–97

Curtis GM (1918) The morphology of the mammalian seminiferous tubule. Am J Anat 24(3):339–394

Ehling J, Bábíčková J, Gremse F, Klinkhammer BM, Baetke S, Knuechel R, Kiessling F, Floege J, Lammers T, Boor P (2016) Quantitative Micro-Computed Tomography Imaging of Vascular Dysfunction in Progressive Kidney Diseases. J Am Soc Nephrol 27(2):520–532

Engström KG, Taljedäl I-B (1987) Altered shape and size of red blood cells in obese hyperglycaemic mice. Acta Physiol Scand 130(4):535–543

Fatemi SH, Halt AR, Realmuto G, Earle J, Kist DA, Thuras P, Merz A (2002) Purkinje cell size is reduced in cerebellum of patients with autism. Cell Mol Neurobiol 22(2):171–175

Haschek WM, Rousseaux CG, Wallig MA (eds) (2010) Fundamentals of toxicologic pathology, chapter 18–male reproductive system, 2nd edn. Academic Press, San Diego, CA, USA

Hlushchuk R, Zubler C, Barré S, Correa Shokiche C, Schaad L, Röthlisberger R, Wnuk M, Daniel C, Khoma O, Tschanz SA, Reyes M, Djonov V (2018) Cutting-edge microangio-CT: new dimensions in vascular imaging and kidney morphometry. Am J Physiol Renal Physiol 314(3):F493–F499

Jeste DV, Barban L, Parisi J (1984) Reduced Purkinje cell density in Huntington's disease. Exp Neurol 85(1):78–86

Kak AC, Slaney M (2001) Principles of computerized tomographic imaging. Society for Industrial and Applied Mathematics, Philadelphia, PA, USA

Kemp KC, Cook AJ, Redondo J, Kurian KM, Scolding NJ, Wilkins A (2016) Purkinje cell injury, structural plasticity and fusion in patients with Friedreich's ataxia. Acta Neuropathol Commun 4(1):53

Khimchenko A, Bikis C, Pacureanu A, Hieber SE, Thalmann P, Deyhle H, Schweighauser G, Hench J, Frank S, Müller-Gerbl M, Schulz G, Cloetens P, Müller B (2018) Hard x-ray nanoholotomography: large-scale, label-free, 3D neuroimaging beyond optical limit. Adv Sci 5(6):1700694

Khimchenko A, Deyhle H, Schulz G, Schweighauser G, Hench J, Chicherova N, Bikis C, Hieber SE, Müller B (2016a) Extending two-dimensional histology into the third dimension through conventional micro computed tomography. Neuro Image 139:26–36

Khimchenko A, Schulz G, Deyhle H, Thalmann P, Zanette I, Zdora M-C, Bikis C, Hipp A, Hieber SE, Schweighauser G, Hench J, Müller B (2016b) X-ray micro-tomography for investigations of brain tissues on cellular level. Proc SPIE 9967:996703

Ko DC, Milenkovic L, Beier SM, Manuel H, Buchanan J, Scott MP (2005) Cell-Autonomous death of cerebellar Purkinje neurons with autophagy in Niemann-pick type C disease. PLoS Genet 1(1):e7

Kottler C, David C, Pfeiffer F, Bunk O (2007) A two-directional approach for grating-based differential phase contrast-imaging using hard x-rays. Opt Express 15(3):1175–1181

Layton AT (2012) Modeling transport and flow regulatory mechanisms of the kidney. ISRN Biomath 2012:170594

Llinas R, Negrello MN (2015) Cerebellum. Scholarpedia 10(1):4606

Lundström U, Larsson DH, Burvall A, Takman PAC, Scott L, Brismar H, Hertz HM (2012) X-ray phase contrast for CO2 microangiography. Phys Med Biol 57(9):2603

Martins J, de e Silva S, Zanette I, Noël PB, Cardoso MB, Kimm MA, Pfeiffer F (2015) Three-dimensional non-destructive soft-tissue visualization with X-ray staining micro-tomography. Sci Rep 5:14088

McInnes E (2005) Artefacts in histopathology. Comp Clin Path 13(3):100–108

Metscher BD (2009a) MicroCT for comparative morphology: simple staining methods allow high-contrast 3D imaging of diverse non-mineralized animal tissues. BMC Physiol 9(1):11

Metscher BD (2009b) MicroCT for developmental biology: a versatile tool for high-contrast 3D imaging at histological resolutions. Dev Dyn 238(3):632–640

Missbach-Guentner J, Pinkert-Leetsch D, Dullin C, Ufartes R, Hornung D, Tampe B, Zeisberg M, Alves F (2018) 3D virtual histology of murine kidneys—high resolution visualization of pathological alterations by micro computed tomography. Sci Rep 8(1):1407

Modregger P, Lübbert D, Schäfer P, Köhler R (2007) Spatial resolution in Bragg-magnified x-ray images as determined by Fourier analysis. Phys Status Solidi A 204(8):2746–2752

Morgan KS, Paganin DM, Siu KKW (2012) X-ray phase imaging with a paper analyzer. Appl Phys Lett 100(12):124102

Müller B, Riedel M, Thurner PJ (2006) Three-Dimensional characterization of cell clusters using synchrotron-radiation-based micro-computed tomography. Microsc Microanal 12(2):97–105

Müller M, de Sena Oliveira I, Allner S, Ferstl S, Bidola P, Mechlem K, Fehringer A, Hehn L, Dierolf M, Achterhold K, Gleich B, Hammel JU, Jahn H, Mayer G, Pfeiffer F (2017) Myoanatomy of the velvet worm leg revealed by laboratory-based nanofocus x-ray source tomography. Proc Natl Acad Sci USA 114(47):12378–12383

Müller M, Kimm MA, Ferstl S, Pfeiffer F, Busse M (2018) Nucleus-specific x-ray stain for 3D virtual histology. Sci Rep 8(1):17855

Nakata H (2019) Morphology of mouse seminiferous tubules. Anat Sci Int 94(1):1–10

Nakata H, Sonomura T, Iseki S (2017) Three-dimensional analysis of seminiferous tubules and spermatogenic waves in mice. Reproduction 154(5):2569–579

Nakata H, Wakayama T, Sonomura T, Honma S, Hatta T, Iseki S (2015) Three-dimensional structure of seminiferous tubules in the adult mouse. J Anat 227(5):686–694

Olgac U, Kurtcuoglu V (2016) The Bohr effect is not a likely promoter of renal preglomerular Oxygen shunting. Front Physiol 7:482

Ourselin S, Roche A, Subsol G, Pennec X, Ayache N (2001) Reconstructing a 3D structure from serial histological sections. Image Vis Comput 19(1):25–31

Pauwels E, van Loo D, Cornillie P, Brabant L, van Hoorebeke L (2013) An exploratory study of contrast agents for soft tissue visualization by means of high resolution x-ray computed tomography imaging. J Microsc 250(1):21–31

Pichat J, Iglesias JE, Yousry T, Ourselin S, Modat M (2018) A survey of methods for 3D histology reconstruction. Med Image Anal 46:73–105

Ray D, Pitts PB, Hogarth CA, Whitmore LS, Griswold MD, Ye P (2015) Computer simulations of the mouse spermatogenic cycle. Biol Open 4(1):1–12

Redondo J, Kemp K, Hares K, Rice C, Scolding N, Wilkins A (2015) Purkinje cell pathology and loss in multiple sclerosis cerebellum. Brain Pathol 25(6):692–700

Schulz G, Weitkamp T, Zanette I, Pfeiffer F, Beckmann F, David C, Rutishauser S, Reznikova E, Müller B (2010) High-resolution tomographic imaging of a human cerebellum: comparison of absorption and grating-based phase contrast. J R Soc Interface 7(53):1665–1676

Shearer T, Bradley RS, Hidalgo-Bastida LA, Sherratt MJ, Cartmell SH (2016) Three-dimensional visualisation of soft biological structures by x-ray computed micro-tomography. J Cell Sci 129(13):2483–2492

Shirai R, Kunii T, Yoneyama A, Ooizumi T, Maruyama H, Lwin T-T, Hyodo K, Takeda T (2014) Enhanced renal image contrast by ethanol fixation in phase-contrast x-ray computed tomography. J Synchrotron Rad 21(4):795–800

Smithies O (2003) Why the kidney glomerulus does not clog: a gel permeation/diffusion hypothesis of renal function. Proc Natl Acad Sci 100(7):4108–4113

Stock SR (2008) Microcomputed tomography: methodology and applications. CRC Press, Boca Raton, FL, USA

Töpperwien M, van der Meer F, Stadelmann C, Salditt T (2018) Three-dimensional virtual histology of human cerebellum by x-ray phase-contrast tomography. Proc Natl Acad Sci 115(27):6940–6945

Tran KD, Smutzer GS, Doty RL, Arnold SE (1998) Reduced Purkinje cell size in the cerebellar vermis of elderly patients with schizophrenia. Am J Psychiatry 155(9):1288–1290

Vasquez SX, Gao F, Su F, Grijalva V, Pope J, Martin B, Stinstra J, Masner M, Shah N, Weinstein DM, Farias-Eisner R, Reddy ST (2011) Optimization of MicroCT imaging and blood vessel diameter quantitation of preclinical specimen vasculature with radiopaque polymer injection medium. PLoS One 6(4):1–6

Velroyen A, Bech M, Zanette I, Schwarz J, Rack A, Tympner C, Herrler T, Staab-Weijnitz C, Braunagel M, Reiser M, Bamberg F, Pfeiffer F, Notohamiprodjo M (2014) X-ray phase-contrast tomography of renal ischemia-reperfusion damage. PLoS One 9(10):e109562

Wilkins SW, Nesterets YI, Gureyev TE, Mayo SC, Pogany A, Stevenson AW (2014) On the evolution and relative merits of hard x-ray phase-contrast imaging methods. Philos Trans R Soc A 372(2010):20130021

Wu J, Takeda T, Lwin TT, Momose A, Sunaguchi N, Fukami T, Yuasa T, Akatsuka T (2009) Imaging renal structures by x-ray phase-contrast microtomography. Kidney Int 75(9):945–951

Zagorchev L, Oses P, Zhuang ZW, Moodie K, Mulligan-Kehoe MJ, Simons M, Couffinhal T (2010) Micro computed tomography for vascular exploration. J Angiogenes Res 2(1):7

Zanette I, Weitkamp T, Le Duc G, Pfeiffer F (2013) X-ray grating-based phase tomography for 3D histology. RSC Adv 3(43):19816–19819

Zanette I, Zdora M-C, Zhou T, Burvall A, Larsson DH, Thibault P, Hertz HM, Pfeiffer F (2015) X-ray microtomography using correlation of near-field speckles for material characterization. Proc Natl Acad Sci USA 112(41):12569–12573

Zanette I, Zhou T, Burvall A, Lundström U, Larsson DH, Zdora M-C, Thibault P, Pfeiffer F, Hertz HM (2014) Speckle-Based x-ray phase-contrast and dark-field imaging with a laboratory source. Phys Rev Lett 112(25):253903

Zdora M-C (2018) State of the art of x-ray speckle-based phase-contrast and dark-field imaging. J Imaging 4(5):60

Zdora M-C, Thibault P, Kuo W, Fernandez V, Deyhle H, Vila-Comamala J, Olbinado MP, Rack A, Lackie PM, Katsamenis OL, Lawson MJ, Kurtcuoglu V, Rau C, Pfeiffer F, Zanette I (2020) X-ray phase tomography with near-field speckles for three-dimensional virtual histology. Optica 7(9):1221–1227

Zdora M-C, Thibault P, Zhou T, Koch FJ, Romell J, Sala S, Last A, Rau C, Zanette I (2017) X-ray phase-contrast imaging and metrology through unified modulated pattern analysis. Phys Rev Lett 118(20):203903

Zhai XY, Birn H, Jensen KB, Thomsen JS, Andreasen A, Christensen EI (2003) Digital three-dimensional reconstruction and ultrastructure of the mouse proximal tubule. J Am Soc Nephrol 14(3):611–619

Chapter 9
Recent Developments and Ongoing Work in X-ray Speckle-Based Imaging

9.1 Introductory Remarks

As outlined in the previous chapters, X-ray speckle-based imaging has already seen rapid development since its first demonstration less than a decade ago. The rising interest and uptake of the technique calls for its further optimisation and customisation in order to widen its user community and explore new applications.

Some of the recent progresses achieved within the framework of this Ph.D. project and reported by other groups were already discussed in Chap. 5, Sect. 5.10. In this chapter, preliminary results of the latest and ongoing work carried out as part of this Ph.D. project are presented. These include:

- the development of customisable diffusers that can be fabricated according to specific setup parameters and imaging conditions;
- the further development of high-energy X-ray speckle-based imaging using the UMPA approach, in particular for geology and materials science applications;
- the first implementation of UMPA speckle-based imaging at an X-ray laboratory source.

9.2 Development of Customised Phase Modulators

It was outlined in Chap. 5, Sect. 5.3.1 that the quality of the speckle pattern, in particular its visibility and feature size, strongly influences the reconstruction results. As the visibility and size of the speckles are determined by the diffuser properties, it would be beneficial if the latter could be adapted specifically to an experiment to optimise the properties of the speckle pattern for a given setup.

In most implementations of X-ray speckle-based imaging commercially available abrasive paper was used as a diffuser material. It consists of silicon carbide particles

M.-C. Zdora, *X-ray Phase-Contrast Imaging Using Near-Field Speckles*,
Springer Theses, https://doi.org/10.1007/978-3-030-66329-2_9

(glued onto a cellulose backing), which cause scattering of the illuminating X-rays, leading to the creation of a speckle pattern that is recorded by the detector. Abrasive paper can be purchased in different granularities (Federation of European Producers of Abrasives 2019) and can hence be roughly adapted to the pixel size and energy of a certain setup. However, there is only a limited range of grain sizes available and the thickness of structures, which determines the phase shift of the X-rays in the diffuser, cannot be chosen independently from the feature size. Other materials that have been used as phase modulators for X-ray speckle-based imaging include biological filter membranes and fine-grain sand. The latter has been employed as a diffuser for setups with higher-energy X-rays (see also Sect. 9.3 and Chap. 3). Biological filter membranes are made from cellulose acetate and have small pores with a diameter in the range of a micron or less. They are hence suitable for use with smaller pixel sizes but have in previous measurements produced speckle patterns of relatively low visibility (Zdora 2014).

Here, a new type of diffuser is investigated. It has the potential to increase the image quality obtained with speckle-based imaging by actively controlling the lateral size and height of the diffuser features to adapt them to a specific setup and to given experimental conditions. Designing these customised phase modulators, in particular for high-resolution applications, was the aim of a collaboration with Dr. Joan Vila-Comamala at ETH Zurich and Paul Scherrer Institute, Switzerland, who conceived this idea and fabricated the diffusers. The principle and first preliminary results of X-ray speckle-based imaging performed with this type of phase modulator are presented in this section after a short review of previous work.

9.2.1 Previous Work

As mentioned above, the classic diffuser used for most speckle-imaging applications is a piece of sandpaper. Recently, with the push of the X-ray speckle-based technique towards higher X-ray energies and lower-coherence sources (see also following two sections of this chapter), alternative speckle masks have been explored as sandpaper often cannot provide speckles with sufficient visibility for these applications. Another challenge is the realisation of very small speckle sizes, which are currently limited to tens of micrometres for the available diffuser materials.[1]

To overcome the limited visibility of the speckle pattern for some applications such as the implementation with low-coherence X-rays and/or high-energy X-rays, it was proposed to use a random absorption mask rather than a phase mask to produce a reference pattern that does not rely on interference effects (Wang et al. 2016a). The feasibility of this approach was demonstrated with high-energy X-rays from synchrotron sources (Wang et al. 2016a) and at low-coherence laboratory sources (Wang et al. 2016a; Zhou et al. 2018c). Following the first demonstrations with steel

[1]A speckle size of down to 4 μm has been reported in the author's Masters thesis using biological filter membranes (Zdora 2014). This, however, led to a speckle visibility of only 1.2%.

wool, engineered porous materials such as aluminium-copper and magnesium-zinc alloys as well as limestone and mortar were also tested as absorption masks (Wang et al. 2018). They have the potential to be fabricated with customised parameters adjustable to a certain experiment. However, the visibilities reported to date with these materials are still low (Wang et al. 2018). It should, furthermore, be considered that using an absorption mask significantly reduces the photon flux and can hence lead to very long scan times, in particular at low-flux laboratory sources.

In another attempt to control the visibility of the speckle pattern at high X-ray energies, stacks of sandpaper were used as diffusers to produce speckle based on a mixture of absorption and phase effects (Bérujon and Ziegler 2017; Zhou et al. 2018c, b).

Technically, every diffuser leads to both absorption and phase effects, to varying degree, that together lead to the formation of the speckled intensity pattern observed by the detector. For the experiments in this thesis and most previous implementations of X-ray speckle-based imaging, the absorption by the diffuser is low. Hence, we assume the resulting speckle to mainly stem from phase and interference effects and we call the diffuser a phase modulator. For the applications with steel wool and similar, absorption is the dominant interaction process with X-rays and these diffusers are sometimes referred to as random absorption masks. Many diffuser materials will lead to a mixture of phase and absorption effects, such as sand and stacks of sandpaper, which were used in Sect. 9.3 below. If possible, it is, however, beneficial to rely mainly on phase effects, as this reduces the loss in flux associated with absorption masks.

9.2.2 Phase Modulators Fabricated by Metal-Assisted Chemical Etching

The aim of the collaborative work conducted during this Ph.D. project was to develop new customisable diffusers that produce a speckle pattern with a desired feature size and high visibility, which is based mainly on phase effects rather than absorption effects as pursued by other groups (see above) and which can be adapted to the properties of a specific experimental setup.

For the proposed approach, the diffuser is fabricated by metal-assisted chemical etching (MACE) of silicon using gold as a catalyst. MACE is a simple, cost-efficient and controllable technique that can produce high-aspect-ratio features of micro- to nanometre lateral size and over a large substrate area in a relatively straightforward manner (Li and Bohn 2000; Chartier et al. 2008; Hildreth et al. 2009; Huang et al. 2011; Romano et al. 2016).

The process of fabricating a diffuser with MACE using a silicon substrate consists of three main steps: evaporation of a thin gold layer, thermal de-wetting and the MACE process itself. Controlling the parameters during the fabrication, in particular the gold layer thickness, de-wetting temperature, etching time and composition of the MACE solution, allows for adjusting the lateral size and height of the silicon

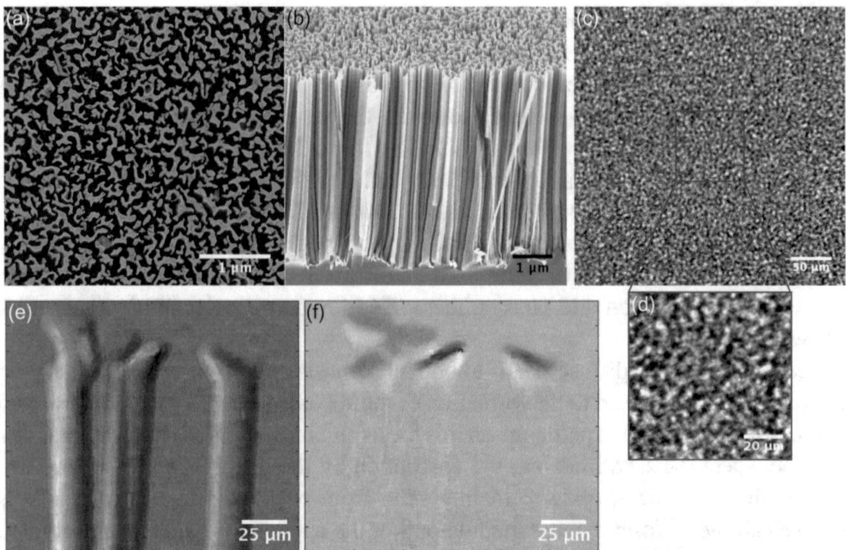

Fig. 9.1 First demonstration of X-ray speckle imaging with diffusers made from silicon nanostructures using metal-assisted chemical etching (MACE). The customisable height and lateral size of the silicon pillars allows for adjusting the speckle size and optimising the visibility and homogeneity of the speckle pattern. **a** Scanning electron micrograph of a diffuser fabricated with MACE in top-down view and **b** side view. **c** Speckle pattern created by the MACE diffuser with a 15 keV X-ray beam and **d** ROI showing the fine speckles of small size. **e**, **f** Horizontal and vertical differential phase images of PET fibres obtained with the XST mode of X-ray speckle-based imaging using the MACE diffuser as a phase modulator. Edge-enhancement fringes around the sample lead to blurring of its edges. Image courtesy of Joan Vila-Comamala, ETH Zurich and PSI, Switzerland

pillars produced by the etching process, which are directly related to the properties of the speckle pattern that is created by the phase modulator. An example of a diffuser produced by MACE is shown in Fig. 9.1a, b. An etching solution of 4.98 mol/l hydrofluoric acid (HF) and 0.43 mol/l hydrogen peroxide (H_2O_2) was used. The etching time was 20 min and the resulting silicon pillars have a height of approximately 6 µm and a sub-micron lateral size.

Using this MACE-fabricated diffuser with very small lateral features, an interference pattern with small speckle size was achieved, as shown in Fig. 9.1c and the ROI in Fig. 9.1d. It can, furthermore, be seen that the resulting speckle pattern is of high homogeneity, which leads to a more uniform visibility map and hence signal sensitivity over the field of view compared to conventional sandpaper diffusers.

Figure 9.1e and f show a first demonstration of applying this diffuser to X-ray speckle-based imaging. The differential phase signals in the horizontal (Fig. 9.1e) and vertical (Fig. 9.1f) direction of a PET fibre (diameter: 23 µm) were obtained with a monochromatic X-ray beam of 15 keV at TOMCAT beamline at the Swiss Light Source (SLS), Paul Scherrer Institute (PSI), Switzerland. The single-shot X-ray speckle-tracking approach was used for this first demonstration. The spatial resolution and sensitivity of the measurement can be improved significantly by using advanced acquisition and processing schemes for speckle imaging, as shown in the following sections.

9.2.3 *X-ray Speckle-Based Projection Imaging with MACE Diffusers*

As a first demonstration of the capabilities of the newly developed diffuser type based on MACE fabrication, the projection of a test sample was collected. In the following, the speckle pattern is inspected and its properties, such as speckle visibility and size, are evaluated. Subsequently, multimodal projections of the sample obtained using the UMPA technique are presented and their quality is assessed.

9.2.3.1 Materials and Methods

The measurements were performed at the I13-2 Imaging branch at Diamond Light Source, UK (see Chap. 3). The X-ray beam coming from the undulator (gap: 5.25 mm) was filtered using 280 µm- and 705 µm-thick pyrolitic graphite filters, 3.2 mm of aluminium and 35 µm of silver. Furthermore, a rhodium mirror was installed. This led to a relatively broad X-ray spectrum with a weighted mean spectral energy of 24.2 keV.

The setup consisted of a test sample made of a silicon sphere (diameter: 480 µm) glued onto a wooden toothpick, the MACE diffuser and a detector system. The latter was a pco.4000 CCD camera (chip size: 4008 $\times$ 2672 pixels, pixel size: 9 µm) coupled via a scintillation screen (europium-doped gallium gadolinium garnet (GGG:Eu) crystal of 17 µm thickness) to a microscope with a 10$\times$ objective in combination with a tube lens and a relay optics system. This led to an effective pixel size of 0.45 µm. The specimen was located between the diffuser and the detector system. The diffuser-sample distance was 4.3 cm and the sample-detector system distance was 39.0 cm.

Data collection was performed following the UMPA scheme with $N = 20$ diffuser positions. 10 reference images without the specimen and a single sample image with the specimen in the beam were acquired at each diffuser position. Before the scan, 10 dark images without X-ray illumination were recorded. The exposure time for each frame was 1 s.

The signal reconstruction was performed following the UMPA reconstruction approach, which is described in more detail in Chap. 6. Prior to the analysis, the averaged dark image was subtracted from all reference and all sample images and the dark-corrected reference images were averaged for each diffuser step to obtain a single reference image per diffuser position. The UMPA processing algorithm was then applied to the reference and sample images using a Hamming analysis window of

9×9 pixels (FWHM of the window $\approx$4.2 pixels). From the reconstructed differential phase signals the integrated phase was retrieved via Fourier phase integration (Kottler et al. 2007).

9.2.3.2 Evaluation of Speckle Visibility and Size

First, the quality of the speckle pattern produced by the MACE diffuser is evaluated. Figure 9.2a shows that fine speckles can be observed over the whole field of view. The low-frequency intensity variations (large horizontal stripes) are caused by the uneven intensity profile of the illuminating beam. The red box in Fig. 9.2a highlights the part of the field of view that was used for the reconstruction of the sphere test sample shown in the next section. A more detailed look at the speckle pattern is provided in the ROI in Fig. 9.2b taken from the centre of the field of view. It reveals speckles of good contrast and small size, suitable for speckle-based imaging at small effective pixel sizes.

As explained in previous chapters, the two key properties of the speckle pattern are its visibility and feature size. The visibility map for the speckle pattern in Fig. 9.2a is shown in Fig. 9.2c for the region in the red box. The visibility was evaluated in ROIs of 150×150 pixels using the definition in Eq. (5.1). It can be seen that the visibility is homogeneous over the field of view, except for some hot spots in the lower third. These show artificially high visibility values due to some dirt on the scintillator (as can be seen in Fig. 9.2a), which leads to a large standard deviation of the intensity values in this region. The mean visibility over the whole field of view is $v = (4.4 \pm 0.5)\%$. The visibility of the speckle pattern was also evaluated for the other 19 diffuser positions of the UMPA scan and the results consistently give an average mean visibility over the field of view of $v = 4.4\%$ (standard deviation of the values of all measurements is 0.01%). While this is not as high as the visibility values reported for the use of sandpaper, it is a promising first result and was sufficient to obtain high-contrast multimodal images using UMPA, as demonstrated in the following.

The speckle size can be estimated from a 2D autocorrelation analysis of the speckle pattern, which was computed here for the region in the red box in Fig. 9.2a. The central area of the 2D autocorrelation is shown in Fig. 9.2d and the horizontal (green) and vertical (orange) line profiles through the centre are plotted in Fig. 9.2e. The FWHM of the autocorrelation peak is typically taken as a measure for the speckle size (see Chap. 5). It was determined here by fitting Gaussians to the horizontal and vertical 1D line profiles through the peak of the 2D autocorrelation, see Fig. 9.2e. This gives a speckle size of $\Delta_x = 11.7$ pixels $\approx 5.3\,\mu$m in the horizontal and $\Delta_y =$ 10.5 pixels $\approx$4.7 μm in the vertical direction. The speckle dimensions are slightly larger in the horizontal due to the lower coherence in this direction caused by the larger horizontal source size and bending of the rhodium mirror in the horizontal direction. The speckle size of approximately 10 pixels across is optimal for X-ray speckle-based imaging which requires a fine but resolvable speckle pattern. The sizes

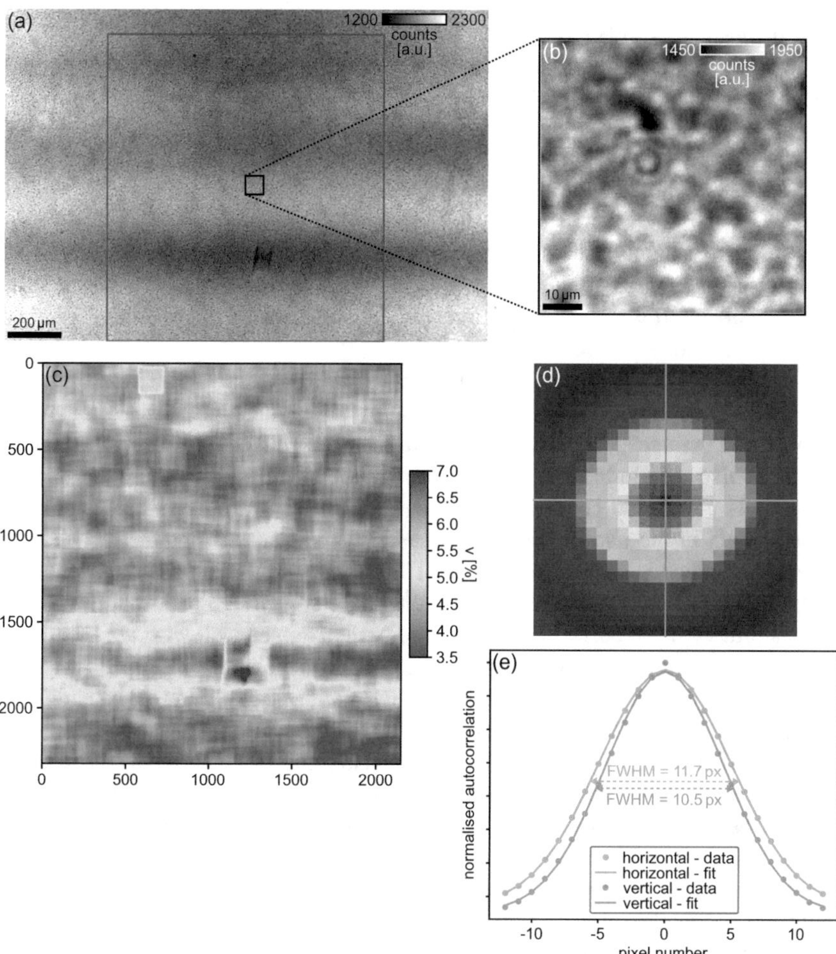

Fig. 9.2 Evaluation of the quality of the speckle pattern created by a MACE-fabricated diffuser (platinum catalyst, 200 μm-thick silicon wafer). **a** Speckle pattern obtained with an X-ray beam of mean spectral energy 24.2 keV. Low-frequency intensity variations are caused by the beam profile (no flat-field correction applied). The red box indicates the area used for the measurements of the sphere sample in Fig. 9.3. **b** ROI of 150 × 150 pixels in the centre of the field of view. **c** Visibility map in the region of the red box in panel (**a**). The visibility was evaluated in ROIs of 150 × 150 pixels around each pixel. **d** Central region of the 2D autocorrelation of the speckle pattern. The width of the peak gives an indication of the speckle size. **e** Horizontal (green) and vertical (red) line profiles through the centre of the autocorrelation peak. The FWHM is used as a measure of the speckle size and was determined by fitting a Gaussian to the measured line profile

of the diffuser structures were chosen deliberately to match this requirement and the results confirm that the speckle size can be tuned to a desired value by adapting the conditions in the MACE fabrication process to achieve a certain feature size.

9.2.3.3 2D Imaging of a Test Sample

After first assessment of the properties of the speckle pattern, a test sample consisting of a silicon sphere glued onto a wooden toothpick was imaged with the setup described above. It is the same specimen as shown in Sect. 6.4.2 of Chap. 6, but not scanned under the same conditions.

The multimodal signals reconstructed with the UMPA analysis method are shown in Fig. 9.3. The differential phase signals in the horizontal and vertical directions in Fig. 9.3a and 9.3b, respectively, clearly show the expected refraction behaviour of the sphere, as also seen in the line profiles through the centre of the sphere in Fig. 9.3d. The signals have good contrast and show a low noise level. From the differential phase signals in the two directions, a phase shift signal was obtained via Fourier integration, which is shown in Fig. 9.3c. The corresponding line profile through the sphere in Fig. 9.3e allows for estimating the phase shift in the centre of the sphere. The maximum of the curve is $\Phi = 49.3$ rad, which agrees well with the theoretical value[2] $\Phi_{\text{theo}} = 48.6$ rad, calculated for the centre of the sphere, i.e. for 480 µm of silicon. The slight overestimation in the measured value is likely caused by additional glue on the surface of the sphere, which increases the phase shift.

The angular sensitivity of the differential phase signal is quantified by the standard deviation of the refraction angle signal in a homogeneous background region, see Chap. 5, Sect. 5.4.4. It was evaluated in a 150 × 150 ROI in the left upper part of the field of view and is $\sigma_x \approx 49.8$ nrad in the horizontal and $\sigma_y \approx 44.8$ nrad in the vertical. The better sensitivity along the vertical axis is in agreement with the higher coherence in this direction as already observed for the speckle size.

The transmission and dark-field signals that were obtained simultaneously from the same data set are presented in Fig. 9.3f and g, respectively. The transmission signal (combination of X-ray absorption and edge-enhancement effects) clearly shows the presence of glue on the surface of the sphere. The dark-field signal is particularly strong in the region of the wooden toothpick where small-angle scattering from features beyond the spatial resolution limit occurs.

The results on the test sample confirm that the MACE-fabricated diffuser is suitable for high-contrast UMPA speckle-based imaging at sub-micrometre pixel sizes delivering high-resolution quantitative information on the specimen.

---

[2]The theoretical value was calculated using the relation $\Phi = (2\pi/\lambda)\delta t$, where λ is the X-ray wavelength, δ the refraction index decrement and t the thickness of the sample. Values for δ of silicon at an energy of 24.2 keV were taken from (Henke et al. 1993).

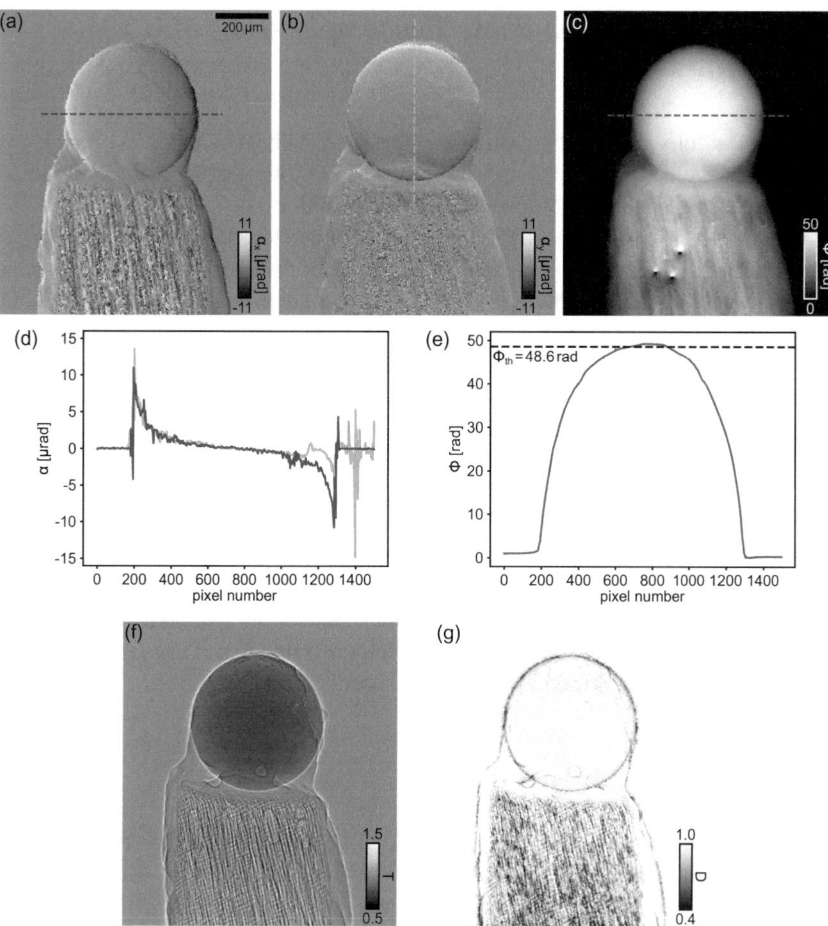

Fig. 9.3 Multimodal signals of a silicon sphere (diameter: 480 μm) glued onto a wooden toothpick obtained with X-ray speckle-based imaging using a diffuser produced with MACE. The UMPA mode for speckle imaging was used with $N = 20$ diffuser positions and a Hamming-type analysis window of size $w = 9 \times 9$ pixels (FWHM of window: ≈ 4.2 pixels). **a** Refraction angle signal in the horizontal and **b** the vertical direction. **c** Phase shift integrated from the two differential phase signals in panels (**a**) and (**b**). **d** Line profile through the centre of the sphere for the refraction angle signals (blue: horizontal, green: vertical). **e** Line profile through the phase shift signal and theoretically expected phase shift value (black dashed line) for 480 μm of silicon. **f** Transmission and **g** dark-field signal. It should be noted that the sphere is partly covered in glue, which slightly increases the phase shift and absorption in the sample and can also be observed in the dark-field signal

9.2.4 High-Resolution Phase Tomography Using MACE Diffusers

After successfully employing a diffuser fabricated with MACE for 2D imaging, it is straightforward to apply the approach also to 3D tomography, which allows for quantitative studies on the sample's density distribution.

To date, X-ray speckle-based tomography is difficult to achieve at sub-micrometre spatial resolution, mainly due to the limited choice of diffusers. As discussed above, diffusers with small enough feature sizes that lead to a high-visibility and fine speckle pattern are not readily available. The proposed MACE fabrication of diffusers provides a solution for this, amongst other benefits, and is able to produce diffusers with features sizes down to tens of nanometres in lateral size in a controllable manner.

In the first proof-of-principle demonstration of speckle imaging with a MACE diffuser in Sect. 9.2.3.3, the properties of the diffuser were chosen to match an effective pixel size of 0.45 µm. Sandpaper diffusers of fine granularity have been used previously with effective pixel sizes in this range (Zdora et al. 2018). One of the benefits of MACE diffuser production is that it can achieve small enough speckle to allow for the use of even smaller effective pixel sizes for high-resolution applications. In the following, it is shown that a pixel size as small as 0.16 µm can be used. A suitable diffuser was fabricated by MACE and it is demonstrated that it can be successfully employed for high-resolution X-ray speckle-based tomography.

9.2.4.1 Materials and Methods

The experiment was carried out at the TOMCAT beamline at the SLS (PSI), Switzerland. A monochromatic X-ray beam of 20 keV was selected from the bending magnet spectrum using a double-crystal multilayer monochromator (DCMM).

The experimental setup was located approximately 25 m from the source. It consisted of the specimen on a tomographic stage, the MACE diffuser on stepping stages located 10 cm further downstream and the detector system positioned 20 cm downstream of the diffuser.

The sample was a tube filled with a mix of pristine and deactivated catalyst particles used for fluid catalytic cracking (FCC) (Boerefijn et al. 2000; Meirer et al. 2015; Vogt and Weckhuysen 2015). It was provided by Dr. Johannes Ihli, PSI, Switzerland and initially supplied by Dr. Yuying Shu and Dr. Wu-Cheng Cheng, W. R. Grace Refining Technologies, USA. More details about the FCC catalyst particles and their composition and fabrication can be found by Ihli et al. (2018). The specimen was chosen due to its great interest for the petroleum industry. Catalyst particles used for FCC deactivate over time and their catalytic activity is reduced. The processes behind this deactivation are subject of intense research. Studies with X-ray ptychographic tomography have previously shown that the deactivation leads to changes in the electron density of the particles (Ihli et al. 2017, 2018). The same particles are studied here at lower resolution using X-ray speckle-based tomography.

The detector system was a pco.edge 5.5 sCMOS camera (chip size: 2560×2160 pixels, pixel size: 6.5 μm) coupled via a 5.9 μm-thick terbium-doped lutetium oxy-orthosilicate (LSO:Tb) scintillator to a microscope system (Optique Peter, Lentilly, France) with a UPLAPO40X objective lens (Olympus, Tokyo, Japan) with numerical aperture 0.9. This led to an effective pixel size of 0.16 μm.

The X-ray speckle-based imaging measurements were performed using the UMPA approach. A total number of 901 evenly spaced rotation angles of the sample were used for the tomography scan and at each of them the diffuser was moved to $N = 24$ positions. 10 dark images were taken before the scan and reference speckle images (without the sample) were acquired in regular intervals after every 50 projections of the tomography scan, i.e. 19 sets of reference images were taken throughout the whole scan. The exposure time was 350 ms for each recorded frame.

For the signal reconstruction, first the average dark image was subtracted from all reference and sample images. Then, the 24 reference images were aligned to the corresponding 24 sample images for each projection via cross-correlation in an empty ROI in the background, in order to correct for slight drifting of the diffuser. The extraction of the multimodal signals was subsequently performed with the UMPA algorithm using a Hamming analysis window of size 7×7 pixels (FWHM of the window ≈ 3.2 pixels). Each set of reference images was used for the reconstruction of the following block of 50 projections. Fourier phase integration was performed to obtain the phase shift signal from the two differential phase signals in the horizontal and vertical directions (Kottler et al. 2007).

The phase and transmission projections were then processed using the Savu tomography reconstruction pipeline at Diamond Light Source (Atwood et al. 2015; Wadeson and Basham 2016). Filtered back-projection (using the Astra Toolbox van Aarle et al. 2016, 2015; Palenstijn et al. 2011) was used to reconstruct the phase (distribution of refractive index decrement δ) and attenuation (distribution of attenuation coefficient μ) tomograms. The δ-values of the phase tomogram were then converted to electron densities using Eq. (2.31) in Chap. 2, Sect. 2.5.4.

9.2.4.2 Speckle Visibility and Size

First, the quality of the speckle pattern that was created by the MACE diffuser is evaluated. The first reference interference pattern (out of the 19 sets) at the first diffuser position is presented in Fig. 9.4a and in more detail in the enlarged 150×150 pixels region from the centre of panel (a) shown in Fig. 9.4b.

It reveals the fine speckles created by the MACE-fabricated diffuser that are less than 5 μm in size. The corresponding sample image at the same diffuser position can be found in Fig. 9.4d for the first projection of the tomography scan and its ROI is shown in Fig. 9.4c. Speckle displacement due to X-ray refraction, intensity reduction due to X-ray absorption and edge-enhancement fringes around the sample due to propagation effects can be observed.

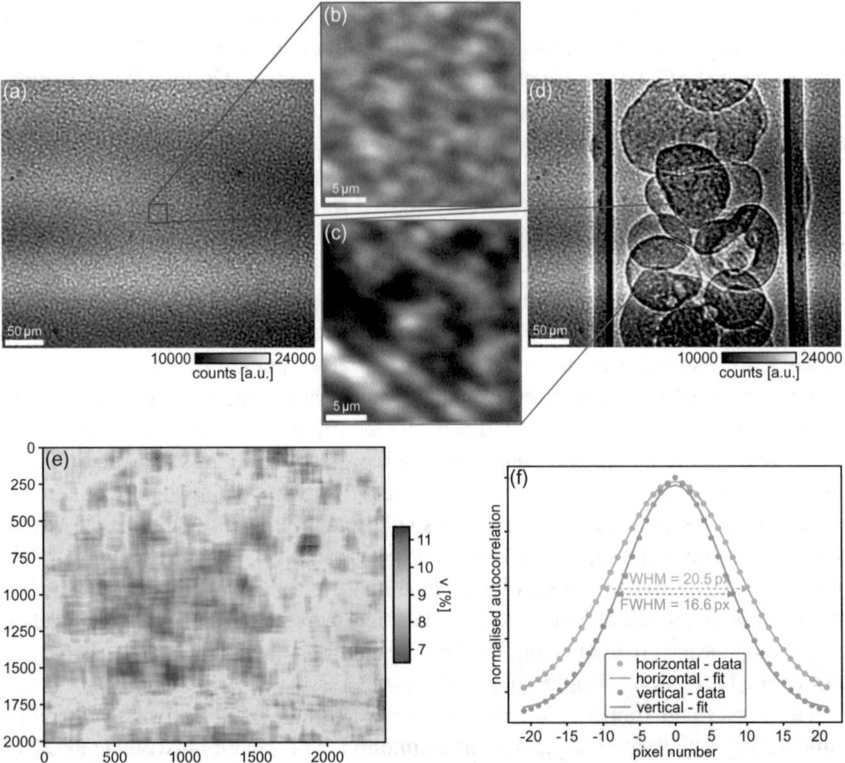

Fig. 9.4 Speckle pattern created by a MACE-fabricated diffuser for high-resolution X-ray speckle-based tomography using the UMPA mode at an effective pixel size of 0.16 μm. **a** Reference speckle pattern at the first diffuser position of the UMPA scan and **b** ROI in the centre of the field of view. **c** and **d** Corresponding ROI and full field of view of the speckle pattern with the sample (FCC catalyst particles in tube) in the beam. **e** Visibility map of the speckle pattern in panel (**a**). **f** Horizontal and vertical line plots through the 2D autocorrelation of the reference speckle pattern in panel (**a**) for estimation of the speckle size. The FWHM of the peak is 20.5 pixels ≈ 3.3 μm in the horizontal and 16.6 pixels ≈ 2.7 μm in the vertical direction

The visibility of the speckle pattern was evaluated using regions of 150 × 150 pixels around each pixel in the field of view and the resulting visibility map for the speckle pattern in Fig. 9.4a is shown in Fig. 9.4e. The mean visibility over the whole field of view is $v = (8.6 \pm 0.8)\%$.

The small size of the speckles was already observed visually in Fig. 9.4b and is assessed quantitatively by calculating the 2D autocorrelation of the reference speckle pattern. As described previously, the FWHM of the autocorrelation peak is taken as a measure for the speckle size. This gives a horizontal speckle size of $\Delta_x = 20.5$ pixels ≈ 3.3 μm and a vertical speckle size of $\Delta_y = 16.6$ pixels ≈ 2.7 μm for the reference pattern in Fig. 9.4a. The speckle size was also evaluated for the other 23 reference images of the UMPA scan and a consistent result of $\Delta_{x;\mathrm{avg}} = (20.4 \pm 0.1)$

pixels $\approx$ 3.3 μm in the horizontal and $\Delta_{y;avg}$ = (16.6 $\pm$ 0.1) pixels $\approx$ 2.7 μm in the vertical is obtained (standard deviations of 0.01 μm). Such small speckle sizes have not been reported to date for speckle-based imaging with any other type of diffusers. In the author's Masters thesis, small speckle sizes of down to 8 μm in the horizontal and 4 μm in the vertical direction were achieved with a similar setup (same energy and propagation distance) at beamline P05 at DESY, PETRAIII, Germany, using a sheet of biological filter membrane of 0.8 μm pore size (Zdora 2014). However, the visibility of the speckle pattern was only 1.2%. Here, a much higher visibility and smaller speckle size are achieved with customised MACE-fabricated diffusers, which opens up a whole new range of possibilities for high-resolution X-ray speckle-based imaging.

9.2.4.3 Tomography of a Catalyst Particle

The fine speckle pattern observed above was used for tomographic imaging of an FCC catalyst particle using a pixel size of 0.16 μm. Preliminary results are presented in Fig. 9.5. The reconstructed signals for the first projection of the tomography scan can be found in Fig. 9.5a–d. The differential phase projections in the horizontal and vertical directions in Fig. 9.5a and b, respectively, show good contrast and low background noise. The angular sensitivity of the differential phase signals was analysed using a small ROI in the air background area (150 $\times$ 150 pixels), indicated by the white box in the upper left corner of the images. It is $\sigma_x = 98.4$ nrad in the horizontal and $\sigma_y = 65.5$ nrad in the vertical. The differential phase signals were integrated to obtain the phase shift signal in Fig. 9.5c.

It can be observed that some of the edges of sample features appear noisy in the differential phase projections. This is due to strong edge-enhancement fringes that appear at the interfaces due to propagation effects (see Chap. 2, Sect. 2.3.3). These can also be observed clearly in the transmission signal in Fig. 9.5d. They are particularly prominent for this measurement due to the relatively low X-ray energy, the small pixel size and the long propagation distance.

As mentioned in Sect. 2.4.1.1, the critical distance to fulfil near-field conditions for the smallest resolvable features in the sample, assuming full coherence of the X-ray beam, is $z_c = (2p_{\text{eff}})^2/\lambda$, where p_{eff} is the effective pixel size and λ the X-ray wavelength. This would require a maximum propagation distance of $z_c = 1.65$ mm in the present case. Due to practical limitations of the setup and to increase the angular sensitivity of the measurement, a significantly larger propagation distance was chosen and hence the holographic regime is reached, for which multiple fringes are observed. In the future, these artefacts can be reduced by shortening the propagation distance, albeit leading to a reduced angular sensitivity, by immersing the sample in a refractive-index matching medium to reduce sharp changes of the latter, or by deliberately reducing the spatial coherence of the illumination, which, however, also affects the speckle visibility and size.

Although the propagation-induced fringes lead to artefacts in the projections that also transfer to the tomogram slices, see Fig. 9.5e and f, the catalyst particles are

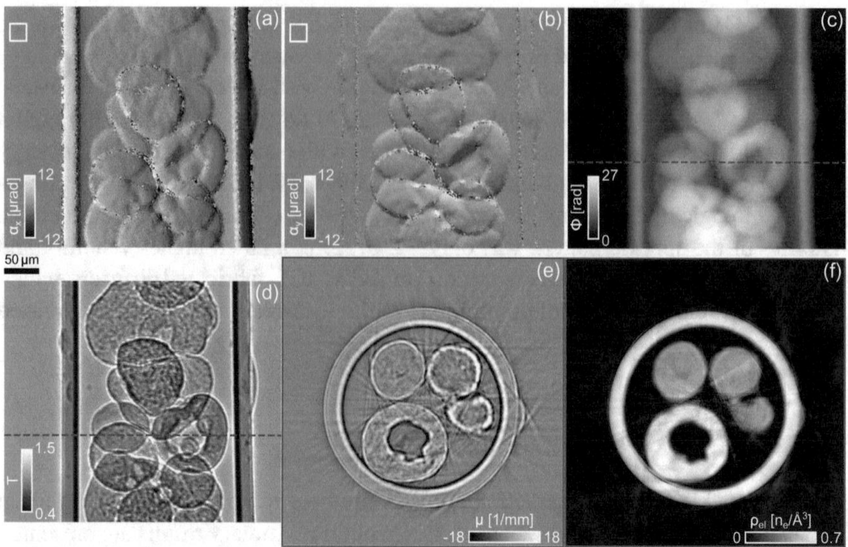

Fig. 9.5 Reconstructed projections and tomogram slices of an FCC catalyst particle sample obtained with UMPA speckle imaging using a MACE diffuser (preliminary results). Analysis was performed with $N = 24$ diffuser positions and a Hamming-type analysis window of size $w = 7 \times 7$ pixels (FWHM of window ≈ 3.2 pixels). **a** Refraction angle signal in the horizontal direction and **b** the vertical direction for one of the 901 projections of the tomography scan. The angular sensitivity (determined in the 150×150 pixels area of the white box) is approximately $\sigma_x \approx 98$ nrad in the horizontal and $\sigma_y \approx 66$ nrad in the vertical direction. **c** Phase-shift projection integrated from the differential phase signal in panels (**a**) and (**b**). **d** Transmission signal. **e** Transverse slice through the attenuation tomogram and **f** corresponding slice through the phase tomogram. Slice positions are indicated by dashed lines in panels (**c**) and (**d**)

still clearly visualised. Artefacts are less pronounced for the phase tomogram in Fig. 9.5f than for the attenuation tomogram in Fig. 9.5e, which is strongly dominated by edge-enhancement fringes leading to streaking.

The phase tomogram allows for the visualisation of the 3D electron density distribution in the specimen, which is studied in more detail in Fig. 9.6. The histogram of the slice in Fig. 9.6b shows three distinct peaks. The first peak is created by air, empty pores and artefacts in the slice. The other two peaks correspond to the catalyst particles: some with lower electron density (labelled with an orange asterisk) and one with higher electron density (labelled with a green dot). The capillary tube wall also contributes to the higher-density peak.

The values can be compared to other electron density measurements reported in the literature for the same kind of catalyst particles. The samples shown here are from the same batch as in the studies by (Ihli et al. 2018, 2017), in which quantitative electron density measurements were performed using X-ray ptycho-tomography. In the first publication, the authors report electron densities in the range of 0.40–0.75 electrons per Å^3 for a deactivated catalyst particle (from core to outer shell), while

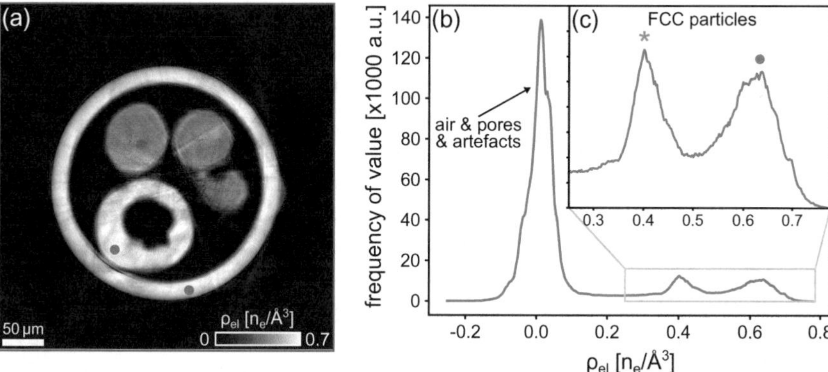

Fig. 9.6 Analysis of the electron density distribution in the sample. **a** Phase tomogram slice from Fig. 9.5f and **b** corresponding histogram showing the distribution of electron density values in the slice. The large peak at low electron density represents the air, empty pores and some of the artefacts in the slice. The other two peaks correspond to the catalyst particles. **c** ROI of the histogram showing the catalyst particle peaks in more detail. The left peak can be attributed to the lower-density FCC particles (orange asterisks) and the right peak to the higher-density FCC particle and the tube wall (green dots). The features are labelled correspondingly in panel (**a**)

they state an electron density of 0.42 electrons per Å^3 for pristine catalyst particles. These numbers compare very well with the peaks in Fig. 9.6c. It can be concluded that the higher-density particle (green dot) in Fig. 9.6a is a deactivated catalyst particle while the other ones (orange asterisks) are pristine particles. As one can mainly see the outer region of the first particle, the measured electron density corresponds to the high-density shell region.

9.2.5 Conclusions and Perspectives

The results in this section demonstrate that custom-made diffusers fabricated with MACE techno-logy can be used to control the properties of the speckle pattern and adapt them to setup parameters such as the effective pixel size. It was shown that MACE diffusers can produce suitable speckles of a few micrometres in size enabling high-resolution speckle-based imaging. Furthermore, the visibility of the speckles was higher than for previously reported speckle patterns of comparable speckle size and it was more homogeneous over the field of view. However, it should be noted that the visibility values were approximately four times higher when using P5000 sandpaper with the same setup. The speckle size, on the other hand, was about 20% larger for P5000 sandpaper[3] than for the MACE diffuser.

[3]Sandpaper of granularity P5000 has the finest commercially available grit size.

A MACE-type diffuser was successfully used for high-resolution X-ray speckle-based tomography of catalyst particles allowing for the reconstruction of quantitative electron density maps in a preliminary analysis. While speckle imaging does not reach the spatial resolution provided by X-ray ptychography, which was used to analyse the electron density of this type of specimen in previous reports (Ihli et al. 2017, 2018), it has the advantages of a much larger field of view, a simple setup, a much shorter scan time and lower requirements on the experimental setup and X-ray beam properties. Speckle-based imaging could in the future be used as a complementary method for studying catalyst particles that allows for classifying the deactivation grade of different catalysts in a distribution of particles of unknown deactivation state, amongst others. The detailed structure of the individual classified catalysts could then subsequently be investigated at higher spatial resolution using X-ray ptychography.

While the demonstration on the catalyst particle illustrated the promising potential of speckle-based imaging for the characterisation of materials at high spatial resolution on the order of hundreds of nanometres, it should be noted that in this regime propagation-based phase-contrast imaging is often used for high-contrast imaging. In particular at conditions of high transverse coherence, where Fresnel fringes are clearly visible, as given for the measurements in this section, the propagation-based method performs well. The obtained images typically reach a higher spatial resolution than results from speckle-based imaging and the propagation-based technique should hence be the choice if resolution is the main concern. However, quantitative information cannot be extracted from single-distance propagation-based data. More elaborate techniques requiring measurements at different sample-detector distances, such as holotomography (see Sect. 2.4.1.2, Chap. 2), are needed for this. Speckle-based imaging, on the other hand, does deliver quantitative information on the electron density distribution in the sample from a single data set and with a simple setup. It should hence be the preferred method for quantitative imaging applications. An extension of the UMPA reconstruction algorithm that includes the Laplacian of the phase to increase information and eliminate artefacts from Fresnel fringes, which are often encountered for high-resolution data under good transverse coherence conditions, will further improve the image quality for these applications, see also Chap. 10.

MACE-fabrication allows for customising the properties of the diffuser to optimise speckle size, visibility and homogeneity, enabling a high angular sensitivity that is uniform across the field of view. However, it should be considered that, similar to X-ray grating interferometry, the fabrication process will make these diffusers more costly than simple sandpaper and limits their maximum size (resulting in a limited field of view compared to sandpaper). Moreover, the need for a silicon wafer as a substrate leads to a higher absorption of low-energy X-rays, which may increase the exposure time and reduce the number of coherent X-rays, in particular at laboratory sources and for low-energy applications. A 200 µm-thick silicon wafer leads to 82.5% transmission of 20 keV X-rays, but its absorption increases rapidly with energy resulting in only 22.4% transmission at 10 keV. In comparison, a diffuser modelled as a 300 µm-thick layer of cellulose $(C_6H_{10}O_5)_n$ with a density of 1.5 g/cm^3 (Zanette et al. 2014) transmits 97.5% of photons at 20 keV and still 84.3% at 10 keV. These aspects should be considered alongside the benefits of customisation and optimised speckle properties.

The results in this section present the first applications of MACE diffusers and promising results have been achieved. In the future, more testing needs to be performed to further optimise the production process and find the best fabrication parameters for specific lateral diameter and height of the features. To date, pillar heights of 10–20 µm have been achieved for the production of speckle masks with MACE. This could be increased in the future once the important parameters and steps for the production process are identified. Moreover, different materials could be tested for fabricating diffusers designed for applications at different X-ray energies. Further reducing the size of the diffuser features will also be investigated, aiming at the use of yet smaller effective pixel sizes. It has recently been shown for the fabrication of gratings for X-ray grating interferometry that coating with iridium using atomic layer deposition can lead to sub-µm-sized features (Vila-Comamala et al. 2018). This approach could be directly transferred to the MACE production of speckle diffusers.

The custom production of speckle masks tailored to specific experimental conditions is a promising step towards optimised high-contrast, high-sensitivity speckle-based imaging under special conditions, such as high-energy X-ray beams, laboratory setups and small effective pixel sizes down to tens of nanometres.

9.3 High-Energy Speckle-Based Imaging for Geology and Materials Science Applications

9.3.1 Previous Work

The two main fields of applications of the X-ray speckle-based technique to date are X-ray optics characterisation and high-contrast imaging of biomedical specimens, both presented in this thesis. Recently, and in particular with the exploration of high-energy speckle imaging, the method is starting to find applications in new areas like materials science and geology.

First demonstrations of these were already briefly mentioned in Chap. 5 (Sect. 5.10.3) on high-energy speckle imaging and include 2D projection imaging of thin slices of volcanic rocks using the 1D XSS method (Wang et al. 2018; Zhou et al. 2018a) and 3D XST tomography of a mortar specimen (Zhou et al. 2018c).

Within the scope of this Ph.D. project, high-contrast 3D tomographies of different volcanic rocks and a mortar specimen were performed at an X-ray energy of 53 keV. The UMPA speckle-based imaging mode was used for this, which enabled the high-contrast 3D visualisation and quantitative analysis of the specimens with a robust and flexible data acquisition scheme and at a relatively short scan time.

During the preparation of this thesis, another report of high-energy X-ray speckle-based imaging of rock samples was published (Wang et al. 2019), mentioned also in

Chap. 5, Sect. 5.10. In this paper, the authors demonstrate the fast phase tomography of similar volcanic rock specimens as shown in this thesis, but using steel wool as a random absorption mask to create a wavefront marker based on absorption effects. Furthermore, the 1D XSS phase tomography of a battery sample at a higher energy of 120 keV is presented. In this study, 1D XSS was used in combination with fly-scan tomography to reduce the acquisition time compared to conventional 1D XSS scanning. It should be noted that fly-scan tomography has been explored already a few years ago for UMPA speckle-based imaging and has been employed for most tomography measurements presented in this thesis. This includes the scans presented in the following sections, for which scan times were similar to the ones reported by Wang et al. (2019).

The work shown here was a collaboration with Drs. Tunhe Zhou and Nghia Vo from Diamond Light Source (UK), who provided beamline access through proposal EE20763-1 and support, Drs. Beverley Coldwell and Matthew Pankhurst from Instituto Tecnològico y de Energìas Renovables (Spain), who provided the rock samples, and Dr. Fei Yang from Excillum (Sweden) who provided the mortar specimen.

9.3.2 Materials and Methods

9.3.2.1 Experimental Setup and Signal Reconstruction

Experiments were carried out at beamline I12-JEEP at Diamond Light Source, UK. Photons with an energy of 53 keV were selected from the X-ray spectrum of the superconducting multipole wiggler source using a silicon (111)-crystal Laue monochromator. The experimental setup for UMPA speckle tomography is shown in Fig. 9.7a.

The sample, here different kinds of volcanic rocks, see Fig. 9.7b, and a mortar specimen, was mounted on a tomography stage located 53 m downstream of the source. The diffuser was placed approximately 1 m upstream of the sample on a table with 2D translational movement for stepping. The detector was located at a distance of 2.2 m downstream of the sample stage for imaging of the rock specimens and 1.0 m downstream of the sample stage for imaging of the mortar specimen. The detector system consisted of a pco.edge 5.5 sCMOS camera (chip size: 2560×2160 pixels, pixel size: 6.5 µm) coupled to a scintillation screen and a visible-light custom-made optical module. Two different modules leading to different magnifications were used for the measurements in this section. The first one resulted in an effective pixel size in the sample plane of 7.59 µm and was used for the volcanic rock samples. The second one gave an effective pixel size in the sample of 3.22 µm and was employed for scanning of the mortar specimen. The effective pixel sizes in the sample plane were measured by translating a test sample by a known amount and detecting the displacement in pixels in the detector image. The effective pixel sizes in the detector plane can be calculated by considering the magnification by the diverging beam. They are 7.91 µm for the rock scans and 3.28 µm for the mortar scan. The diverging nature of the X-ray beam was taken into account here, but not for the previous

Fig. 9.7 UMPA speckle-based phase tomography of volcanic rocks and mortar at an X-ray energy of 53 keV. **a** Experimental setup at Diamond I12 beamline. The X-ray beam enters from the left and passes through the diffuser mounted on an optics table before hitting the sample on the tomography table. The middle two modules of the detector system (white box) were used for the experiment. **b** Examples of some of the investigated volcanic rock samples

measurements at Diamond I13 and ESRF ID19 beamlines, as Diamond I12 is a much shorter beamline and hence the approximation of a parallel beam does not hold well.

Different materials were used as speckle masks, depending on the pixel size: For the scan of the mortar specimen with a smaller pixel size it was a stack of 30 sheets of P400 sandpaper (average particle diameter: 35 µm, see (Federation of European Producers of Abrasives 2019) and for the volcanic rock scans it was about 25 mm of sand (thickness in beam direction) in a polystyrene box (2.5 mm wall thickness), as shown in Fig. 3.3 in Chap. 3.

The UMPA mode (see Chap. 6) for speckle-based imaging was used for data collection, using 7–25 diffuser steps (depending on the scan, see Table 9.1) with step sizes on the order of 30× the effective pixel size. 2401 or 1801 (depending on the scan, see Table 9.1) projections at equidistant viewing angles of the specimen were recorded over 180° of sample rotation in fly-scanning mode. This was repeated for each of the diffuser positions, so that the tomography rotation was the fast axis of the scan. Before each of the tomography acquisitions, 20 reference speckle images were taken without the sample and with just the diffuser at the respective position in the beam. For normalisation of the images, 20 dark images, without beam on the camera, and 50 flat images, with beam but without diffuser or sample, were acquired before the entire scan. The exposure time per frame was 4 ms for all measurements of the volcanic rocks and 10 ms for the scan of the mortar specimen with a smaller effective pixel size.

From the acquired data, the phase as well as the attenuation tomograms of the samples were reconstructed. For this, first the differential phase and transmission signals were retrieved for each projection of the tomography scan. All acquired sample

Table 9.1 Scan and reconstruction parameters used for the UMPA speckle-based tomography scans of volcanic rocks and a mortar specimen. p_{eff}: effective pixel size in the sample plane, z: sample-detector distance, t_{exp}: exposure time per frame, N: number of diffuser positions, w: analysis window size, n_p: number of projections of the tomography scan

Sample	Ignimbrite pumice (Fig. 9.8)	Basalt 1 (Fig. 9.9)	Basalt 2 (Fig. 9.10)	Picrite basaltic rock (Fig. 9.11)	Mortar (Fig. 9.12)
p_{eff} [μm]	7.59	7.59	7.59	7.59	3.22
z [m]	2.2	2.2	2.2	2.2	1.0
t_{exp} [ms]	4	4	4	4	10
Diffuser	Sand	Sand	Sand	Sand	Sandpaper
N	20	15	7	20	25
w [pixels]	5 × 5	5 × 5	5 × 5	5 × 5	5 × 5
n_p	1801	2401	2401	1801	2401

and reference speckle patterns were corrected for the detector dark current (average over all dark frames) and subsequently normalised by the dark-corrected flat field (average over all flat frames). The 20 reference speckle images for each diffuser position were then averaged to obtain a single reference per step. For each diffuser position, references were realigned to the corresponding sample images by performing a cross-correlation in a small area in the air background, in order to correct for slight drifts and instabilities of the speckle pattern. Multimodal image reconstruction following the UMPA formalism was then conducted for each projection using a Hamming window of size 5×5 pixels.

The projections obtained this way were subsequently processed using the Savu tomography reconstruction pipeline at Diamond Light Source (Atwood et al. 2015; Wadeson and Basham 2016). Using Savu, first ring filtering of the sinograms was performed applying the methods proposed by Vo et al. (2018) and Raven (1998). For the rock scans, both of these methods were applied, while for the mortar scan only the latter was used. The ring removal was followed by filtered back-projection using the Astra Toolbox (van Aarle et al. 2016, 2015; Palenstijn et al. 2011). This delivered the 3D attenuation and phase tomograms of the specimen showing the distribution of the attenuation coefficient μ and refractive index decrement δ, respectively. The phase tomogram values were, furthermore, converted to electron and mass densities using Eqs. (2.31) and (2.32) in Chap. 2, Sect. 2.5.4.

The various scan and reconstruction parameters are summarised in Table 9.1.

9.3.2.2 Samples

To demonstrate the potential of UMPA for geological research, different types of volcanic rock samples were investigated: ignimbrite from Tenerife, Spain, alkali basalts from La Palma island, Spain, and a picrite rock sample from Iceland.

The first sample, which is shown on the left in Fig. 9.7b, is an ignimbrite rock. It is the product of an explosive gas-driven eruption and subsequent pyroclastic flow, i.e. fast flow of hot gas and volcanic matter (Coldwell 2019). The bulk consists of silica-rich material from welded ash. Additionally, pieces of foreign rocks destroyed in the explosion are present.

The alkali basalt samples contain large olivine crystals (Coldwell 2019). Alkali basalt often occurs in the lava of the ocean basins at ocean islands.

The picrite rock sample was created as the result of rapid cooling of basaltic lava and accumulation of olivine crystals in part of the magma chamber or in a lava lake (Coldwell 2019). It is abundant in olivine crystals and high in magnesium.

The mortar sample was prepared as a mixture of water, cement and sand particles. It was cast with a 0.51 water-to-cement ratio (w/c) by mass. A mixture of 50% sand (average grain size: 0.3–0.4 mm, maximum grain size: 1 mm) by volume and 19% CEM I 52.5 N Portland cement by volume were used. The specimen was cured for 91 d and subsequently a cylinder of 5 mm diameter was cored out while flushing with de-ionised water. The sample was then cured for five days at 50 °C, wrapped with a 70-μm-thick polyimide film around its lateral surface and mounted on the sample holder. It was then left in air for several months before the UMPA tomography scan, which was performed in dry state.

9.3.3 Speckle Visibility, Speckle Size and Angular Sensitivity

A major challenge in X-ray speckle-based imaging at higher X-ray energies is to achieve a sufficient visibility of the speckle pattern. This is essential for a successful operation of the reconstruction algorithm and high angular sensitivity of the reconstructed differential phase projections.

A previous work on high-energy X-ray speckle-based imaging with a similar setup geometry at the same beamline reports a speckle visibility (determined as the ratio of standard deviation and mean of the speckle pattern) of 6% with a diffuser made from an aluminium-copper alloy (Wang et al. 2018). In the measurements presented in this thesis, fine sand was used for the scans of the volcanic rocks and a stack of 30 sheets of P400 sandpaper for the mortar sample scan. For the rock data presented here, which were acquired under almost the same conditions as the measurements by (Wang et al. 2018), speckle visibilities in the range of $(13.0-14.8)\%$ were achieved, calculated as the ratio of standard deviation and mean intensity in a ROI of 150×150 pixels in the centre of the field of view, see Eq. (5.1) in Chap. 5, Sect. 5.3.1. This is more than double the visibility obtained with the piece of aluminium-copper alloy by (Wang et al. 2018). It can, hence, be concluded that fine sand is a suitable material to produce a high-visibility speckle pattern at high X-ray energies and can, at the given conditions, supersede masks that are purely based on absorption effects.

For the mortar specimen, stacked sandpaper was used as a diffuser to achieve a speckle size suitable for the effective pixel size of the scan. The visibility of the speckle pattern in this case was 5.4% measured in a 150×150 pixels ROI and averaged over all reference images at the different diffuser positions. Although lower than for the sand diffuser, this visibility was sufficient to achieve a high angular sensitivity of the differential phase projections, see Table 9.2. The size of the speckles (average over all diffuser positions) was about 8.0 pixels, corresponding to 26.0 μm, in the horizontal and 7.9 pixels, corresponding to 25.5 μm, in the vertical. The standard deviation of the values determined for the different diffuser positions was 0.01 pixels for both directions. It was determined as the FWHM of the peak of the 2D autocorrelation of the reference speckle pattern (see Chap. 5, Sect. 5.3.1).

For the rock samples, sand was chosen as a diffuser with coarser grains, required to match the larger effective pixel size. For the different rock scans, the speckle size ranged between 9.4–9.6 pixels, corresponding to approximately (71.1−72.9) μm, in the horizontal and 9.1–9.7 pixels, corresponding to approximately (69.3−73.3) μm, in the vertical direction. The slight differences in speckle size for the various scans, despite using the same diffuser material, are caused by the differences in the illuminated part of the sand, which contains a range of grain sizes that are likely not homogeneously distributed.

As outlined in Sect. 5.4.4 of Chap. 5, the angular sensitivity depends on a number of factors, including the visibility of the speckle pattern and the propagation distance. For UMPA, the number of steps and the size of the analysis window also play important roles. The angular sensitivities of the scans presented in this section were determined in small ROIs in the air background of the differential phase projections, as described previously.

The best achievable sensitivity values as well as their average over all projections of the tomography scan are listed in Table 9.2. It can be seen that the best sensitivities were achieved for the mortar scan (see Fig. 9.12) despite the smaller pixel size and shorter propagation distance compared to the rock scans. For this measurement, the largest number of diffuser positions ($N = 25$) was used, which has a strong influence on the noise properties in the projections. The sensitivity of the second basalt sample (see Fig. 9.10) is the lowest as only $N = 7$ diffuser steps were used. The angular sensitivity achieved in the presented scans with UMPA is very high compared to other demonstrations of high-energy speckle imaging. For example, as mentioned above, (Wang et al. 2018) report angular sensitivities in the range of (210−450) nrad for a similar setup with an effective pixel size of 3.8 μm (comparable to the mortar scan here) at an average beam energy of 55 keV. In their latest publication, the same authors performed 1D XSS speckle tomography of volcanic rocks under almost identical conditions as reported here and the angular sensitivity achieved for $N = 20$ diffuser positions was 260 nrad (Wang et al. 2019).

This shows that the UMPA approach is capable of delivering high-sensitivity images at high X-ray energies with a relatively low number of diffuser positions, making it favourable over other speckle acquisition and processing modes.

Table 9.2 Angular sensitivities obtained for the UMPA phase tomography scans of the rock and mortar specimens

Sample	Ignimbrite pumice (Fig. 9.8)	Basalt 1 (Fig. 9.9)	Basalt 2 (Fig. 9.10)	Picrite basaltic rock (Fig. 9.11)	Mortar (Fig. 9.12)
$\sigma_{x,\min}$ [nrad]	164	147	242	204	125
$\sigma_{y,\min}$ [nrad]	156	135	212	197	126
$\sigma_{x,\mathrm{avg}}$ [nrad]	174	183	287	217	185
$\sigma_{y,\mathrm{avg}}$ [nrad]	165	169	251	208	187

9.3.4 Multimodal Tomography of Volcanic Rocks

9.3.4.1 Background and Motivation

In the last decade, X-ray computed tomography has gained popularity for studying volcanic rock samples. Up until recently, this was, however, mainly limited to conventional absorption-based CT (Zandomeneghi et al. 2010; Stevenson et al. 2010; Voltolini et al. 2011; Baker et al. 2012; Cnudde and Boone 2013; Madonna et al. 2013; Couves et al. 2016; Pankhurst et al. 2018). Other absorption-based methods such as dual-energy (also known as spectral or differential absorption) CT, for which images are acquired at different X-ray energies to exploit the energy-dependent attenuation behaviour of materials for improved visualisation, have also been reported (Gualda et al. 2010; Baker et al. 2012). Quantitative density information can be derived from absorption-CT data using additional information input (Pankhurst et al. 2018). Speckle-based phase tomography, however, gives direct access to the electron density distribution without requiring any *a priori* information on the sample. X-ray phase-contrast imaging is also advantageous to increase contrast between grain boundaries in the rock that have similar attenuation properties. Moreover, speckle-based imaging is a multimodal technique, delivering the phase-contrast, attenuation and dark-field signals from the same data set, which can help to obtain 3D compositional information on the specimen by combining the complementary image signals. This, for example, allows for drawing conclusions on the growth rates of crystals such as olivines.

Here, some preliminary results on multimodal imaging of rock samples is presented. Further in-depth analysis of e.g. the composition, porosity, pore morphology and connectivity or grain sizes is subject of ongoing and future work.

9.3.4.2 Phase Tomograms and Density Measurement

Slices through the phase tomograms of the different volcanic rock samples are shown in Figs. 9.8, 9.9, 9.10 and 9.11. It is immediately striking that the specimens, despite all being volcanic rocks, have a rather different inner structure.

Figure 9.8a–c shows transverse and longitudinal slices through the phase volume of the ignimbrite rock from Tenerife, reconstructed with $N = 20$ diffuser steps and a window size of 5×5 pixels.

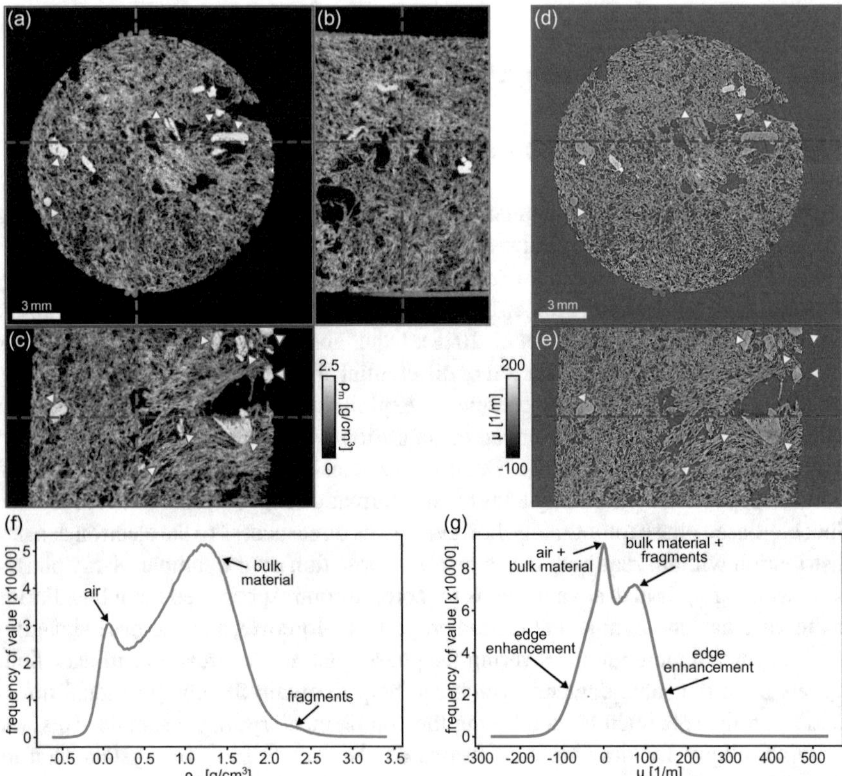

Fig. 9.8 Slices through the UMPA phase and attenuation volumes of a volcanic ignimbrite pumice sample (Tenerife). **a–c** Slices through the phase and **d**, **e** the attenuation volume. The complementary character of phase and attenuation signals allows for distinguishing the different types of high-density foreign rock fragments labelled with orange arrows and magenta arrow heads from each other and from the background bulk of the sample. **f** Histogram of the transverse phase slice in panel (**a**) showing the distribution of mass density ρ_m values and **g** histogram of the corresponding attenuation slice in panel (**d**) showing the distribution of the attenuation coefficient μ. The distribution of grey values is broader for the phase volume and the high-density rock fragments show as a small peak to the right. The lower-density parts of the rock are better separated from the air background for the phase slice

The magma bulk mass of the rock contains a large number of bubbles. The wavy structure of the rock that can be observed in Fig. 9.8c is due to the process of rock formation by deposition of extremely hot magma along the path of pyroclastic flow following the volcanic eruption. Furthermore, small pieces of higher-density material can be identified in the specimen. They originate from foreign rocks shattered in the explosion. The type of these inclusions is unknown. The information from the attenuation and phase-contrast volumes could help in the identification of their composition in a future in-depth analysis. As can be seen by comparing the phase slices in Fig. 9.8a and c with the corresponding attenuation slices in Fig. 9.8d and e, some of the pieces have a significantly higher density and attenuation coefficient than the bulk material (pointed out by orange arrows). Furthermore, there are high-density fragments that stand out clearly in the phase slices (pointed out by magenta-lined arrow heads), but seem to have a similar attenuation coefficient to the bulk background material. This underlines the complementary character of the attenuation and phase-contrast signals. Combining this information will be essential for characterising these fragments that are hard to distinguish from the other bright pieces (orange arrows) in the phase slices only, and at the same time hard to separate from the bulk materials based on the attenuation signal only.

Furthermore, in the phase signal one can observe even small density differences in the bulk material, which consists of volcanic ash and pumice, see e.g. the brighter areas in the centre and top left regions of Fig. 9.8a. This can also be seen from the histogram of the transverse slice (Fig. 9.8a) presented in Fig. 9.8f, which shows a much broader peak for the grey values in the bulk material than the corresponding attenuation histogram in Fig. 9.8g. Furthermore, the low-density parts of the rock are better separated from the air background in the phase histogram, while for the attenuation histogram parts of the rock contribute to the first (air) peak. Moreover, the high-density rock fragments can be identified in the phase histogram in Fig. 9.8f as a small peak, which is not present in the attenuation histogram in Fig. 9.8g. Negative values in the histograms stem from reconstruction artefacts and for the attenuation histogram in Fig. 9.8g also from edge-enhancement fringes.

Figures 9.9 and 9.10 show two different specimens of alkali basalts from La Palma island. They are of interest to study the large crystals of olivine and rock-forming minerals (pyroxenes) within them.

In the phase slices of the first specimen in Fig. 9.9a–e, these two main types of inclusions can be distinguished from the bulk by their higher density. It can, furthermore, be seen that the olivine crystals (black and white arrow heads in Fig. 9.9b, c) have a higher density than the pyroxenes (orange arrows in Fig. 9.9b, c). The nominal density of olivine is in the range of (3.27−4.39) g/cm^3 (iron-rich to magnesium-rich endmembers), while the typical density of the type of pyroxene expected to be present in the specimen is (3.2−3.56) g/cm^3 (Coldwell 2019).

The olivine crystals also show a high attenuation coefficient and are clearly visible in the corresponding attenuation slices in Fig. 9.9f and g. On the other hand, the pyroxenes, which were observed with good contrast in the phase slices, are almost invisible in the attenuation signal as their absorption properties are similar to the bulk

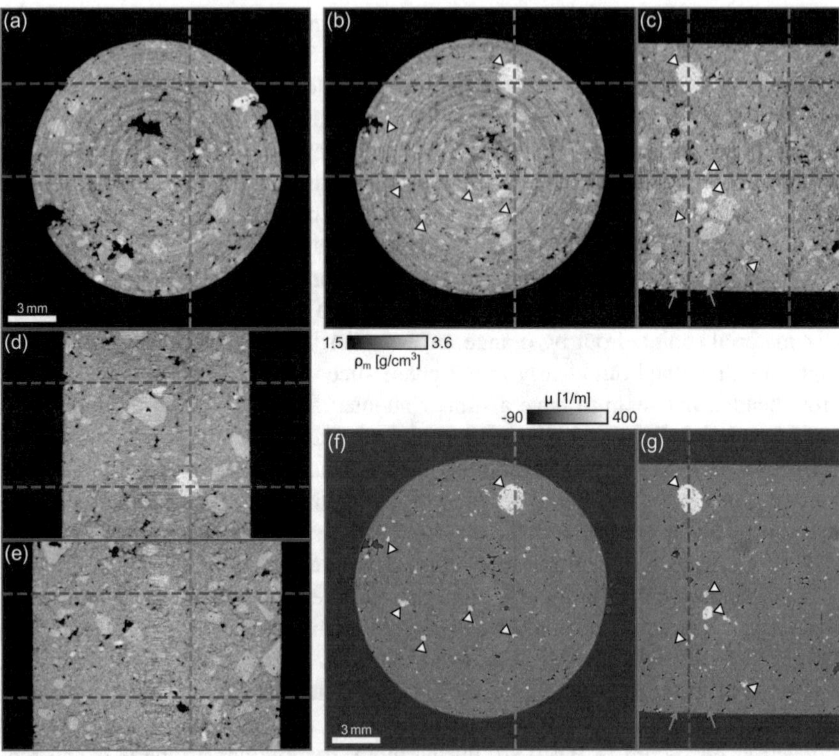

Fig. 9.9 Slices through the UMPA phase and attenuation volumes of a volcanic basalt sample (La Palma island). **a–e** Slices through the phase and **f**, **g** corresponding slices through the attenuation volume. Olivine (black and white arrow heads) and pyroxene (orange arrows) crystals can be distinguished by their different densities in the phase slices. The attenuation slices clearly show olivines, but the attenuation coefficient of the pyroxenes is similar to the surrounding matrix, making them almost indiscernible

background material. Upon close inspection, the outlines of these crystals and some internal cracks can be recognised thanks to the propagation-induced enhancement of edges.

The phase slices are slightly noisier here than for the previous scan as only $N = 15$ diffuser steps and the same window size of 5×5 pixels were used in the UMPA reconstruction. Furthermore, the sample was denser and hence photon statistics are lower. However, the image quality is still sufficient to identify the different types of crystals in the sample. Ring artefacts can be observed in the phase tomogram slices, which are not as obvious in the attenuation volume. This is due to the additional phase retrieval step for the former, which results in broadening of the rings. Ring removal was performed after the phase retrieval step and with the same parameters for the phase and the attenuation tomograms, leading to some remains of the wider rings in the phase slices. A larger filter kernel could have been applied, but this would have

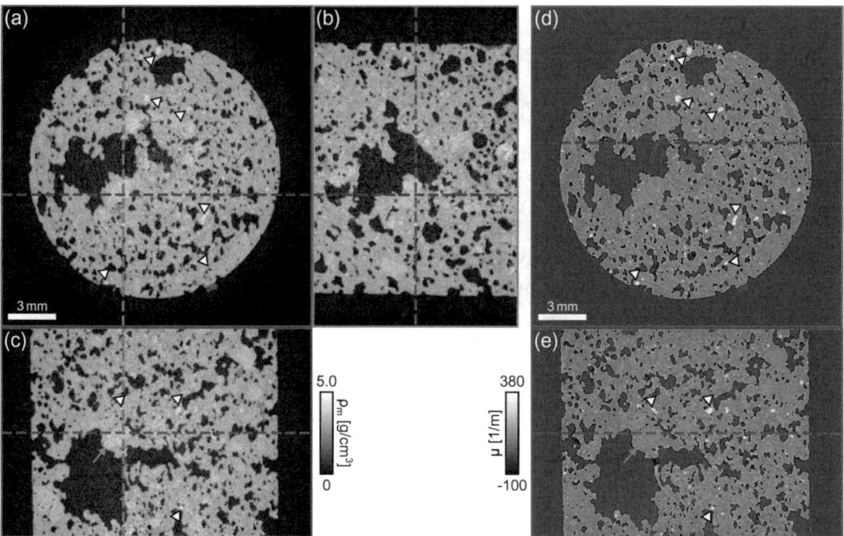

Fig. 9.10 Slices through the UMPA phase and attenuation volumes of another volcanic basalt sample (La Palma island). **a–c** Slices through the phase volume and **d**, **e** corresponding slices through the attenuation volume. Olivine (black and white arrow heads) and pyroxene (orange arrows) crystals can again be distinguished in the phase slices, while in the attenuation slices only olivines show up clearly against the bulk of the rock

also led to blurring of some sample features. Compared to the other rock samples, the tomogram of the alkali basalt in Fig. 9.9 exhibits more ringing, which is most likely due to intensity fluctuations occurring over the longer total scan time (scan was interrupted and continued after some time).

The second alkali basalt sample comes from the same geographical location, however the difference in overall appearance of the rock in Fig. 9.10 compared to Fig. 9.9 is remarkable. The tomogram slices in Fig. 9.10 reveal that the specimen contains a much larger number of pores. The olivine pieces (black and white arrow heads) are smaller than in the previous sample. As observed for the other specimen, olivines can be identified in both the attenuation and the phase maps, but the lower-density pyroxenes (orange arrows) are visualised more clearly in the phase slices. The pyroxenes appear darker in Fig. 9.10a–c compared to Fig. 9.9a–e due to the larger intensity window. The density of the olivines is higher, which leads to the conclusion that they might be richer in magnesium than the olivines in the sample in Fig. 9.9. The density of the surrounding matrix is also higher in Fig. 9.10 and it is closer to the density of the olivines. For both samples, the enhanced edges in the attenuation signal reveal some fine-structure of the crystals that is not visible in the phase slices.

The tomograms presented in Fig. 9.10 were reconstructed with UMPA using only $N = 7$ diffuser steps at a window size of $w = 5 \times 5$ pixels. Despite the small number of steps, the results allow for the discrimination of the olivine and pyroxene crystals. This demonstrates that UMPA can be used for high-contrast and quantitative

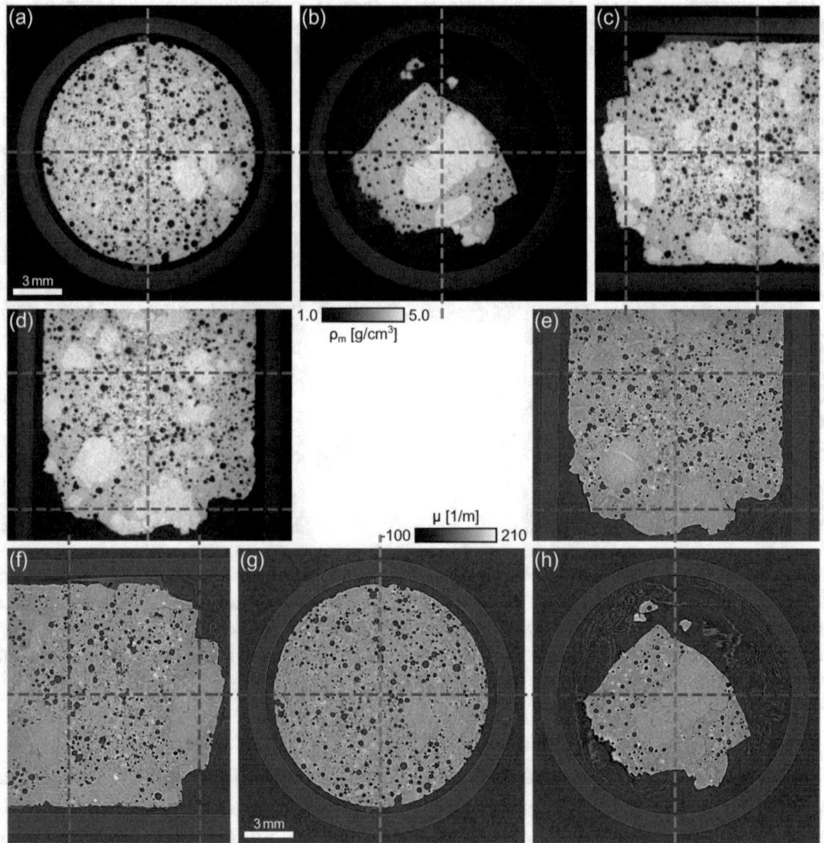

Fig. 9.11 Slices through the UMPA phase and attenuation volumes of a picrite basaltic rock sample (Iceland). **a–d** Slices through the phase and **e–h** the attenuation volume. Large olivine and clinopyroxene crystals can clearly be distinguished from the bulk in the phase slices. The two types of crystals show similar densities. Furthermore, their attenuation coefficients seem similar to the surrounding bulk and they are therefore hard to identify in the attenuation slices. The latter, however, reveal the fine cracks within the crystals

tomography at short imaging times even at high X-ray energies, making it a suitable candidate for high-throughput studies on the properties of geological samples.

Olivine crystals are also present in the high-magnesium picrite basalt sample in Fig. 9.11.

It shows a vesicular structure, which is caused by bubbles from dissolved gas trapped in the lava. The phase slices in Fig. 9.11a–d reveal a large number of crystals, some of which are olivines and some clinopryroxenes. As these types of crystals are of similar densities, they have similar grey values in the phase slices and are hard to distinguish. As can be seen in the attenuation tomogram slices in Fig. 9.11e–h, both types of crystals, furthermore, have similar attenuation coefficients to the bulk material, which consists of clinopyroxene, olivine and plagioclase feldspar. While the phase tomogram can achieve a better separation of the crystals from the bulk, the attenuation signal visualises fine cracks within the crystals. The combination of the two image modalities can hence give information on the location and shape of crystals in the rock, as well as their internal structure.

9.3.5 *Phase Tomography of a Mortar Sample*

9.3.5.1 Motivation and Background

Materials science is another promising area of application for high-energy X-ray speckle-based imaging. High X-ray energies are required for such studies due to the high density of most of the specimens. As an example, UMPA tomography of a piece of mortar is presented here. Mortar is of major importance for the building industry and there is ongoing research on understanding the mechanisms and parameters that determine its strength and other physical properties. High-contrast, high-resolution imaging can provide a pathway to this kind of information by visualising the micro-structure of the material. X-ray computed tomography has proven to be promising for this purpose as it is a non-destructive method capable of delivering undistorted 3D information at isotropic spatial resolution. X-ray absorption-based microCT has successfully been explored for the characterisation of mortar samples for over two decades (Bentz et al. 1994; Chotard et al. 2003; Carrara et al. 2018). More recently, it has been recognised in the materials science community that X-ray phase-contrast imaging can often deliver images with superior contrast than conventional absorption CT and multi-modal imaging, exploiting the absorption-, phase-contrast and dark-field signals, can provide complementary information as demonstrated with X-ray grating interferometry (Sarapata et al. 2015; Yang et al. 2016, 2014; Prade et al. 2016).

X-ray speckle-based imaging delivers multi-modal signals with a simple and robust setup and is therefore a promising alternative to grating interferometry for this kind of study. Here, the 3D density and attenuation maps reconstructed from UMPA speckle scans are used for a preliminary analysis of the composition of a mortar sample.

9.3.5.2 Phase and Attenuation Tomograms

Transverse and longitudinal slices through the phase and attenuation volumes are shown in Fig. 9.12a–e and f–i, respectively. The different components of the mortar specimen can clearly be distinguished in the phase slices, which visualise its internal density distribution. The attenuation slices show fewer variations between the components, but their edges are enhanced aiding the localisation of features.

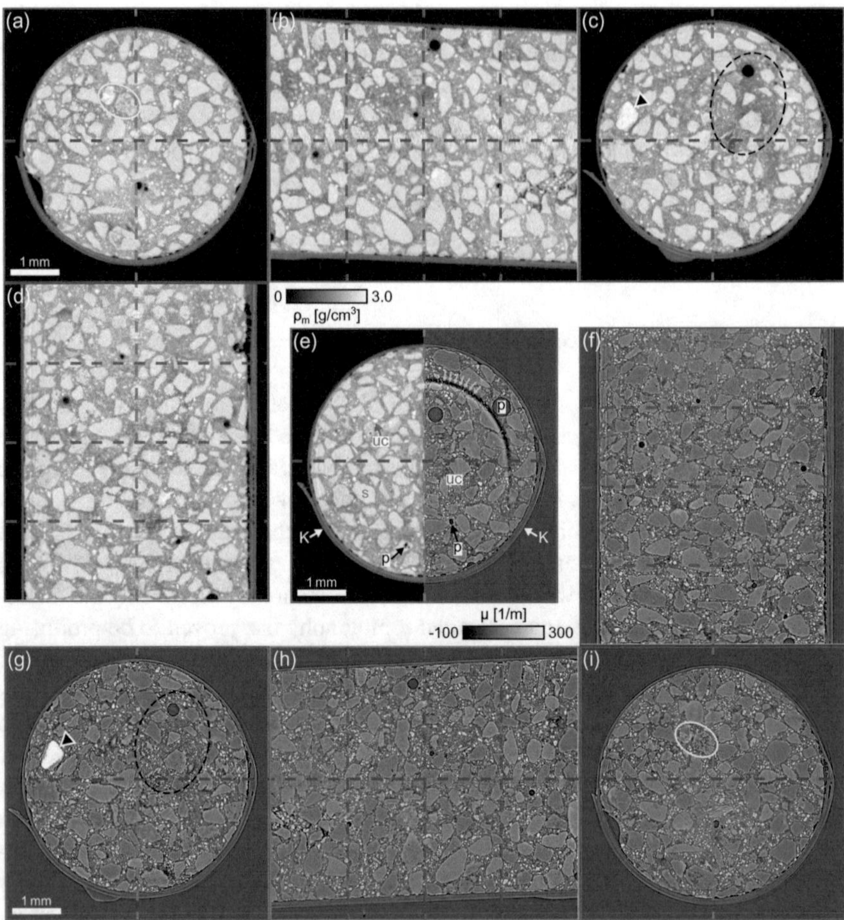

Fig. 9.12 Slices through the UMPA phase and attenuation volumes of a mortar sample. **a**–**d** Slices through the phase and **f**–**i** the attenuation volume of the sample. **e** Central transverse slice of the phase volume (left) and attenuation volume (right) in direct comparison. The components of the mortar specimen can clearly be separated from the background paste: p: pores, s: sand, uc: unhydrated cement, K: Kapton foil. Artefacts are visible in the attenuation slices in panels (**e**) (turquoise arrows) and (**i**) (light red arrows). Areas of lower density due to unresolved pores can be seen clearly in panel (**c**) (black circle). A bright unhydrated cement particle shows up in the phase and attenuation signals (arrow head in panels **c** and **g**). The detailed internal structure of some sand particles is observed in the slice in panels (**a**) and (**i**) (yellow circle)

The main components of the sample are labelled in Fig. 9.12e, which shows a transverse slice, half phase slice on the left and half attenuation slice on the right. In the phase signal, one can clearly differentiate between the large pieces of high-density sand (s, orange label) and the bulk background cement paste with lower density. This separation is not clear in the attenuation slices, which indicates that sand and cement paste have similar absorption properties. However, edge enhancement highlights the outlines of the sand crystals allowing for their distinction from the bulk. Within the cement paste, small high-density particles show up bright in the phase as well as the attenuation signal. These are unhydrated cement pieces (uc, magenta label), which have a high density and attenuation coefficient. They are denser than the hydrated cement in the paste because of the absence of water and pores. Larger air-filled pores (p, black label), which vary in size, are clearly visible as dark areas in both modalities. The Kapton (polyimide) foil that was wrapped around the specimen can also be seen in the slices (K, white arrow).

Some artefacts such as rings (see small turquoise arrows in Fig. 9.12e) and streaking (see light red arrows in Fig. 9.12i) arise in the attenuation signal, but are not visible in the phase slices. The rings are caused by background features such as dirt on the monochromator or similar in the flat-field image, giving rise to ring artefacts. In the phase modality, this effect is less prominent as the signal extraction step is less sensitive to background features. The streaking artefact, on the other hand, is a result of the highly elongated shape of the sand grain containing strongly absorbing particles.

When looking at the areas containing mainly cement paste, some regions of lower density can be identified (see black dashed circle in Fig. 9.12c). They are due to unresolved clusters of pores and hydrated cement that reduce the average density in this region. Brighter areas can be attributed to pieces of high-density unhydrated cement that cannot be resolved by the imaging system but increase the average density in the area. Upon close inspection the regions of reduced density can also be observed in the attenuation slices (see black dashed circle in Fig. 9.12g), however the difference in contrast to the rest of the paste is much less pronounced than in the phase slices.

Also visible in the slice in Fig. 9.12c and g is a bright particle of very high density as well as attenuation coefficient (see black and white arrow head). This could be a large piece of unhydrated cement. Here, it can be distinguished more clearly from the sand particles in the attenuation signal due to the large difference in absorption properties.

As a last notable feature, attention is drawn to the fine structure inside the two sand particles indicated by yellow circles in Fig. 9.12a and i. While one of them contains some lower-density spots, the other one shows higher-density inclusions. Although this is visualised in both imaging modalities, the detailed structure of these inclusions is better visible in the attenuation signal, which renders the edges of the small features visible.

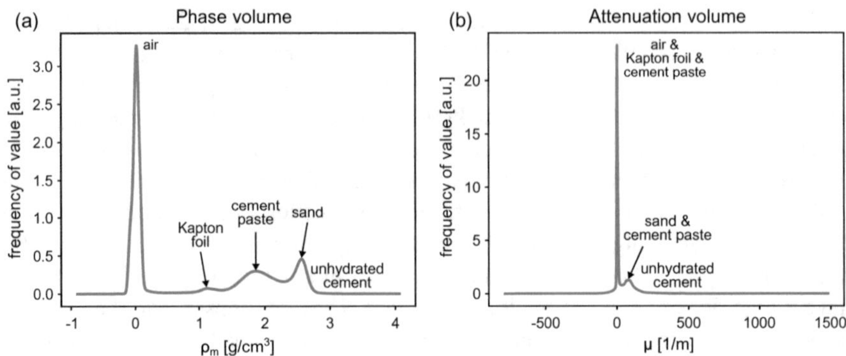

Fig. 9.13 Histograms of the entire UMPA **a** phase and **b** attenuation volumes of the mortar specimen (every fifth slice of the tomogram included). The phase histogram shows several distinct peaks and a wider distribution of values than the attenuation histogram

9.3.5.3 Quantitative Density Measurement

The high contrast in the phase tomogram slices in Fig. 9.12a–e indicates a good density resolution of the data set. The density resolution was here determined in the same manner as in Sect. 8.4.4.3 of Chap. 8 by calculating the standard deviation in a $50 \times 50 \times 50$ pixels sub-volume of the phase tomogram in a homogeneous background region (here air). It was evaluated for four different volumes of interest in the 50 central slices of the tomogram.[4] The values range from $31.2\,\text{mg/cm}^3$ to $33.6\,\text{mg/cm}^3$, with an average of $(32.3 \pm 0.9)\,\text{mg/cm}^3$. This is a similar value to the density variations reported for X-ray grating interferometry at the same energy of 53 keV performed at ESRF ID19 beamline (Gradl et al. 2016).[5] However, the effective pixel size in the experiment presented here is less than half of the pixel size in this publication.

The good density resolution allows for a clear separation of different components in the mortar specimen, which can be illustrated by looking at the histograms of the phase tomogram. It is shown in Fig. 9.13a for the whole volume (including every fifth slice in the analysis). The histogram of the attenuation volume is presented in Fig. 9.13b for comparison. Negative values in the histograms are caused by artefacts in the tomographic reconstruction and noise in the background regions, and for the attenuation volume also by edge-enhancement fringes. It can be seen that the distribution of values is much broader in the phase than in the attenuation volume and there are distinct peaks that correspond to the different components of the sample. This is explored in more detail in Fig. 9.14.

[4]In a top-down view, see transverse slice in Fig. 9.12e, these are areas in the left middle, right middle, top centre, and bottom centre regions of the slice.

[5]The measurement of the standard deviation was performed in a $60 \times 60 \times 10$ pixels region in this publication.

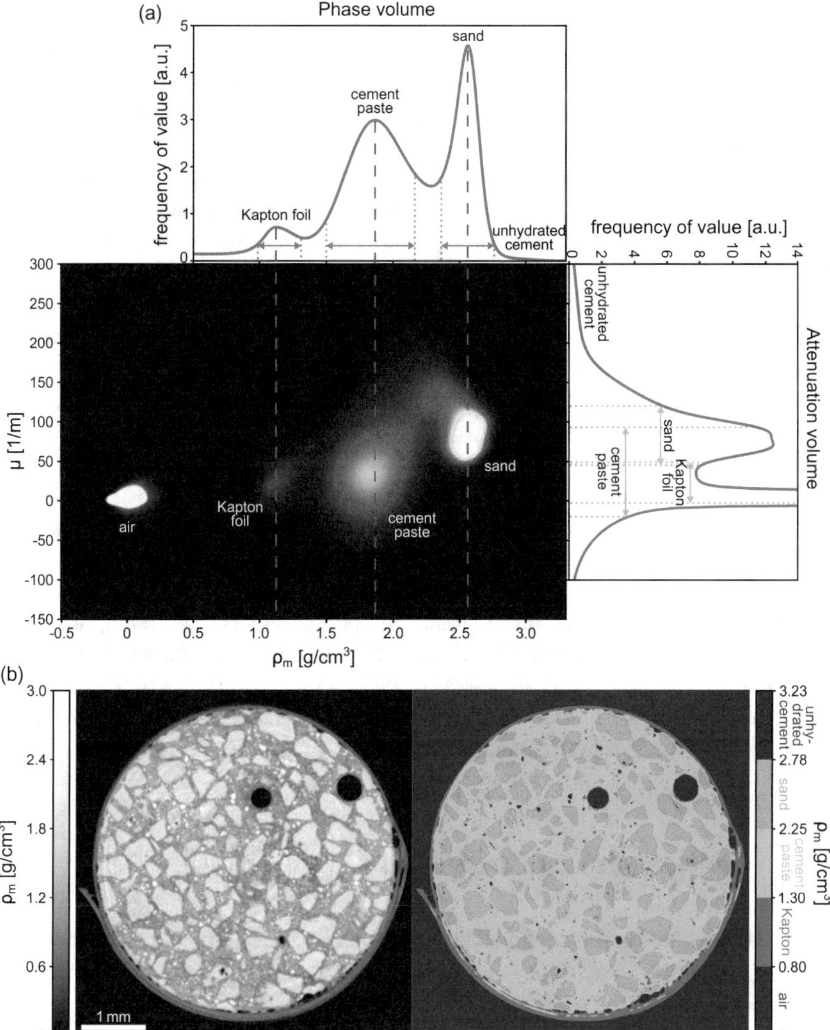

Fig. 9.14 2D histogram including the UMPA phase and attenuation tomogram values and threshold-based segmentation. **a** 2D histogram combining the phase and attenuation information. The corresponding 1D histograms are shown for value ranges of interest. The peaks for Kapton foil (and low-density paste regions), cement paste and sand are clearly separated in the phase volume and show as distinct spots in the 2D histogram. Dashed lines indicate the peak positions of the mortar components. Arrows in the 1D histograms show the width of the peaks in the 2D histogram. The peaks overlap in the 1D attenuation histogram, while they are well separated in the phase histogram. **b** Threshold-based segmentation of the different components in the specimen (right) based on their mass-density values in the phase slice (left)

Smaller regions of interest of the histograms in Fig. 9.13 are shown in in Fig. 9.14a and attenuation and phase information are combined together in a 2D histogram. The 2D histogram enables the clear distinction of the different components in the mortar specimen, which show as well separated patches. The reliable identification of components from the phase volume, as evidenced by the separation of peaks in the 1D phase histogram, indicates that this signal modality allows for threshold-based automated segmentation of the volume. This is demonstrated in Fig. 9.14b on the central slice of the phase tomogram. By simply setting different density thresholds, the various constituents of the sample can be separated. It can, furthermore, be noted that the measured density values in Fig. 9.14 agree well with the nominal mass densities. For example, the peak for the sand grains is found at $\rho_{m;\,\mathrm{s}} = 2.57\,\mathrm{g/cm^3}$ and the unhydrated (and partly hydrated) cement particles correspond to density values between approximately $2.9\,\mathrm{g/cm^3}$ and $4\,\mathrm{g/cm^3}$. The nominal values are $\rho_{m;\,\mathrm{s;\,theo}} = 2.65\,\mathrm{g/cm^3}$ for the sand grains and $\rho_{m;\,\mathrm{uc;\,theo}} = 3.13\,\mathrm{g/cm^3}$ for the unhydrated cement particles (Yang 2017).

Upon close inspection it can be seen that surrounding the unhydrated cement particles (red) some lower-density rings (orange) appear, which can be explained by the fact that the particles are not completely unhydrated on the outside where they are in contact with the paste and hence have slightly lower density values. The higher-density rings (lighter blue) around the pores, on the other hand, are due to strong edge enhancement at the interface of air and cement paste, which can cause artefacts in the phase-contrast signal, as discussed in previous chapters. Some areas within the paste appear light blue. They have a lower density due to unresolved pores and more water uptake in these regions, reducing the average density value.

9.3.6 Conclusions and Perspectives

It was demonstrated in this section that X-ray speckle-based imaging can be extended to the high-energy X-ray regime without major efforts, which opens up a range of new applications for the technique.

Here, it was employed for geology and materials science studies, investigating different types of volcanic rocks and a mortar sample. For both applications, the UMPA technique allowed for the visualisation of the 3D detailed inner structure of the samples and the identification of their different components based on density differences extracted from the phase volume. The complementary information from the attenuation volume reconstructed from the same data set can aid in the classification of the components. It, furthermore, highlights fine structural details, such as cracks and inclusions, due to the edge-enhancement fringes arising upon propagation. In the future, phase retrieval could be performed on the attenuation projections by using approaches commonly applied to propagation-based phase-contrast imaging data, e.g. the single-distance algorithm proposed by Paganin et al. (2002). This is particularly interesting for high-energy measurements, for which the absorption of X-rays is low and the transmission images are mainly dominated by edge enhancement effects. Phase retrieval from the transmission signal could provide additional qualitative information and could, furthermore, be exploited for an iterative refinement of the UMPA phase reconstruction.

The high sensitivity to small density differences achieved by UMPA speckle imaging even at high energies (here density resolution of better than 32 mg/cm^3 at 3.22 μm effective pixel size at 53 keV) made it possible to perform a threshold-based segmentation of the compound mortar specimen. A broader and in-depth analysis of the results presented in this section including a full 3D segmentation and the extraction of various properties like porosity, pore sizes and the relative fraction and distribution of different components, are subject of ongoing work.

With future optimisations of the setup, such as a higher speckle visibility, and further developments in the analysis code, it is expected that high-energy speckle-based UMPA will yield even better density resolutions, which will enable detailed quantitative studies on the 3D density distribution in geological and material science specimens.

After the first demonstrations in the high-energy X-ray regime presented in this section, the UMPA technique is also anticipated to find applications in other areas of research that benefit from a high density resolution and require high X-ray penetration power such as archaeology, palaeontology and industrial applications.

9.4 Speckle-Based Imaging with the Unified Modulated Pattern Analysis at a Laboratory Source

X-ray speckle-based imaging does not impose very strict requirements on the coherence properties of the X-ray source and is therefore a promising candidate for laboratory-based implementations. In the last few years, increased efforts have been invested into translating the technique from the synchrotron to the laboratory, as outlined in Chap. 5, Sect. 5.6. This is an important step towards opening up the method to a wider user community and to routine and high-throughput applications.

9.4.1 Previous Work

The first demonstration of X-ray speckle-based imaging at a laboratory source dates back to 2014 when single-shot X-ray speckle tracking (XST) was implemented at a liquid-metal-jet source (Excillum, Sweden) (Zanette et al. 2014). A year later, XST tomography (Zanette et al. 2015) as well as projection imaging with the 2D speckle scanning (2D XSS) method (Zhou et al. 2015) were successfully performed at the same source.

The liquid-metal-jet X-ray source (Hemberg et al. 2003) is well suited for the adaptation of speckle-based phase-contrast imaging thanks to its high brilliance compared to other laboratory sources. A small spot size of a few micrometres diameter can be achieved at a higher power than for conventional sources, leading to a higher degree of spatial coherence and shorter exposure times. However, scan times are many orders of magnitude longer than for imaging at synchrotron sources and the implementation of the multi-frame approaches for speckle-based imaging, such as 1D and 2D XSS and UMPA, therefore remains a technical challenge, in particular for tomography.

In a few demonstrations, speckle imaging has also been implemented with conventional (solid target) microfocus X-ray sources, which typically allow for significantly less power at small source sizes and cannot reach the very small spot diameters available at the liquid-metal-jet source. Projection imaging was performed in these cases using the 1D XSS mode (Wang et al. 2016a, b) and[6] the most recently introduced optical flow method (Labriet et al. 2019). Furthermore, tomographic imaging has been reported with the single-shot XST mode at a microfocus source (Zhou et al. 2018c). In these demonstrations, absorption masks of large-grain sandpaper or steel wool were used and speckles are generated mainly by absorption effects.[7] This greatly relaxes the requirements on the coherence properties of the source, but, on the other hand, also leads to a loss in flux.

To date, speckle imaging with the UMPA approach has not yet been implemented with laboratory sources. UMPA offers many advantages over other methods for speckle-based imaging, as discussed in detail in Chap. 6. In particular its tunable character, robustness to instabilities in the diffuser position and low number of required diffuser steps compared to other scanning-based approaches for speckle imaging, make UMPA an ideal candidate for the implementation at laboratory X-ray sources.

In the following, a proof-of-principle demonstration of UMPA speckle-based imaging at a liquid-metal-jet X-ray laboratory source is presented, which was implemented at the Nanoimaging Group at the University of Southampton (UK). The experiments were conceived and planned by the author of this thesis together with her supervisors, Dr. Irene Zanette and Prof. Pierre Thibault and performed with the support of Toby Walker, Department of Physics and Astronomy, University of Southampton (UK), Dr. Nicholas W. Phillips, Department of Engineering Science, University of Oxford (UK) and Ronan Smith, Department of Physics and Astronomy, University of Southampton (UK). Some of the results presented in the following were later (after submission of the thesis) prepared in a journal article published as (Zdora et al. 2020): M.-C. Zdora, I. Zanette, T. Walker, N. W. Phillips, R. Smith, H. Deyhle, S. Ahmed, and P. Thibault, "X-ray phase imaging with the unified modulated pattern analysis of near-field speckles at a laboratory source," Appl. Opt. **59**(8), 2270–2275 (2020), licensed under the OSA Open Access Publishing Agreement.

Several theoretical and experimental aspects for the implementation of X-ray speckle-based imaging at the liquid-metal-jet source have been studied previously

[6] Spot sizes of $16 \times 16\,\mu m^2$ and $20 \times 20\,\mu m^2$ were used by Wang et al. (2016a) and Wang et al. (2016b), respectively.

[7] In the study by Labriet et al. (2019) it is not mentioned what kind of speckle mask was used.

by Walker (2019). The further optimisation of the setup, in particular for reducing the scan time, improving the speckle visibility and dealing with the limited spatial and temporal coherence conditions, is currently underway.

9.4.2 Materials and Methods

9.4.2.1 Experimental Setup and Data Acquisition

The measurements were performed at a liquid-metal-jet X-ray source by Excillum (Sweden), model MetalJet D2 with ExAlloy I1 anode material, which is comprised of 68% (by weight) gallium, 22% (by weight) indium and 10% (by weight) tin (Excillum 2019).

The source was operated at an accelerating voltage of 50 kV and with a current of 0.200 mA giving a power of 10 W. The electron spot size varied during the scan between 5.6 µm × 5.6 µm and 5.9 µm × 5.9 µm.

The source spectrum and detected spectrum at the given setup conditions have not been measured yet. For some preliminary estimation, it is assumed here that the average energy of the detected spectrum is similar to the one reported by (Zanette et al. 2014), where a liquid-metal-jet source of similar anode alloy was operated at the same accelerating voltage. However, it should be noted that the anode alloy used by (Zanette et al. 2014) had a higher relative fraction of gallium and the experimental setup was not exactly the same (in particular a longer propagation distance was used).

The experimental arrangement for the measurements is shown in Fig. 9.15. The X-ray beam coming from the source passed through an aperture and illuminated the diffuser mounted on a linear translation stage at a distance of approximately 56 cm from the source. The diffuser was a sheet of P800 grade sandpaper with a nominal mean grain size of (21.8 ± 1) µm (Federation of European Producers of Abrasives 2019). The sample was placed 14 cm downstream of the diffuser (source-sample distance: ~70 cm). The detector system was comprised of a scintillation screen, here a 600 µm-thick cerium-doped gadolinium aluminium gallium garnet (GAGG:Ce) crystal, for conversion of X-rays to visible light, coupled to an optical microscope and an sCMOS camera. The latter was a pco.edge 4.2 camera with a chip of 2048 × 2048 pixels and a pixel size of 6.5 µm. The microscope system was produced by Optique Peter (Lentilly, France) and contained a 2× magnifying objective, leading to an effective detector pixel size of $p_{\mathrm{eff;det}} = 3.25$ µm. The source-scintillator distance was approximately 81 cm, giving a sample-scintillator distance of 11 cm. This configuration led to a geometric magnification of the sample of approximately 1.16 and an effective pixel size in the sample plane of approximately $p_{\mathrm{eff;sam}} = 2.8$ µm. Furthermore, features in the diffuser plane are magnified by a factor 1.45 in the detector plane and by a factor of 1.25 in the sample plane.

Two measurements were performed during the experiment: an initial test scan of a demonstration sample and a longer scan of a biological specimen, a small bug.

Scanning was performed following the UMPA scheme, see Chap. 6, with step sizes significantly larger than the size of the analysis window used for UMPA processing.

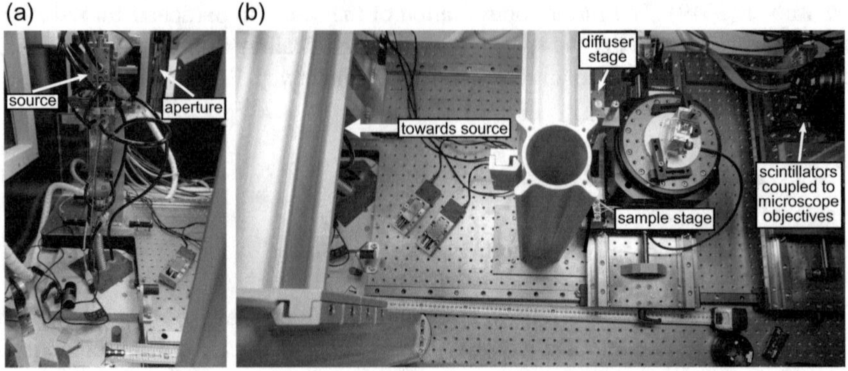

Fig. 9.15 Experimental setup for UMPA speckle-based imaging with a liquid-metal-jet laboratory source. **a** Side view of the source and aperture used for cutting air scatter. **b** Top-down view of the setup showing the arrangement of the diffuser, sample and detector. The distances from the source are 56 cm for the diffuser, 70 cm for the sample and 81 cm for the detector system (scintillator position)

Although regular unidirectional stepping is not required for UMPA, the diffuser was stepped vertically with a fixed step size of 75.95 μm for the presented measurements due to temporary experimental constraints. For the first measurement of the test sample, the diffuser was moved to $N = 15$ different positions. For the second scan of the bug specimen, the data was acquired at $N = 80$ diffuser positions. The large number of positions was chosen here to ensure optimum image quality for this first proof-of-principle demonstration. It will be shown below that images of good quality can be achieved with the current setup by using data from only $N = 20$ diffuser positions. The exposure time per frame for each of the reference and sample images was 750 s for the test measurement and 1100 s for the bug scan. For the latter, sample and diffuser were moved out of the beam and three flat-field images were acquired every three hours of the scan (corresponding to every seven diffuser positions). For the test measurement of the sphere, one flat-field image was recorded prior to each set of reference and sample images. Dark images without X-ray illumination were recorded after both scans, 44 dark images for the sphere data and 30 dark images for the bug.

9.4.2.2 Signal Reconstruction

In a pre-processing step, the reference and sample images were dark-current and flat-field corrected. For this, the mean of the dark images and of the flat-field images[8] was determined. The mean dark signal was subtracted from all frames and subsequently

[8]For the sphere scan only one set of flat-field images was acquired and hence no averaging was performed.

the reference and sample images were divided by the dark-corrected flat-field images. For the test scan, a new set of flat-field images was acquired at each diffuser position. For the scan of the bug, each flat-field image was used to correct the previous four and the following three reference and sample images.

In a second pre-processing step, dead and bright pixels, originating from detector defects and direct hits of the sensor by scattered X-rays, respectively, were removed from the images using a mask. They were replaced by the median value in a 5×5 pixels area around the affected pixels. Furthermore, slight drifts of the diffuser between sample and reference frames were eliminated by aligning the reference to the corresponding sample image at each diffuser position by means of cross-correlation in a homogeneous background region (air).

The reconstruction of the multimodal image signals, i.e. differential phase, transmission and dark-field images, was performed using the UMPA analysis code. Phase integration from the two differential phase signals was carried out using a Fourier integration routine (Kottler et al. 2007). A fourth-order polynomial was fitted to the air background region of the integrated phase signal and subtracted to eliminate low-frequency artefacts that can arise during the integration step in particular in the presence of noise.[9]

9.4.2.3 Samples

The first measurement was performed on a test sample made from a silicon sphere (diameter: 480 µm) glued onto a wooden toothpick, which is the same specimen as presented in Sect. 9.2.3.3 above and in Sect. 6.4.2 of Chap. 6. After this first test, a biological sample was scanned. It was a true bug, a specimen of the species *Orius Niger* of the family of Anthocoridae known as minute pirate bugs or flower bugs. It was approximately 3.8 mm long and 1.8 mm wide. Bugs from the *Orius* genus are often used for biological pest control and are hence of interest for the agriculture sector (de Veire and Degheele 1992; Deligeorgidis 2002).

9.4.3 Scan of a Test Sample

First, a scan of a known demonstration sample was performed to assess the potential of the setup. The exposure time per frame and number of steps of the UMPA scan were kept low to reduce the scan time for this first evaluation. Hence, a limited image quality was to be expected.

---

[9]Subtracting a first-order polynomial from the integrated phase signal as done for the data presented in the previous chapters of this thesis is mathematically justified as this corresponds to an offset in the differential phase that might stem from remaining effects of diffuser instabilities. However, in the presence of increased noise, as encountered at laboratory setups, further low-frequency artefacts can arise in the integrated phase signal which are difficult to control. They were eliminated here empirically by subtracting a fourth-order polynomial.

The visibility of the reference interference pattern for this initial test was $(12.5 \pm 0.1)\%$ over the field of view used for signal reconstruction (average visibility value over all 15 diffuser positions). The speckle size was on average (2.74 ± 0.06) pixels in the horizontal and (2.78 ± 0.08) pixels in the vertical direction, corresponding to approximately (8.3 ± 0.2) µm and (9.1 ± 0.3) µm, respectively, in the detector plane (more specifically the scintillator plane). The speckle visibility and size are discussed in more detail in the next section.

The multimodal image signals of the sample, reconstructed using a window size of $w = 9 \times 9$ pixels (Hamming-type window), are shown in Fig. 9.16. The angular sensitivity of the differential phase measurement was approximately $\sigma_{x,y} = (0.59 \pm 0.03)$ µrad in both the horizontal and the vertical direction, averaged over 14 regions of 150×150 pixels each in the air background area.

As expected, the differential phase signals in Fig. 9.16a and b show significantly more noise than for the synchrotron measurement of the same specimen in Sect. 6.4.2 of Chap. 6 due to the much lower photon statistics at laboratory sources. However, the sample is still clearly visible. The transmission signal in Fig. 9.16d is less affected by image noise as it is based on a change in mean local intensity only. Edge-enhancement fringes can be observed around the sample, see ROI in Fig. 9.16h, as described in previous chapters. The presence of these fringes indicates that a level of transverse coherence is given at the setup.

The ROIs of the differential phase signals centred around the silicon sphere in Fig. 9.16e and f reveal that the noise level is high and the signal in the central part of the sphere is lower than expected, which can also be observed in the integrated phase in Fig. 9.16g. Furthermore, the phase shift in the sphere appears to be lower than in the upper part of the toothpick, which is an unexpected observation and contrary to the results in Sect. 6.4.2 of Chap. 6 on the same sample. Looking at the dark-field signal in Fig. 9.16i and the ROI in Fig. 9.16j, this can be understood better. Small-angle scattering is expected to be negligible in the sphere that is made of a homogeneous material (see also Fig. 6.16 in Sect. 6.4.2 of Chap. 6), but a strong dark-field signal is detected here. This can be explained by beam-hardening effects leading to a reduction in speckle visibility, as discussed e.g. by Zdora et al. (2015). The strong dark-field signal in the wooden toothpick, on the other hand, is to be expected due to the highly scattering nature of the wooden fibres. As the thicker part of the toothpick contains more scattering material, the signal increases towards the bottom of the sample. However, also beam hardening partly contributes to the dark-field signal in the toothpick. Beam hardening in the toothpick is, furthermore, visible in the integrated phase shift in Fig. 9.16c, where the signal is weaker in the thicker lower part of the toothpick contrary to what would be expected for the case of a monochromatic illumination. It can be observed in the dark-field image that the background is not unity, which is caused by noise in the reference and sample images. The dark-field signal hence contains contributions from small-angle scattering from unresolved features and edges, but also beam hardening and noise, i.e. any factor that reduces the local visibility of the speckles.

In general, beam hardening affects the multimodal contrast signals in different ways. Looking at the horizontal line plots through the centre of the sphere, the effec-

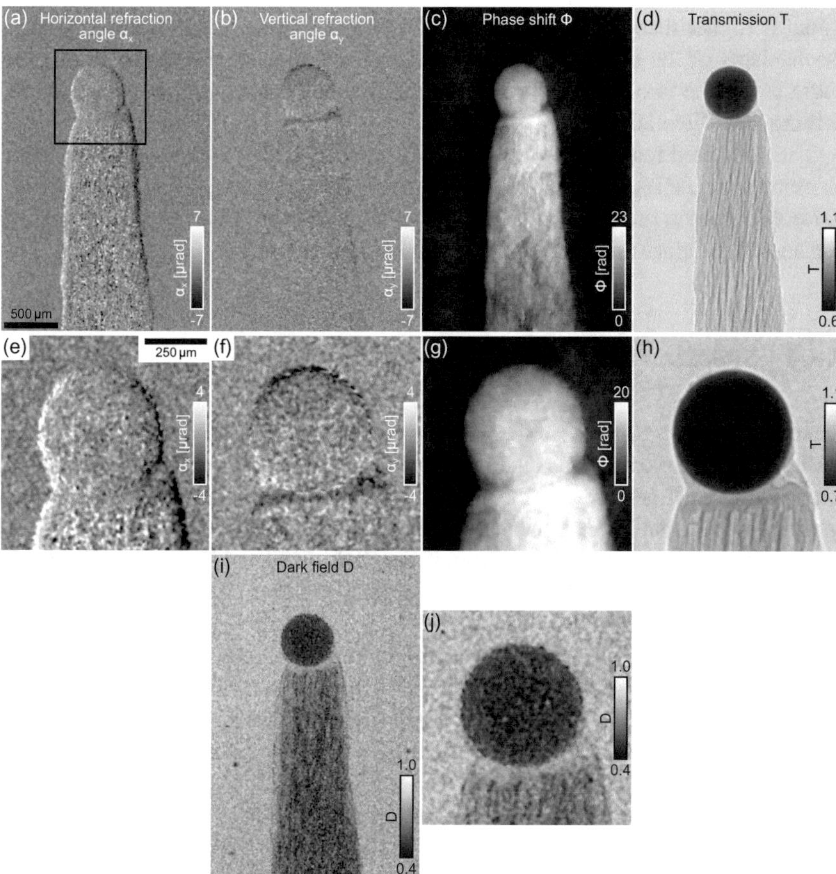

Fig. 9.16 Multimodal signals of a test sample (silicon sphere of 480 µm diameter glued onto a wooden toothpick) imaged using UMPA at a liquid-metal-jet laboratory source. **a** Refraction angle in the horizontal, **b** refraction angle in the vertical, **c** integrated phase shift, **d** transmission. **e–h** ROIs (as indicated by the black box in panel (**a**)) of the horizontal and vertical differential phase, integrated phase shift and transmission signals. The phase shift in the centre of the sphere is lower than expected. **i, j** Dark-field signal and corresponding ROI. The strong signal suggests significant beam hardening in the silicon sphere. The data was reconstructed with $N = 15$ diffuser steps and a window size of 9×9 pixels (Hamming window). The scale bars indicate the sizes in the sample plane

tive energy after beam hardening by the thickest part of the sphere (480 µm of silicon) was determined for the transmission and phase signals separately. The transmission in the centre of the sphere is approximately $T \approx 63.9\%$, which corresponds to an effective energy of $E_{\text{eff};\,T} \approx 20\,\text{keV}$. The phase shift in the centre leads to the assumption of an effective energy of $E_{\text{eff};\,\Phi} \approx 30\,\text{keV}$, giving a phase shift of $\Phi \approx 39\,\text{rad}$. This indicates that significant hardening from the estimated average spectrum energy of 16 keV occurs within the sphere. The fact that the effective energy for the phase signal is higher than for the transmission is expected due to the different contrast mechanisms of the two signals. It has been shown that the difference in effective energies for the two signal modalities increases with increasing sample thickness (Munro and Olivo 2013).

The presented results illustrate that the effects of beam hardening on the values obtained with UMPA speckle imaging can be significant when using a polychromatic X-ray spectrum as produced by the liquid-metal-jet source. They need to be carefully considered in future quantitative studies at laboratory sources.

9.4.4 Speckle Visibility and Size

The characteristics of the speckle pattern were studied in detail on the data acquired with the bug sample that was taken following the test scan. It was conducted with a longer exposure time to reduce the noise in the images. The reference speckle pattern created by illumination of a sheet of P800 sandpaper is shown in Fig. 9.17a for the first diffuser position and in more detail in the ROI of 150 × 150 pixels in Fig. 9.17b. In a first qualitative visual assessment, the speckle pattern appears to be uniform and the speckles are of good visibility.

The speckle visibility was analysed for the reference speckle pattern at each of the $N = 80$ diffuser positions using Eq. (5.1) in Chap. 5. The visibility map for the pattern in Fig. 9.17a is shown in Fig. 9.17c. It was determined by evaluating the visibility in a 150 × 150 pixels region around each pixel. The mean speckle visibility over the field of view at this diffuser position is $\nu_0 = (10.6 \pm 0.5)\%$, which is comparable to, but slightly lower than, the speckle visibility in the range $(15 - 20\%$ achieved at a similar setup (Zanette et al. 2015) with better transverse coherence conditions but lower spatial resolution. The visibility map shows different average values for the left and right half. This is most likely caused by tilted mounting of the scintillation screen in the detector system and slight inhomogeneities in the diffuser material. The average of all mean visibility values at the different diffuser positions is $\nu_{\text{avg}} = (10.7 \pm 0.1)\%$.

The size of the speckles was determined for each diffuser position by 2D autocorrelation of the reference speckle pattern and the FWHM is taken as a measure for the speckle size. Horizontal and vertical line profiles through the centre of the 2D autocorrelation of the speckle pattern in Fig. 9.17a are shown in Fig. 9.17d. The data points of the line plots were interpolated and fitted by Gaussians (solid lines) to determine the FWHM of the peaks. This gives a speckle size of approximately 3.20 pixels in the horizontal and 3.24 pixels in the vertical direction, corresponding to 10.4 µm and 10.5 µm, respectively, in the detector plane. The mean speckle size averaged over all diffuser positions is (3.21 ± 0.06) pixels in the horizontal and (3.25 ± 0.07) pixels in the vertical direction, i.e. 10.4 µm and 10.6 µm, respectively, in the detector plane.

The results indicate that the speckle pattern created by the sandpaper using the liquid-metal-jet source is of good visibility and the speckles are very small, covering less than five pixels.

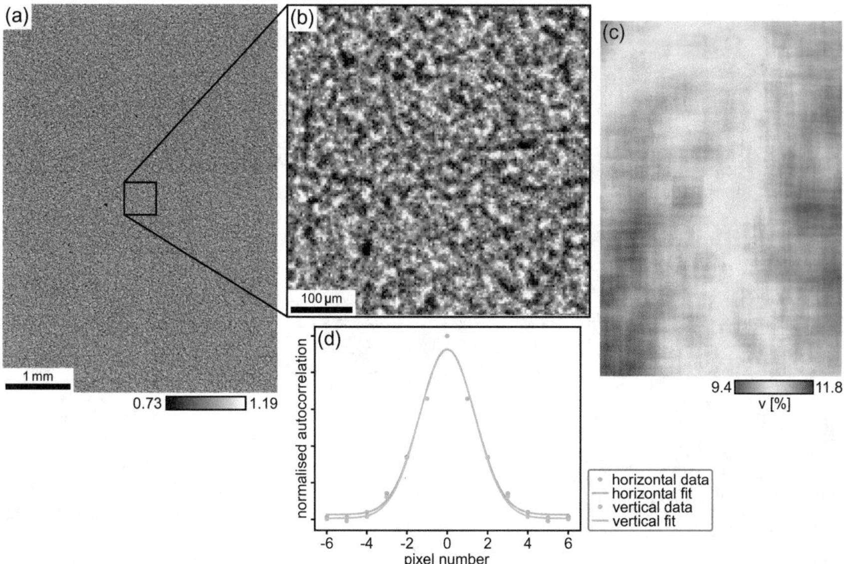

Fig. 9.17 Characteristics of the speckle interference pattern created by illumination of a piece of P800 sandpaper at the liquid-metal-jet source. **a** Reference speckle pattern at the first diffuser position of the UMPA scan and **b** ROI in the centre of the field of view. **c** Visibility map of the speckle pattern in panel (**a**). The mean visibility over the field of view is $(10.6 \pm 0.5)\%$. Bright areas are due to spikes in the speckle pattern due to direct hits of the sCMOS sensor by scattered X-rays. **d** Horizontal (green) and vertical (orange) line profiles through the centre of the 2D autocorrelation of the speckle pattern in panel (**a**). The data points were fitted with Gaussian curves to determine the FWHM of the peaks, which are taken as a measure for the speckle size, here 3.20 pixels in the horizontal and 3.24 pixels in the vertical direction. The scale bars indicate the sizes in the detector (scintillator) plane. Adapted with permission from Zdora et al. (2020) © The Optical Society

The slight differences in the characteristics of the speckle pattern compared to the test sample measurements in the previous section are due to the facts that the latter showed a higher noise level due to the shorter exposure time, the analysed field of view was smaller (as the sample was smaller) and a different region of the diffuser was illuminated.

9.4.5 Multimodal Imaging of a Bug

The measurement of the bug was performed with improved statistics by increasing the exposure time and number of diffuser steps compared to the test scan, albeit at the cost of a longer scan time.

The multimodal signals of the bug specimen are presented in Fig. 9.18.

As expected, while mostly vertical features are visible in the horizontal refraction angle signal in Fig. 9.18a, the vertical refraction angle in Fig. 9.18b highlights hori-

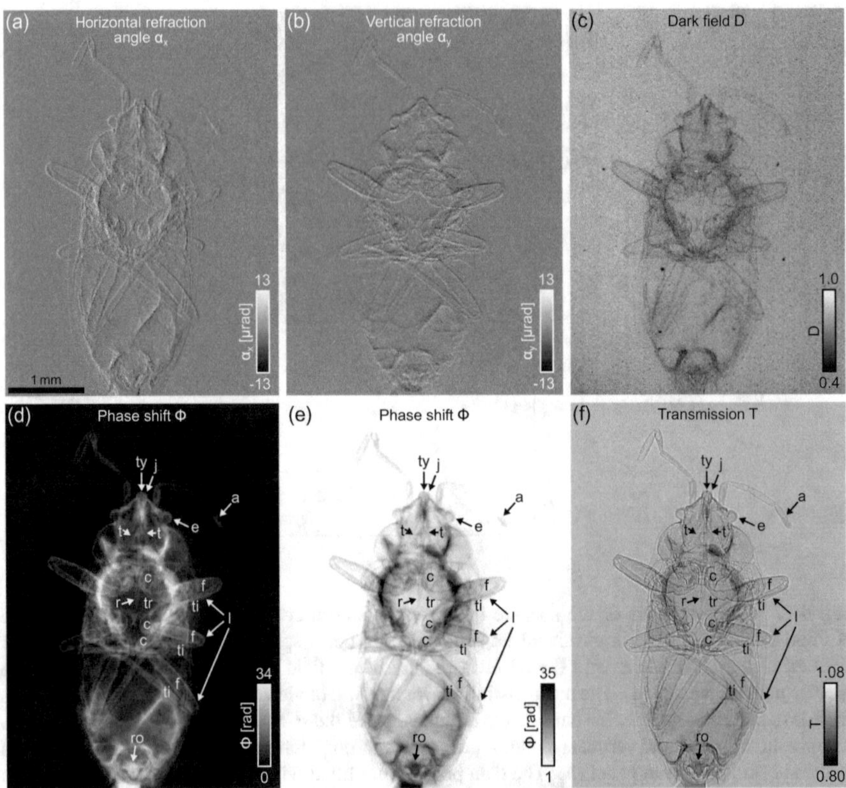

Fig. 9.18 Multimodal signals of a small bug (*Orius Niger*) imaged using UMPA at the liquid-metal-jet laboratory source. **a** Refraction angle in the horizontal, **b** refraction angle in the vertical, **c** dark-field signal, **d** integrated phase shift, **e** integrated phase shift shown with inverted contrast for better comparison with the transmission, and **f** transmission signal. Detailed features of the bug can be observed in the transmission and phase modalities, as labelled: a: antenna, ty: tylus, j: jugum, e: eye, t: tracheae, r: rostrum, l: legs with c: coxa, tr: trochanter, f: femur and ti: tibia, ro: reproductive organ. The dark-field image shows less beam hardening than for the test sample due to the lower absorption of the bug. The data was reconstructed with $N = 80$ diffuser steps and a window size of 5×5 pixels. The scale bar indicates the size in the sample plane. Adapted with permission from Zdora et al. (2020) © The Optical Society

zontal features. The information from the differential phase signals was combined in the integrated phase shift, shown in Fig. 9.18d. The same signal modality is presented in Fig. 9.18e with inverted contrast for better comparison with the transmission signal. The latter is shown in Fig. 9.18f. It contains some edge enhancement, as observed for the sphere previously, due to Fresnel propagation. The absorption by the body of the insect is low. Both modalities, phase and transmission, clearly visualise small features of the specimen, as labelled for the right side of the bug in Fig. 9.18d–f. These include details of the head such as the tylus (ty), the forward-projecting part at the front of the head, and the juga (j), the outer lobes at the front of the head. The mouth part contains a long rostrum (r) that extends downwards and is a long beak with a sharp tip that the bug uses for sucking plants or insect prey after injecting them with digestive enzymes. The antennae (a) consist of four segments, as typical for bugs of the Anthocoridae family. Variations in density can be observed for the different segments, which are visualised most prominently in the phase signal. These are likely due to drying of the specimen, which left the right antenna almost hollow. It can be seen that the compound eyes (e) are denser on the outside than the inside. Interestingly, even two of the head tracheae (t) can be observed. These tubes that run through the entire body of the bug deliver oxygen and are the essential part of the respiratory system. They are best visible in the phase signal due to the low density inside the air-filled tubes. Details of the structure of the legs (l) with its various segments can be seen. The leg is attached to the bug's body at the coxa (c), which is followed by the trochanter (tr), the thick femur (f) and the tibia (ti). The last segment of the leg, here partly covered by overlaying structures, is the tarsus. At the bottom of the body, the reproductive organ (ro) can be seen as well as some glue used to mount the specimen.

As already observed for the sphere sample, the dark-field signal in Fig. 9.18c shows some background signal due to the effects of noise, which is, however, less pronounced than for the test scan taken with shorter exposure time. A non-negligible dark-field signal can be observed also inside the sample and mainly in thicker areas of the specimen, which can be explained by beam hardening, as previously investigated on the test sample. The beam hardening is not as strong for the bug sample as for the sphere due to its much lower absorption of X-rays. As common for X-ray speckle- and grating-based imaging, the dark-field image also highlights the edges of the specimen.

To achieve a good angular sensitivity, the data presented in Fig. 9.18 were reconstructed with the entire set of the images acquired at $N = 80$ diffuser positions. It is now investigated how reducing the number of steps, while maintaining the analysis window size and the exposure time per frame, affects the angular sensitivity.

Figure 9.19b–i show ROIs (as indicated in Fig. 9.19a) of the horizontal and vertical refraction angle signals of the bug reconstructed with different numbers of diffuser steps. While the images in Fig. 9.19b, c were obtained by using the whole data set, panels (d) and (e) show the result of including only the last 40 diffuser positions in the UMPA analysis. A total of 40 positions was used also for the images in Fig. 9.19f, g, but in this case every second step of the whole data set was included. This effectively doubles the step size and keeps the overall time over which the bug was imaged the same as for $N = 80$. By comparing changes between the signals in panels (d), (e) and (f), (g), the effect of different step sizes and varying overall time of the sample in the beam can be studied. Figure 9.19h, i, on the other hand, show the differential phase signals retrieved from $N = 20$ steps with the same step size and overall time as in panels (d), (e), i.e. every second of the last 40 steps was included in the analysis.

It can be observed that, despite the decrease in image quality with decreasing number of steps, most of the features of the bug's head are still discernible in detail even when using only a fourth of the data ($N = 20$). Good image quality is also confirmed for the integrated phase and transmission signals, see Fig. 9.20.

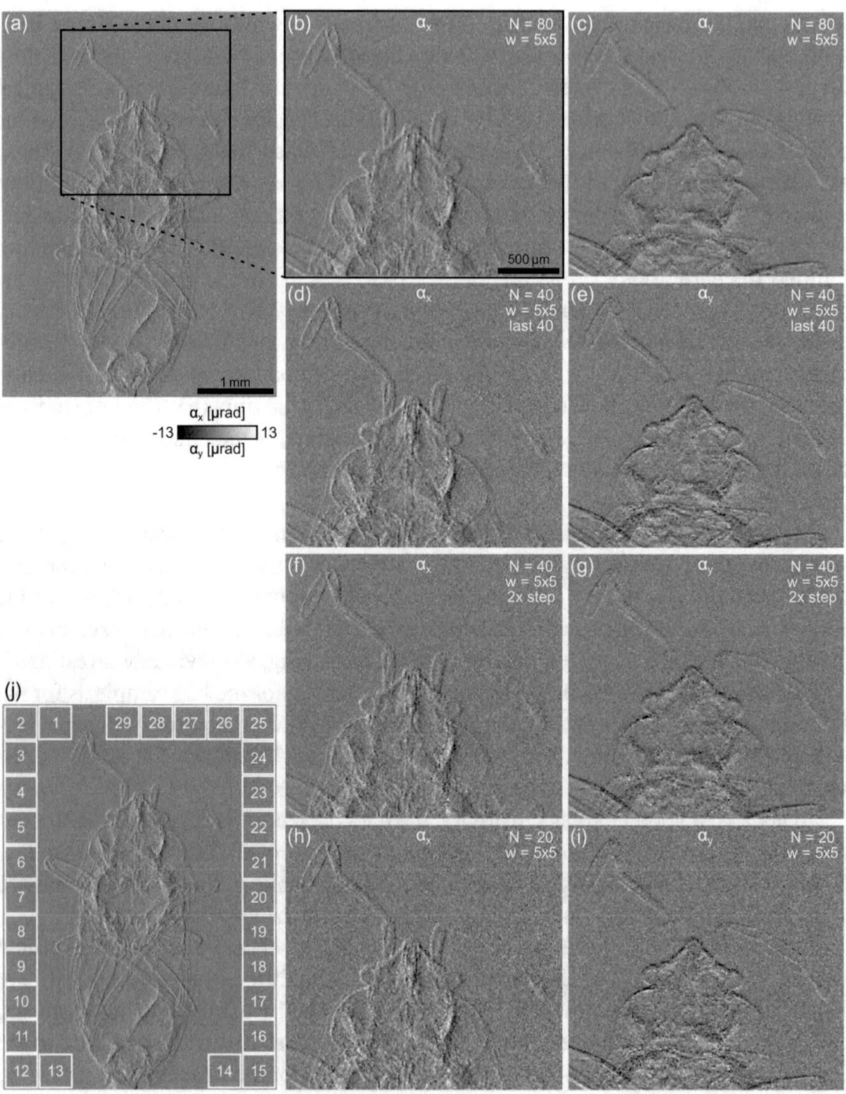

Fig. 9.19 Comparison of UMPA reconstructions of the bug with different numbers N of diffuser positions. The analysis window size ($w = 5 \times 5$ pixels) and exposure time per frame (1100 s) were kept constant. **a** Full frame of the horizontal refraction angle signal reconstructed with $N = 80$ (ROI in panels (**b**)–(**i**) indicated by the black box). **b**, **c** ROIs of the horizontal and vertical refraction angle signals, respectively, reconstructed with $N = 80$. **d**, **e** Corresponding ROIs reconstructed with $N = 40$, keeping the step size constant and halving the total time that the sample is observed by using the last 40 steps of the scan. **f**, **g** Corresponding ROIs reconstructed with $N = 40$, with double the step size and keeping the total overall time by taking every second step of the scan. **h**, **i** Corresponding ROIs reconstructed with $N = 20$, with the same step size as in panels (**f**), (**g**) by taking every second of the last 40 steps of the scan. **j** Horizontal refraction angle image highlighting the ROIs (150 × 150 pixels) used for the determination of the angular sensitivity values listed in Table 9.3. The scale bars indicate the sizes in the sample plane

Upon closer inspection it can, furthermore, be seen that the spatial resolution for the images reconstructed from only the second half of the data set, see Fig. 9.19d, e and h, i, seems to be better than for the reconstruction using data acquired over the full length of the scan, see Fig. 9.19b, c and f, g. This is also observable in the integrated phase and transmission signals in Fig. 9.20. Although Fig. 9.20a and Fig. 9.20b were reconstructed with the same number of diffuser positions ($N = 40$) and analysis window size (5×5 pixels), the features of the bug appear sharper when only the second half of the data set is used (Fig. 9.20a). This is also the case for the corresponding transmission signals shown in Fig. 9.20d, e. The reason for this is that the specimen, which was collected only a day before the scan and frozen overnight, underwent slight changes in shape and inner structure over the course of the scan due to progressive drying. The longer the observation time, the more changes in the sample can be seen, leading to more pronounced blurring for Fig. 9.20b, e compared to Fig. 9.20a, d.

The reconstructions with $N = 20$ in Fig. 9.20c, f, for which every second diffuser step of the last 40 steps was included in the analysis, were also reconstructed from the data acquired over a shorter time. The level of details observed in the sample is comparable to Fig. 9.20a, d, as expected. Moreover, it can be seen that the integrated phase and transmission signals are still of good image quality even at $N = 20$, indicating the potential for much shorter scan times.

As a last step, the effect of pixel binning on the image quality is studied. It is planned to perform UMPA at the liquid-metal-jet source with a different detector with larger pixels in the future. This will allow for decreasing the exposure time and overall scan time, as well as extending the field of view to scan larger samples. Here, the effect of larger pixels is studied by reconstructing the same data set as analysed previously after 2×2 binning. Figure 9.21 compares the vertical differential phase signals of the bug's head for UMPA reconstructions of the unbinned data (Fig. 9.21a and b) and of the corresponding binned data (Fig. 9.21c and d) reconstructed with adapted window sizes to include a similar total area in the detector plane and hence total number of photons in the analysis. Similar image quality is expected for the images in Fig. 9.21a and c and Fig. 9.21b and d. While by visual comparison the spatial resolution seems comparable for the reconstruction from the binned and unbinned data, the former seems slightly more noisy.

The quality of the signals reconstructed with the different parameters was assessed quantitatively by measuring the horizontal and vertical angular sensitivities in the differential phase images. They were determined as the average standard deviation of the refraction angle in 29 ROIs of 150×150 pixels in the air region around the bug, as indicated in Fig. 9.19j. The values for the different reconstruction parameters (average over all regions) are listed in Table 9.3.

It can be seen that the sensitivity is slightly better in the vertical than in the horizontal. This is most likely due to the mounting of the diffuser on a vertical holder, which makes the diffuser more stable in the vertical direction. As expected, the sensitivity deteriorates with decreasing number of steps when keeping the window size constant. Furthermore, the sensitivity values for $N = 40$ using the different step sizes are the same, confirming the fact that the step size is not relevant for UMPA as

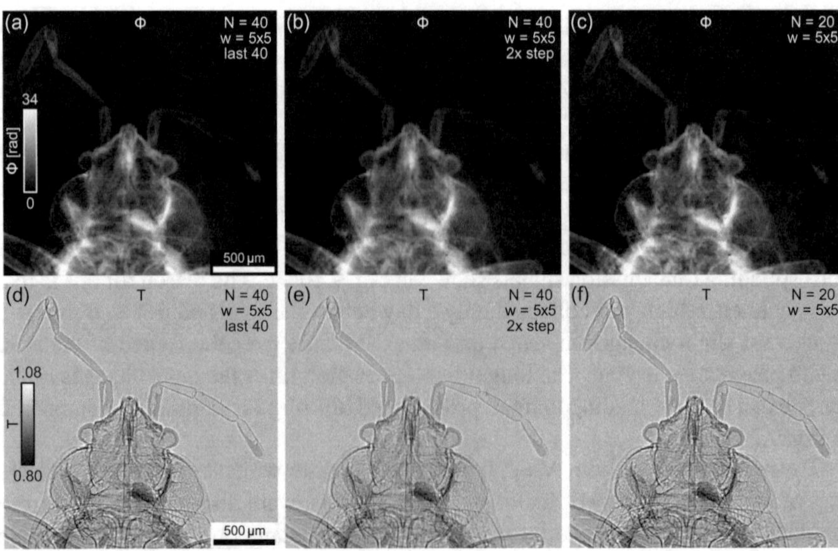

Fig. 9.20 UMPA phase and transmission signals of the bug reconstructed with $N = 40$ and $N = 20$ diffuser positions. The analysis window size ($w = 5 \times 5$ pixels) and exposure time per frame (1100 s) were kept constant. Integrated phase shift signal of the bug reconstructed with **a** $N = 40$, using the last 40 steps of the scan, **b** $N = 40$, using every second step of the scan and **c** $N = 20$, using every second of the last 40 steps of the scan. Even at $N = 20$, fine details of the bug can be observed. The images show better resolution for panels (**a**), (**d**) and (**c**), (**f**) than for panels (**b**), (**e**) due to drying of the sample occurring during the scan that is more pronounced with increasing period of time. **d–f** Corresponding transmission signals. Panel (**e**) also shows the sample blurring for data acquired over a longer period of time, as observed in panel (**b**). The scale bars indicate the sizes in the sample plane

long as it is larger than the analysis window size. Increasing the window size improves the sensitivity significantly to less than 0.5 μrad, in accordance with the sensitivity model introduced for UMPA in Chap. 6, albeit at the cost of spatial resolution, as can be observed by comparing Fig. 9.21a and b. The total number of photons and effective window size are similar for the reconstruction of the unbinned data with $N = 80$, $w = 5 \times 5$ pixels and the binned data (2×2 pixels binning) with $N = 80$, $w = 3 \times 3$ pixels, as well as for the reconstruction of the unbinned data with $N = 80$, $w = 9 \times 9$ pixels and the binned data (2×2 pixels binning) with $N = 80$, $w = 5 \times 5$ pixels. However, the values listed in Table 9.3 lead to the conclusion that binning seems to deteriorate the angular sensitivity for the presented data set. The effect is more pronounced for smaller analysis windows. A reason for this might be the small size of the speckles that only cover a few pixels and are hence on the order of less than 2 pixels in size after binning. For a future setup with larger pixels, larger speckles

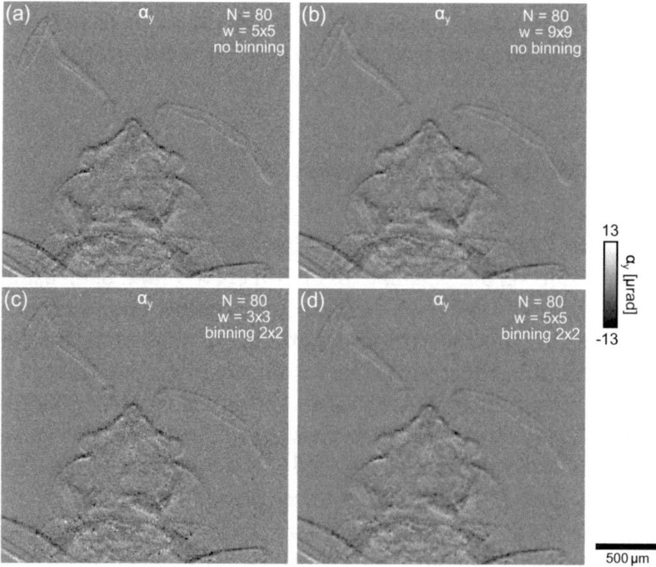

Fig. 9.21 Comparison of UMPA reconstructions of the bug from original unbinned and binned data. **a** Vertical refraction angle signal reconstructed from unbinned data with $N = 80$, $w = 5 \times 5$ pixels and **b** from unbinned data with $N = 80$, $w = 9 \times 9$ pixels. **c** Vertical refraction angle signal reconstructed from binned data (2×2 binning) with $N = 80$, $w = 3 \times 3$ pixels and (**b**) from binned data (2×2 binning) with $N = 80$, $w = 5 \times 5$ pixels. The spatial resolution in panels (**a**) and (**c**) and panels (**b**) and (**d**) seems comparable, as expected. However, the angular sensitivity is decreased for the binned data, see Table 9.3, most likely due to the small speckle size

Table 9.3 Horizontal and vertical angular sensitivities, σ_x and σ_y, for different UMPA reconstruction parameters (number N and size of diffuser steps, analysis window size w, binning factor) of the bug scan. The values are the average ($\pm$ standard deviation) over all ROIs indicated in Fig. 9.19j. The last two columns give the figure numbers and panels in which the horizontal and vertical refraction angle signals, respectively, are shown

N	Step size [μm]	w [pixels]	Binning	σ_x [μrad]	σ_y [μrad]	Figs. α_x	Figs. α_y
80	75.95	5×5	–	0.92 ± 0.16	0.90 ± 0.11	9.19b	9.19c
40	75.95	5×5	–	1.33 ± 0.24	1.30 ± 0.17	9.19d	9.19e
40	151.90	5×5	–	1.33 ± 0.25	1.30 ± 0.17	9.19f	9.19g
20	151.90	5×5	–	2.05 ± 0.66	1.91 ± 0.27	9.19h	9.19i
80	75.95	9×9	–	0.45 ± 0.07	0.44 ± 0.05	–	9.21b
80	75.95	5×5	2×2	0.50 ± 0.13	0.45 ± 0.08	–	9.21d
80	75.95	3×3	2×2	1.27 ± 0.31	1.13 ± 0.20	–	9.21c

should be chosen. Then, larger pixel sizes will provide a simple solution to reduce exposure times and increase the field of view, albeit at the cost of spatial resolution.[10]

9.4.6 Conclusions and Perspectives

In this section, first demonstrations of UMPA using a laboratory X-ray source were presented. The results on a test sample of known composition demonstrate the feasibility and potential of UMPA in the lab. They, furthermore, showed that beam hardening can play a significant role for speckle-based multimodal imaging at polychromatic sources. This effect should be investigated in more detail in the future, in particular for possible quantitative studies. One could consider filtering the spectrum, as evaluated theoretically in some first simulation studies by Walker (2019), which could, moreover, increase the temporal coherence of the X-rays for improved speckle visibility, albeit at the cost of a reduced photon flux.

The second scan performed on a small bug demonstrates the promising potential of UMPA multimodal X-ray imaging for real-world samples at easily accessible laboratory sources and fine details of the insect were observed. The number of diffuser steps for this data set was high, in order to explore the angular sensitivity limits of the given setup. However, details of the specimen were still clearly visible when using only $N = 20$ diffuser steps for the reconstruction, which is the number of steps commonly used for UMPA measurements at the synchrotron. This indicates that the total scan time can be significantly reduced in the future, in particular with further optimisation of the setup, which will allow for shorter exposure times. It was observed that a shorter overall scan time is particularly important for specimens that might be susceptible to changes over time, e.g. hydrated biological tissue, in order to avoid blurring of features.

With the given setup configuration, angular sensitivities of better than 1 μrad were achieved at a small window size of 5×5 pixels. Although sensitivities of almost an order of magnitude better have been reported for similar measurements conducted at synchrotron sources (see previous chapters), it will be possible to significantly increase the angular sensitivity achievable at the liquid-metal-jet source setup in the future. It should be noted that the propagation distance (sample-detector) was only 11 cm for the data presented here, which is approximately 5–10 times less than usually used at synchrotrons. A longer propagation distance could greatly improve the angular sensitivity. However, the setup geometry limits the total distance between source and detector for the given laboratory setup. The best performing setup configurations considering the relative positions of diffuser, sample and detector and the associated implications on the speckle visibility, speckle size, sample magnification and other factors will be studied systematically in the future. Furthermore, the influ-

[10] Although the spatial resolution can be adjusted by the window size in the UMPA analysis and can hence be decreased to improve the spatial resolution, the minimum window size and best achievable spatial resolution is limited by the effective pixel size.

ence of the spot size on the speckle characteristics will be investigated in more detail. A larger spot size would allow for a higher power and hence reduced exposure time. Lastly, improvements and optimisations of the detector system will contribute to a shorter scan time. Different scintillators with higher efficiency and different types of detectors will be tested. In particular, larger pixel sizes will be used to increase photon statistics at shorter scan times and allow for a larger field of view.

This was first investigated in the previous section by applying pixel binning. It was observed that for small speckle sizes binning can deteriorate the image quality compared to the corresponding reconstructions from unbinned data. It was concluded that larger speckles, realised by using different diffusers or placing the diffuser further upstream to exploit the geometric magnification, are needed to overcome this.

The presented results are first feasibility demonstrations of UMPA at a polychromatic laboratory source. Detailed theoretical and experimental studies to optimise the setup for UMPA are subject of currently ongoing work. The main challenge that needs to be tackled is to achieve a good image quality, in terms of angular sensitivity and spatial resolution, at reasonable scan times. Once this has been accomplished, sensitivities and overall acquisition times will be similar to lab-based X-ray grating interferometry, while using a simpler and more robust setup. UMPA, furthermore, has advantages over propagation-based phase-contrast imaging for lab-based setups, including its multimodal and quantitative character and high phase sensitivity. Furthermore, while propagation-based imaging requires relatively small detector pixels in order to resolve the Fresnel fringes, this is not a limitation for UMPA. The use of large pixels will in the future allow for UMPA imaging of larger specimens and will reduce scan times. UMPA tomography scans with an optimised setup will pave the way for high-throughput applications, e.g. for biomedical and pre-clinical imaging.

9.5 Concluding Remarks

In this chapter, some of the ongoing work on X-ray speckle-based imaging using UMPA that was investigated within the scope of this Ph.D. project was presented. These projects all aim at making speckle-based imaging accessible for a broader range of applications and a wider user community.

It was shown that customised phase modulators can be fabricated using MACE technology, in order to adapt the properties of the speckle pattern to specific setups and experimental conditions. It was demonstrated that these diffusers are particularly promising for high-resolution speckle-based phase tomography. This was illustrated on the example of catalyst particles, for which a quantitative analysis of the electron density distribution allowed to draw conclusions on their deactivation status.

It was, furthermore, shown in this chapter that UMPA can be performed at high X-ray energies delivering high-sensitivity phase information with angular sensitivities better than 200 μrad at 53 keV. This was applied to geology and materials science samples, which are fields of research that were previously not accessible by speckle-based imaging. Combining the information from the phase and transmission tomograms allowed for the identification of various inclusions in volcanic rock samples, for which quantitative density values could be determined. For a mortar specimen, the information from the phase tomogram made it possible to distinguish its components based on their density differences, enabling simple threshold-based segmentation. In-depth analysis of the preliminary results presented here, e.g. full segmentation, density measurement and extraction of information on the porosity, particle size and other properties, will be performed in the near future. Work on further high-energy X-ray imaging applications such as palaeontology and industrial testing is also currently being undertaken by the Ph.D. candidate.

In the last section of the chapter, it was demonstrated that UMPA can be translated to X-ray laboratory sources. First results on a small bug obtained with a liquid-metal-jet source illustrate the potential of UMPA in laboratory settings. It was shown that $N = 20$ diffuser steps can be sufficient to obtain detailed information on the specimen even with a laboratory source. It is anticipated that with the currently ongoing optimisation of the setup, scan times can be significantly reduced, the angular sensitivity will be improved and the field of view increased, enabling tunable multimodal imaging that is readily accessible for a larger user community.

It can be concluded from the results in this chapter that X-ray speckle-based imaging with UMPA can be implemented at most existing X-ray imaging setups and for a wide range of applications that might not have been considered previously.

Work is ongoing on optimising the diffuser fabrication, speckle-imaging setups and acquisitions protocols and the UMPA analysis routine to achieve the best possible image quality at short scan times. Following the studies of this Ph.D. project, wider user uptake of UMPA speckle-based imaging is currently progressing at laboratory sources and at several synchrotron beamlines such as Diamond I13 and Diamond I12.

References

Atwood RC, Bodey AJ, Price SWT, Basham M, Drakopoulos M (2015) A high-throughput system for high-quality tomographic reconstruction of large datasets at Diamond Light Source. Phil Trans R Soc A 373(2043):20140398

Baker D, Mancini L, Polacci M, Higgins M, Gualda G, Hill R, Rivers M (2012) An introduction to the application of X-ray microtomography to the three-dimensional study of igneous rocks. Lithos 148:262–276

Bentz D, Martys NS, Stutzman P, Levenson MS, Garboczi E, Dunsmuir J, Schwartz LM (1994) X-ray microtomography of an ASTM C109 mortar exposed to sulfate attack. Mater Res Soc Symp Proc 370:77

Bérujon S, Ziegler E (2017) Near-field speckle-scanning-based x-ray tomography. Phys Rev A 95(6):063822

Boerefijn R, Gudde N, Ghadiri M (2000) A review of attrition of fluid cracking catalyst particles. Adv Powder Technol 11(2):145–174

Carrara P, Kruse R, Bentz D, Lunardelli M, Leusmann T, Varady P, Lorenzis LD (2018) Improved mesoscale segmentation of concrete from 3D X-ray images using contrast enhancers. Cement Concrete Comp 93:30–42

Chartier C, Bastide S, Lévy-Clément C (2008) Metal-assisted chemical etching of silicon in HF-H2O2. Electrochim Acta 53(17):5509–5516

Chotard T, Boncoeur-Martel M, Smith A, Dupuy J, Gault C (2003) Application of X-ray computed tomography to characterise the early hydration of calcium aluminate cement. Cement Concrete Comp 25(1):145–152

Cnudde V, Boone M (2013) High-resolution X-ray computed tomography in geosciences: A review of the current technology and applications. Earth-Sci Rev 123:1–17

Coldwell B (2019) Instituto Tecnológico y de Energías Renovables. (Santa Cruz de Tenerife, Spain), Private Communication

Couves C, Roberts S, Racey A, Troth I, Best A (2016) Use of X-ray computed tomography to quantify the petrophysical properties of volcanic rocks: a case study from Tenerife, Canary Islands. J Petroleum Geol 39(1):79–94

de Veire MV, Degheele D (1992) Biological control of the western flower thrips, Frankliniella occidentalis (Pergande) (Thysanoptera: Thripidae), in glasshouse sweet peppers with Orius spp. (Hemiptera: Anthocoridae). A comparative study between O. niger (Wolff) and O. insidiosus (Say). Biocontrol Sci Technol 2(4):281–283

Deligeorgidis PN (2002) Predatory effect of Orius niger (Wolff) (Hem., Anthocoridae) on Frankliniella occidentalis (Pergande) and Thrips tabaci Lindeman (Thysan., Thripidae). J Appl Entomol 126(2–3):82–85

Excillum (2019) Metal Jet D2 Data Sheet. https://www.excillum.com/products/metaljet-sources/metaljet-d2-160-kv/. Accessed 22 July 2019

Federation of European Producers of Abrasives (2019) FEPA P-grit sizes coated abrasives. https://www.fepa-abrasives.com/abrasive-products/grains. Accessed 21 June 2019

Gradl R, Zanette I, Ruiz-Yaniz M, Dierolf M, Rack A, Zaslansky P, Pfeiffer F (2016) Mass density measurement of mineralized tissue with grating-based x-ray phase tomography. PLoS ONE 11(12):1–13

Gualda GAR, Pamukcu AS, Claiborne LL, Rivers ML (2010) Quantitative 3D petrography using x-ray tomography 3: documenting accessory phases with differential absorption tomography. Geosphere 6(6):782–792

Hemberg O, Otendal M, Hertz HM (2003) Liquid-metal-jet anode electron-impact x-ray source. Appl Phys Lett 83(7):1483–1485

Henke BL, Gullikson EM, Davis JC (1993) X-ray interactions: photoabsorption, scattering, transmission, and reflection at E = 50–30,000 eV, Z = 1–92. Atomic Data Nucl Data Tables 54(2):181–342

Hildreth OJ, Lin W, Wong CP (2009) Effect of catalyst shape and etchant composition on etching direction in metal-assisted chemical etching of silicon to fabricate 3D nanostructures. ACS Nano 3(12):4033–4042

Huang Z, Geyer N, Werner P, de Boor J, Gösele U (2011) Metal-assisted chemical etching of silicon: a review. Adv Mater 23(2):285–308

Ihli J, Diaz A, Shu Y, Guizar-Sicairos M, Holler M, Wakonig K, Odstrcil M, Li T, Krumeich F, Müller E, Cheng W-C, Anton van Bokhoven J, Menzel A (2018) Resonant ptychographic tomography facilitates three-dimensional quantitative colocalization of catalyst components and chemical elements. J Phys Chem C 122(40):22920–22929

Ihli J, Jacob RR, Holler M, Guizar-Sicairos M, Diaz A, da Silva JC, Ferreira Sanchez D, Krumeich F, Grolimund D, Taddei M, Cheng W-C, Shu Y, Menzel A, van Bokhoven JA (2017) A three-dimensional view of structural changes caused by deactivation of fluid catalytic cracking catalysts. Nat Commun **8**(1), 809

Kottler C, David C, Pfeiffer F, Bunk O (2007) A two-directional approach for grating-based differential phase contrast-imaging using hard x-rays. Opt Express 15(3):1175–1181

Labriet H, Bérujon S, Broche L, Fayard B, Bohic S, Stephanov O, Paganin DM, Lhuissier P, Salvo L, Bayat S, Brun E (2019) 3D histopathology speckle phase contrast imaging: from synchrotron to conventional sources. Proc SPIE 10948:109481S

Li X, Bohn PW (2000) Metal-assisted chemical etching in HF/H2O2 produces porous silicon. Appl Phys Lett 77(16):2572–2574

Madonna C, Quintal B, Frehner M, Almqvist BSG, Tisato N, Pistone M, Marone F, Saenger EH (2013) Synchrotron-based X-ray tomographic microscopy for rock physics investigations. Geophysics 78(1):D53–D64

Meirer F, Kalirai S, Morris D, Soparawalla S, Liu Y, Mesu G, Andrews JC, Weckhuysen BM (2015) Life and death of a single catalytic cracking particle. Sci Adv 1(3)

Munro PRT, Olivo A (2013) X-ray phase-contrast imaging with polychromatic sources and the concept of effective energy. Phys Rev A 87(5):053838

Paganin D, Mayo SC, Gureyev TE, Miller PR, Wilkins SW (2002) Simultaneous phase and amplitude extraction from a single defocused image of a homogeneous object. J Microsc 206(1):33–40

Palenstijn W, Batenburg K, Sijbers J (2011) Performance improvements for iterative electron tomography reconstruction using graphics processing units (GPUs). J Struct Biol 176(2):250–253

Pankhurst MJ, Vo NT, Butcher AR, Long H, Wang H, Nonni S, Harvey J, Gudfinnsson G, Fowler R, Atwood R, Walshaw R, Lee PD (2018) Quantitative measurement of olivine composition in three dimensions using helical-scan X-ray micro-tomography. Am Mineral 103(11):1800–1811

Prade F, Fischer K, Heinz D, Meyer P, Mohr J, Pfeiffer F (2016) Time resolved X-ray dark-field tomography revealing water transport in a fresh cement sample. Sci Rep 6:29108

Raven C (1998) Numerical removal of ring artifacts in microtomography. Rev Sci Instrum 69(8):2978–2980

Romano L, Kagias M, Jefimovs K, Stampanoni M (2016) Self-assembly nanostructured gold for high aspect ratio silicon microstructures by metal assisted chemical etching. RSC Adv 6(19):16025–16029

Sarapata A, Ruiz-Yaniz M, Zanette I, Rack A, Pfeiffer F, Herzen J (2015) Multi-contrast 3D X-ray imaging of porous and composite materials. Appl Phys Lett 106(15):154102

Stevenson DM, Yin L, Stewart MA (2010) X-ray tomography verification for determining phase proportions in volcanic rocks. Proc SPIE 7804:780416

van Aarle W, Palenstijn WJ, Beenhouwer JD, Altantzis T, Bals S, Batenburg KJ, Sijbers J (2015) The ASTRA toolbox: a platform for advanced algorithm development in electron tomography. Ultramicroscopy 157:35–47

van Aarle W, Palenstijn WJ, Cant J, Janssens E, Bleichrodt F, Dabravolski A, Beenhouwer JD, Batenburg KJ, Sijbers J (2016) Fast and flexible X-ray tomography using the ASTRA toolbox. Opt Express 24(22):25129–25147

Vila-Comamala J, Romano L, Guzenko V, Kagias M, Stampanoni M, Jefimovs K (2018) Towards sub-micrometer high aspect ratio X-ray gratings by atomic layer deposition of iridium. Microelectron Eng 192:19–24

Vo NT, Atwood RC, Drakopoulos M (2018) Superior techniques for eliminating ring artifacts in X-ray micro-tomography. Opt. Express 26(22):28396–28412

Vogt ETC, Weckhuysen BM (2015) Fluid catalytic cracking: recent developments on the grand old lady of zeolite catalysis. Chem Soc Rev 44(20):7342–7370

Voltolini M, Zandomeneghi D, Mancini L, Polacci M (2011) Texture analysis of volcanic rock samples: Quantitative study of crystals and vesicles shape preferred orientation from X-ray microtomography data. J Volcanol Geotherm Res 202(1):83–95

Wadeson N, Basham M (2016) Savu: a Python-based, MPI framework for simultaneous processing of multiple, N-dimensional, large tomography datasets. arXiv:1610.08015

Walker T (2019) Speckle-based phase-contrast imaging with a laboratory source. Master's thesis, University of Southampton, Southampton, United Kingdom

Wang H, Atwood RC, Pankhurst MJ, Kashyap Y, Cai B, Zhou T, Lee PD, Drakopoulos M, Sawhney K (2019) High-energy, high-resolution, fly-scan X-ray phase tomography. Sci Rep 9(1):8913

Wang H, Cai B, Pankhurst MJ, Zhou T, Kashyap Y, Atwood R, Le Gall N, Lee P, Drakopoulos M, Sawhney K (2018) X-ray phase-contrast imaging with engineered porous materials over 50keV. J Synchrotron Rad 25(4):1182–1188

Wang H, Kashyap Y, Cai B, Sawhney K (2016a) High energy X-ray phase and dark-field imaging using a random absorption mask. Sci Rep 6:30581

Wang H, Kashyap Y, Sawhney K (2016b) From synchrotron radiation to lab source: advanced speckle-based X-ray imaging using abrasive paper. Sci Rep 6:20476

Yang F (2017) Multi-contrast X-ray imaging of water transport in cement-based materials. Ph.D. thesis, ETH Zurich, Zurich, Switzerland

Yang F, Griffa M, Hipp A, Derluyn H, Moonen P, Kaufmann R, Boone MN, Beckmann F, Lura P (2016) Advancing the visualization of pure water transport in porous materials by fast, talbot interferometry-based multi-contrast x-ray micro-tomography. Proc SPIE 9967:99670L

Yang F, Prade F, Griffa M, Jerjen I, Di Bella C, Herzen J, Sarapata A, Pfeiffer F, Lura P (2014) Dark-field X-ray imaging of unsaturated water transport in porous materials. Appl Phys Lett 105(15):154105

Zandomeneghi D, Voltolinia M, Mancini L, Brun F, Dreossi D, Polacci M (2010) Quantitative analysis of X-ray microtomography images of geomaterials: application to volcanic rocks. Geosphere 6(6):793–804

Zanette I, Zdora M-C, Zhou T, Burvall A, Larsson DH, Thibault P, Hertz HM, Pfeiffer F (2015) X-ray microtomography using correlation of near-field speckles for material characterization. Proc Natl Acad Sci USA 112(41):12569–12573

Zanette I, Zhou T, Burvall A, Lundström U, Larsson DH, Zdora M-C, Thibault P, Pfeiffer F, Hertz HM (2014) Speckle-based x-ray phase-contrast and dark-field imaging with a laboratory source. Phys Rev Lett 112(25):253903

Zdora M-C (2014) X-ray multimodal imaging using near-field speckles. Master's thesis, Technische Universität München, Munich, Germany

Zdora M-C, Thibault P, Pfeiffer F, Zanette I (2015) Simulations of x-ray speckle-based dark-field and phase-contrast imaging with a polychromatic beam. J Appl Phys 118(11):113105

Zdora M-C, Zanette I, Walker T, Phillips NW, Smith R, Deyhle H, Ahmed S, Thibault P (2020) X-ray phase imaging with the unified modulated pattern analysis of near-field speckles at a laboratory source. Appl Opt 59(8):2270–2275

Zdora M-C, Zanette I, Zhou T, Koch FJ, Romell J, Sala S, Last A, Ohishi Y, Hirao N, Rau C, Thibault P (2018) At-wavelength optics characterisation via X-ray speckle- and grating-based unified modulated pattern analysis. Opt Express 26(4):4989–5004

Zhou T, Wang H, Connolley T, Scott S, Baker N, Sawhney K (2018a) Development of an X-ray imaging system to prevent scintillator degradation for white synchrotron radiation. J Synchrotron Rad 25(3):801–807

Zhou T, Wang H, Sawhney K (2018b) Single-shot X-ray dark-field imaging with omnidirectional sensitivity using random-pattern wavefront modulator. Appl Phys Lett 113(9):091102

Zhou T, Yang F, Kaufmann R, Wang H (2018c) Applications of laboratory-based phase-contrast imaging using speckle tracking Technique towards high energy X-rays. J Imaging 4(5)

Zhou T, Zanette I, Zdora M-C, Lundström U, Larsson DH, Hertz HM, Pfeiffer F, Burvall A (2015) Speckle-based x-ray phase-contrast imaging with a laboratory source and the scanning technique. Opt Lett 40(12):2822–2825

Chapter 10
Summary, Conclusions and Outlook

10.1 Summary and Conclusions of this Ph.D. Project

This Ph.D. project encompasses work on a range of different aspects of X-ray speckle- and single-grating-based phase-contrast imaging. These include theoretical and experimental work on technique development and optimisation on the one hand, and the identification and demonstration of scientifically interesting applications on the other hand. The latter cover a broad range of research fields, giving this work an interdisciplinary character.

X-ray speckle-based imaging is the main focus of this work. It is the most recent X-ray phase-contrast imaging method and was still in its infancy at the start of this Ph.D. project. Conventional X-ray double-grating interferometry (with a phase and an absorption grating) was already an established technique. However, setups using a single phase grating had not been explored extensively. Both speckle- and single-grating-based imaging were studied here and the two approaches were later unified in a single data acquisition and reconstruction method, named unified modulated pattern analysis, which was developed during this Ph.D. project.

In the following sections the main contributions and conclusions of this Ph.D. work are summarised.

10.1.1 Technique Development

One of the principal accomplishments was making X-ray quantitative phase-contrast imaging simpler to implement, more flexible and easily adjustable to the specific requirements and purpose of a particular application.

- For **X-ray grating-based imaging**, the use of a single (phase) grating allowed for a simpler, more robust and more photon- and dose-efficient implementation of

M.-C. Zdora, *X-ray Phase-Contrast Imaging Using Near-Field Speckles*,
Springer Theses, https://doi.org/10.1007/978-3-030-66329-2_10

X-ray grating interferometry capable of achieving a higher spatial resolution, see Chap. 4. The concept of using a single grating had been reported previously in a few publications, but was further pursued, developed and applied to biomedical imaging in this thesis. In particular, a setup was installed at a synchrotron beamline within the scope of this Ph.D. project and optimised for shorter scan times by implementation with a multilayer monochromator.

- For **X-ray speckle-based imaging**, experimental testing and analysis of the initially proposed implementations (single-shot speckle tracking and multi-frame 2D and 1D speckle scanning) led to the conclusion that there was the need for advanced operational modes to overcome some of their inherent limitations, see Chap. 6. To tackle this, the unified modulated pattern analysis (UMPA) approach was developed as a flexible way to perform differential phase-contrast and dark-field imaging. UMPA was designed to be adaptable to and compatible with most existing setups for X-ray phase-sensitive imaging. It is, furthermore, photon-efficient and cost-effective and does not require additional specialised equipment such as high-performance stepping stages. Most importantly, UMPA allows for tuning of the angular sensitivity and spatial resolution of the reconstructed images by adapting the scan and reconstruction parameters accordingly. This had not been possible with the previous speckle-tracking and speckle-scanning modes and UMPA, in fact, encompasses both of these implementations as its extreme cases. Furthermore, UMPA bridges the gap between speckle- and grating-based imaging as it can be operated with any kind of reference pattern, including the periodic line pattern from a phase grating, see Chaps. 6 and 7. This way, the single-grating technique and the various implementations of speckle-based imaging were unified in a single approach.
- A more **technical development** explored as part of this Ph.D. project was the work on alternative customisable phase modulators for X-ray speckle-based imaging, see Chap. 9. These phase modulators were produced using metal-assisted chemical etc.hing (MACE) of silicon. The main aim was to optimise speckle-based imaging, in particular in UMPA implementation, by customising the diffuser to a specific experimental setup. The first UMPA measurements using MACE diffusers confirmed a sufficient speckle visibility and speckle sizes compatible with effective pixel sizes down to 160 nm. This could open up possibilities for new high-resolution applications of speckle-based imaging.
- Further **experimental developments** include the extension of UMPA speckle-based imaging to high X-ray energies and polychromatic laboratory sources, see Chap. 9. For the first demonstrations of high-energy UMPA performed at 53 keV, angular sensitivities down to 125 nrad were achieved, enabling high-contrast imaging of rocks and mortar specimens, see Sect. 10.1.2. A proof-of-principle implementation of UMPA at a liquid-metal-jet laboratory X-ray source allowed for the multimodal visualisation of detailed structures in a bug. The effect of the broad X-ray spectrum, increased scan time and different analysis parameters on the reconstruction results were studied. This lays the foundations for the future optimisation of the setup to make it compatible with UMPA phase tomography studies, in particular for biomedical imaging applications such as virtual histology.

10.1.2 Applications

The second major contribution of this Ph.D. project is the translation of X-ray speckle-based imaging from first proof-of-principle demonstrations to applied science cases. New areas of applications were identified and the potential of UMPA in these areas was demonstrated experimentally.

- As a first application, UMPA was applied to **X-ray optics characterisation** using both a random speckle pattern and a periodic grating pattern as a wavefront marker, see Chap. 7. Making use of the flexible, simple setup and tunable character of UMPA, aberrations in the focussing behaviour of X-ray refractive lenses were identified and quantified with high precision and accuracy.
- One of the most promising applications of the speckle-based technique, in particular using UMPA, is **biomedical imaging**. In Chap. 8, the potential of UMPA phase tomography for 3D virtual histology was demonstrated on various unstained soft-tissue specimens, some of which were, moreover, fully hydrated. The results show that the approach not only allows for the high-contrast 3D visualisation of the inner micro-architecture of biological tissue at unprecedented quality, but is also able to deliver quantitative density values with a density resolution of better than 2 mg/cm^3 measured at a spatial resolution of 8 µm.
  Biomedical imaging was also performed using single-grating interferometry as presented in Chap. 4. The results on a mouse embryo indicate its great potential as an alternative to commonly used destructive methods such as conventional histology and high-resolution episcopic microscopy.
- Further applications of X-ray speckle-based phase tomography explored in this thesis are in the area of **geology** for the visualisation of the inner structure of volcanic rocks and in **materials science** for studying the composition of mortar, see Chap. 9.
- An application of speckle-based phase tomography at higher resolution that was realised using the customised MACE phase modulators is the investigation of **fluid catalytic cracking particles** used in the petroleum industry to draw conclusions on their activation state, see Chap. 9.

10.2 Outlook and Perspectives

The results of this thesis show the promising potential of UMPA speckle-based imaging for a wide range of fields and it can be expected that it will become a powerful tool for 3D investigations. The future directions for the **applications** that were first demonstrated in this thesis include the following aspects:

- For biomedical imaging applications, the combination of high-contrast 3D structural and high-sensitivity quantitative density information provided by UMPA speckle-based virtual histology could in the future be exploited for further medi-

cal studies such as the identification, staging and fundamental research of diseases. In particular the fact that the method can be applied to unstained hydrated tissue raises hopes for future investigations of tissue in its native state. It, furthermore, allows for the combination with other complementary analysis techniques such as conventional histology or serial block-face scanning electron microscopy.
- The first demonstrations of high-energy speckle-based UMPA phase tomography on volcanic rocks and mortar samples indicate its future potential for high-energy 3D imaging with subsequent automated segmentation, structural and compositional analysis. The examples in this thesis pave the way for future applications in other areas of research that require quantitative high-resolution density measurements at high X-ray energies, e.g. palaeontology, archaeology, industrial testing of metal components and many more.
- For applications of speckle-based imaging at sub-micrometre pixel sizes, the first results of UMPA speckle-based phase tomography on catalyst particles promise high potential for the analysis of small samples of tens to hundreds of micrometers in size. This is in particular promising in combination with other methods capable of providing a higher spatial resolution such as X-ray ptychography, for a multi-scale analysis approach.

The applications demonstrated in this thesis cover a wide range of fields, but are, of course, not exhaustive. With the rising awareness of X-ray speckle-based imaging within different user communities, an increasing number of applications is anticipated in the near future.

Furthermore, work on **algorithmic development** of X-ray speckle-based imaging with the UMPA approach will be carried out in the future:

- Further improvements of the UMPA reconstruction approach will be performed. In particular, the implementation of the code will be optimised to increase reconstruction speed, improve the angular sensitivity and spatial resolution and minimise artefacts. Combinations of UMPA with other phase retrieval methods could also be explored in the future.
- Fly-scan UMPA phase tomography, as demonstrated in this thesis, could in the future be combined with real-time online data reconstruction. The reconstruction result could be updated after the acquisition of each diffuser step and the data acquisition terminated when a sufficient image quality is reached.
- Dark-field imaging using UMPA to date delivers the average signal over all directions. This can in the future be extended to (omni-)directional dark-field imaging, also in 3D.

Future work on **technical and experimental developments** will include the following aspects:

- The optimisation of phase modulators for speckle-based imaging is ongoing with the aim of customisation via controlled fabrication using metal-assisted chemical etc.hing, as introduced in Chap. 9.

- The improvement of experimental setups and components will require simulations and theoretical work to determine the optimal imaging arrangement for given experimental conditions, such as the type and feature size of the diffuser, the X-ray beam spectrum, the distances between diffuser, sample and detector and more. This will be particularly important for applications at laboratory sources, as demonstrated in Chap. 9. The optimisation of lab-based UMPA speckle imaging is currently being further investigated.
- Another future plan is the exploration of different data acquisition routines for the UMPA approach with the aim of eliminating artefacts and extending the reconstructed field of view beyond the size of the illuminated area. A combination of UMPA with near-field ptychography measurements, holographic imaging or other propagation-based imaging methods could also be considered in the future.

Following recent publications and this Ph.D. thesis, X-ray speckle-based imaging, especially in UMPA implementation, is anticipated to draw increasing attention in a wide range of user communities. It is expected to be taken up in two **different scientific environments**: for specialised and fundamental research and metrology at synchrotrons and novel XFEL sources and in more accessible laboratory settings for high-throughput and routine applications in different areas of research as well as for clinical and industrial/commercial use.

- At the synchrotron, the speckle-based technique has already become an established method for metrology and optics characterisation. It has also seen increasing interest for imaging applications in the fields of biomedical imaging and geology, which is expected to be expanded in the near future.
- High-energy X-ray imaging will be explored further at synchrotron facilities, which will be pushed forward also by the current upgrade programmes, e.g. ESRF-EBS, designed to deliver a significantly higher coherent flux at high energies. This will improve the speckle visibility, opening up the potential of speckle-based imaging to new applications for thicker and denser samples. However, this will also need to be accompanied by algorithmic developments to cope with the inherent increase in propagation fringes around sample features at higher coherence levels.
- At XFELs, the speckle-based technique has recently found its first applications for metrology purposes and imaging applications are expected to follow. Methods for fast data acquisition will have to be developed for this purpose to deal with the short X-ray pulse duration, its destructive character and the pulse-to-pulse beam fluctuations, which do not allow for diffuser stepping.
- The largest user community is expected to be found for applications in the laboratory. Although the feasibility of speckle-based imaging at X-ray laboratory sources has been demonstrated, including for the UMPA approach as shown in this thesis, there is still work required on increasing imaging speed and optimising setups, especially for tomography implementations. After this has been successfully achieved, the applications presented in this thesis will move towards lab-based environments.

It is anticipated that in the medium- to long-term future, the X-ray speckle-based technique, in particular when implemented with UMPA, will become an indispensable tool for X-ray imaging in the fields of high-throughput biomedical imaging for research and clinical histology, optics characterisation, quality control of materials and food, the study of geological and palaeontological samples, and many more.

Appendix A
X-ray Grating- versus Speckle-Based Methods

Although this Ph.D. project focuses mainly on X-ray speckle-based imaging, both grating- and speckle-based methods were explored. The two approaches were later unified in the Unified Modulated Pattern Analysis, see Chap. 6, to overcome some of the limitations of both techniques.

To illustrate the main advantages and limitations of both imaging techniques, the X-ray grating- and speckle-based methods are here compared. The choice of the most suitable phase modulator and imaging method depends on the particular application and experimental conditions. Some of the considerations discussed in the following can be found throughout the main part of this thesis, in particular in Chaps. 4 and 5, but they are summarised here to provide an overview.

The grating- and speckle-based approaches are in the following compared in terms of the phase modulator used to create a reference pattern as well as the common analysis methods.

A.1 Gratings versus Speckle Diffusers

Costs and Availability

One of the most obvious differences between the gratings used for X-ray grating interferometry and common diffuser materials for X-ray speckle-based imaging is the cost factor.

Gratings are custom-made optical elements tailored to specific experimental setups. They are elaborate and expensive to fabricate, in particular when small periods and high aspect ratios are required. Furthermore, the number of grating producers is small and hence overall availability is limited.

M.-C. Zdora, *X-ray Phase-Contrast Imaging Using Near-Field Speckles*, Springer Theses, https://doi.org/10.1007/978-3-030-66329-2

The most common diffusers used for speckle-based imaging, on the other hand, are commercially available, easily accessible and cost-effective, such as sandpaper, sand or biological filter membrane. The simplicity and wide availability of diffusers are among the main benefits that make the speckle-based technique particularly attractive. A type of customised phase modulator has been explored in this thesis, see Sect. 9.2, Chap. 9, but the production of these is still significantly less costly than for gratings.

Coherence Requirements

First, it should be noted that here true near-field speckle and periodic reference patterns created mainly by phase-modulating diffusers and gratings are considered. For both grating- and speckle-based imaging, it has also been proposed and experimentally demonstrated to use optical elements based on absorption effects, i.e. absorption gratings and random absorption masks such as steel wool, respectively. While this allows for the use of higher X-ray energies and low-brilliance laboratory sources by not relying on coherence effects, the number of photons reaching the detector is significantly reduced by the absorption in the mask, which can greatly increase the scan time. Most diffuser materials for speckle imaging produce speckle by a combination of phase and absorption effects.

For many implementations of X-ray speckle-based imaging, the use of a highly absorbing diffuser is not necessary since the spatial and temporal coherence requirements for creating a speckle pattern based on scattering and interference of X-rays from the diffuser particles are relatively relaxed. As a general rule, one can estimate that the spatial coherence in the diffuser plane should be larger than the size of the scatterers in the diffuser. The coherence requirements for grating-based imaging are similar when only using a single (phase) grating. The drawbacks of a limited spatial coherence can, however, be overcome by introducing a source grating G0 to the setup that creates an array of mutually incoherent, but individually coherent sourcelets, as explained in Chap. 4. This, on the other hand, also leads to the absorption of photons by the mask, which increases the scan time.

Flexibility of the Experimental Setup

For a practical implementation of X-ray phase-contrast imaging, the use of a beam-splitter phase grating as a phase modulator sets some inherent limitations on the setup flexibility as the phase grating-detector distance (or phase grating-analyser grating distance) should correspond to one of the fractional Talbot distances, in order to achieve best possible contrast of the interference pattern. This makes the experimental setup less flexible. However, it has been shown that under polychromatic illumination, the variation of the pattern visibility with the grating-detector (or inter-grating) distance is smoothed out (Engelhardt et al. 2008; Hipp et al. 2014), which relaxes the requirements on the positioning of the setup components at laboratory sources, see also Sect. 2.4.5 in Chap. 2.

A random diffuser, as used for speckle-based imaging, does not impose requirements on the choice of distances and allows for a more flexible setup, even under monochromatic illumination. The positioning of the setup components can be freely chosen to optimise the angular sensitivity, number of photons arriving at the detector, geometrical magnification (if applicable), and other factors. Although the Talbot effect also applies to the case of a random phase modulator, the latter typically contains a range of different spatial frequencies and the observed pattern is a superposition of contributions from all of them. Therefore, the variations in contrast upon propagation of the speckle pattern are small as long as near-field conditions are fulfilled.

Spatial Resolution

The spatial resolution that can be achieved with a given grating- or speckle-based imaging setup depends on a number of factors and is strongly influenced by the analysis method.

For speckle-based imaging in single-shot X-ray speckle-tracking mode, the spatial resolution is determined by the size of the analysis window. It needs to be larger than the speckle size, which hence determines the ultimate limit in resolution. For other implementations using multiple diffuser positions, the spatial resolution is ultimately determined by the resolving power of the detector system or the size of the analysis window used in the signal reconstruction process, which is in these cases significantly smaller than the speckle size. Furthermore, the width of the Fresnel fringes arising around the features of the sample upon propagation can contribute to a reduction of spatial resolution and can lead to artefacts. The width of the first Fresnel zone is $\Delta = \sqrt{\lambda z}$, where λ is the X-ray wavelength and z the distance between the sample and the detector system. If Δ is larger than the resolution limit of the detector system, blurring of features can be observed. This is generally also true for X-ray grating interferometry measurements. However, the pixel sizes are typically larger in this case, see below.

The limit in spatial resolution for grating-based approaches is typically given by the shear of the X-rays induced by the phase grating. The beam-splitting nature of the grating leads to a separation of the beam mainly into the $+1$ and -1 or $+1$, 0, -1 diffraction orders for a π- and a $\pi/2$-shifting phase grating, respectively, which are separated by an angle on the order of sub-μrad. The separation of the diffracted beams in the detector plane sets a limit to the spatial resolution of the data set. For a π-shifting grating, this is given by $s_{+1-1} = 2\frac{\lambda}{p_1}z$, for an X-ray wavelength λ, a grating period p_1 and a grating-detector distance z. In practice, this means that often speckle-based approaches are able to reach a higher spatial resolution of the reconstructed images as they are not limited by this effect. However, on the other hand, the field of view for speckle-based imaging is typically significantly smaller than for grating interferometry in its common two-grating implementation, see below.

The spatial resolution of grating-based imaging when using a double grating interferometer is also influenced by the period of the analyser grating G2. It, generally, cannot be better than twice the period of G2 (Weitkamp et al. 2005).

The ultimate resolution limit of both speckle- and grating-based techniques is given by the resolution of the detector system, which is limited by twice the effective detector pixel size, but is in practice decreased by the point spread function of the detector.

It should be noted that for X-ray grating interferometry in its conventional implementation with a 1D beam-splitter grating, the spatial resolution is not isotropic and strongly reduced in the direction parallel to the grating lines. For speckle-based imaging, which makes use of a 2D reference pattern, this is not the case. However, anisotropy of the spatial resolution can be introduced by certain reconstruction approaches, e.g. for the 1D scanning mode, see Chap. 5, Sect. 5.4.2.2.

Directional Sensitivity

Anisotropy can be observed not only for the spatial resolution, but also, and even more prominently, for the signal sensitivity of grating-based approaches when 1D gratings are used. Data acquired with conventional grating interferometry using one or two 1D gratings (or three gratings for lab applications), shows a sensitivity in only one direction, i.e. perpendicular to the grating lines, for the differential phase, but also the dark-field signal. This is particularly problematic for samples with strong anisotropy in their refraction and scattering behaviour.

The 2D speckles do not impose this kind of limitation and speckle-based imaging can hence detect the refraction and dark-field signals in both directions from the same data set. This can not only provide directional information on the properties of the specimen, but can also significantly reduce artefacts in the integrated phase signal.

It should be noted here that 2D gratings have been developed and, furthermore, measurements can be performed by using two 1D gratings rotated against each other by 90 degrees to create a 2D interference pattern with grating-based setups, see Sect. 4.5 in Chap. 4. This, however, leads to a more elaborate, more costly production for the former and a more complicated data acquisition process that is susceptible to instabilities for the latter.

Requirements on the Detector System

Another point to consider is that X-ray speckle-based imaging in its current implementations requires the ability to directly resolve the fine interference pattern with the detector system. This implies that the effective pixel size needs to be relatively small—a few times smaller than the average speckle size, which is typically a few to a few tens of microns. This excludes the use of flatpanel detectors commonly employed in medical imaging and also leads to a limited field of view. Hence, the technique is restricted to smaller samples or requires stitching or off-axis tomography to enlarge the field of view for imaging of larger specimens. One might argue that a diffuser giving a sufficiently large speckle size could be chosen to overcome this problem. However, for near-field speckle based on phase and interference effects, the limited coherence length will become an issue in this case. For diffuser grain sizes on the order of hundreds of micrometres, sufficient spatial coherence will not be given at most synchrotron and laboratory sources. An alternative for these cases, as currently explored at laboratory sources, is the use of a random absorption mask for the creation of an interference pattern mainly based on absorption effects, see Sect. 2.4.5 in Chap. 2 and Sect. 5.6 in Chap. 5.

For X-ray grating-based imaging with a single grating, similar limitations apply as the line pattern needs to be resolved by the detector system also in this case. The conventional X-ray grating interferometry implementation with two gratings can overcome this by introducing an analyser grating. The relative translation between the two gratings converts the fine interference pattern into intensity variations in the detector pixels, which allows for the use of pixels that are significantly larger than the period of the interference pattern. While this setup is compatible with detectors with larger pixels and hence enables imaging of larger samples, it places further limitations on the spatial resolution, which cannot be better than twice the period of the absorption grating, and makes the experimental procedure less robust to instabilities.

A.2 Fourier versus Real-Space Analysis

There is a fundamental difference in the analysis methods for conventional X-ray grating interferometry and X-ray speckle-based imaging. The analysis of X-ray grating interferometry data is typically performed in Fourier space and the change in the Fourier coefficients between the sample and reference phase-stepping scans is analysed for each pixel. The analysis of speckle-imaging data, on the other hand, is conducted using local windowed cross-correlation or least-squares minimisation procedures applied in real space.

The Fourier-based approach has the advantage of being straightforward to implement, fast and efficient. It is, however, based on the assumption of a sinusoidal shape of the phase-stepping curves. This condition is not always given and deviations from a sinusoidal curve can lead to image artefacts. Moreover, due to the periodic nature of the reference pattern, phase-wrapping artefacts can occur for strongly phase-shifting objects and at sharp edges, which means that the phase shift can reach values exceeding 2π.

Analysis in real space allows for the use of any kind of reference pattern, including speckles and periodic patterns, and eliminates the problem of phase wrapping.

However, artefacts can still arise at the edges of features in the presence of strong Fresnel-diffraction fringes and sharp density changes that are currently not incorporated in the analysis code. Furthermore, real-space reconstruction algorithms are commonly more computationally intensive than Fourier-based methods.

Real-space analysis also provides the possibility of a more flexible, tunable data analysis, as for example demonstrated with the UMPA approach. The sensitivity and spatial resolution of the reconstructed images can be adapted to given conditions and fine-tuned even after data acquisition by modifying the reconstruction parameters such as the size of the analysis window.

Appendix B
Mathematical Description of the UMPA Algorithm

This section contains a mathematical derivation of the original UMPA reconstruction code, which is freely available at https://github.com/pierrethibault/UMPA.

We start the derivation by creating a model of the speckle pattern. We denote the intensity of the reference interference pattern without sample in the beam at a certain diffuser position j as $I_{0;\,j}(\mathbf{r})$, where $\mathbf{r}$ is the pixel position. The intensity $I_j(\mathbf{r})$ of the corresponding sample interference pattern that is measured in a pixel $\mathbf{r}$ at the same diffuser position j when the sample is inserted into the beam can then be modelled as:

$$I_j(\mathbf{r}) = T(\mathbf{r})\left[\langle I\rangle + v(\mathbf{r})(I_{0;\,j}(\mathbf{r}+\mathbf{u}(\mathbf{r})) - \langle I\rangle)\right]. \tag{B.1}$$

Here, $T(\mathbf{r})$ is the transmission through the sample at pixel position $\mathbf{r}$, which leads to a decrease in the measured intensity of the interference pattern. $\langle I\rangle$ describes the mean intensity value of the interference pattern and hence $(I_{0;\,j}(\mathbf{r}+\mathbf{u}) - \langle I\rangle)$ can be seen as the amplitude of the reference speckle pattern. The latter is reduced by a factor $v(\mathbf{r})$ when the sample is in the beam due to small-angle scattering and we call $v(\mathbf{r})$ the dark-field signal, analogous to X-ray grating interferometry. As we have seen in Chap. 6, the presence of the sample in the beam not only leads to a reduction in intensity and a loss in visibility of the reference speckle pattern, but also a displacement due to refraction, which is expressed here as the vector $\mathbf{u}$.

Now, we define $\xi = T(1-v)$ and $\eta = Tv$ and rewrite Eq. (B.1) as:

$$\begin{aligned} I_j(\mathbf{r}) &= T(\mathbf{r})(1-v(\mathbf{r}))\langle I\rangle + T(\mathbf{r})v(\mathbf{r})I_{0;\,j}(\mathbf{r}+\mathbf{u}(\mathbf{r})) \\ &= \xi(\mathbf{r})\langle I\rangle + \eta(\mathbf{r})I_{0;\,j}(\mathbf{r}+\mathbf{u}(\mathbf{r})). \end{aligned} \tag{B.2}$$

The next step is to define a cost function $\mathscr{L}$ as the sum over all diffuser positions j and pixels $\mathbf{r}$ in the window of the squared difference between the measured intensity $I_j(\mathbf{r})$ of the sample interference pattern and the model defined in Eq. (B.2):

M.-C. Zdora, *X-ray Phase-Contrast Imaging Using Near-Field Speckles*, Springer Theses, https://doi.org/10.1007/978-3-030-66329-2

$$\mathscr{L}(\mathbf{r_0}) = \sum_{\mathbf{r},j} w(\mathbf{r}-\mathbf{r_0}) \left| I_j(\mathbf{r}) - \left[\xi(\mathbf{r})\langle I\rangle + \eta(\mathbf{r}) I_{0;\,j}(\mathbf{r}+\mathbf{u}(\mathbf{r}))\right]\right|^2 . \tag{B.3}$$

The function is evaluated in an analysis window $w(\mathbf{r}-\mathbf{r_0})$ of extent $\mathbf{r}-\mathbf{r_0}$, centred around pixel position $\mathbf{r_0}$. Different window types can be used. A Hamming window was chosen for the results presented in this thesis as it gave the best image quality.

Equation (B.3) can be expanded:

$$\begin{aligned}\mathscr{L}(\mathbf{r_0}) = \sum_{\mathbf{r},j} w(\mathbf{r}-\mathbf{r_0}) \big[& I_j^2(\mathbf{r}) - 2\xi(\mathbf{r})\langle I\rangle I_j(\mathbf{r}) - 2\eta(\mathbf{r}) I_j(\mathbf{r}) I_{0;\,j}(\mathbf{r}+\mathbf{u}(\mathbf{r})) \\ & + \xi^2(\mathbf{r})\langle I\rangle^2 + 2\xi(\mathbf{r})\langle I\rangle \eta(\mathbf{r}) I_{0;\,j}(\mathbf{r}+\mathbf{u}(\mathbf{r})) + \eta^2(\mathbf{r}) I_{0;\,j}^2(\mathbf{r}+\mathbf{u}(\mathbf{r}))\big] .\end{aligned} \tag{B.4}$$

Further expansion and rewriting with the definition $\mathbf{s} = \mathbf{r} - \mathbf{r_0}$, which corresponds to the distance from the centre pixel $\mathbf{r_0}$ in the window, leads to the following form of the cost function:

$$\begin{aligned}\mathscr{L}(\mathbf{r_0}) = & \sum_{\mathbf{s},j} w(\mathbf{s}) I_j^2(\mathbf{r_0}+\mathbf{s}) - 2\xi(\mathbf{r_0})\langle I\rangle \sum_{\mathbf{s},j} w(\mathbf{s}) I_j(\mathbf{r_0}+\mathbf{s}) \\ & - 2\eta(\mathbf{r_0}) \sum_{\mathbf{s},j} w(\mathbf{s}) I_j(\mathbf{r_0}+\mathbf{s}) I_{0;\,j}(\mathbf{r_0}+\mathbf{s}+\mathbf{u}(\mathbf{r_0})) \\ & + \xi^2(\mathbf{r_0}) \sum_{\mathbf{s},j} w(\mathbf{s}) \langle I\rangle^2 \\ & + 2\xi(\mathbf{r_0})\langle I\rangle \eta(\mathbf{r_0}) \sum_{\mathbf{s},j} w(\mathbf{s}) I_{0;\,j}(\mathbf{r_0}+\mathbf{s}+\mathbf{u}(\mathbf{r_0})) \\ & + \eta^2(\mathbf{r_0}) \sum_{\mathbf{s},j} w(\mathbf{s}) I_{0;\,j}^2(\mathbf{r_0}+\mathbf{s}+\mathbf{u}(\mathbf{r_0})).\end{aligned} \tag{B.5}$$

Here, the approximations $\xi(\mathbf{r_0}+\mathbf{s}) \approx \xi(\mathbf{r_0})$, $\eta(\mathbf{r_0}+\mathbf{s}) \approx \eta(\mathbf{r_0})$ and $\mathbf{u}(\mathbf{r_0}+\mathbf{s}) \approx \mathbf{u}(\mathbf{r_0})$ were made. Now the following expressions are defined:

$$\begin{aligned} S_1(\mathbf{r_0}+\mathbf{s}) &= \sum_j I_j(\mathbf{r_0}+\mathbf{s}) \\ S_2(\mathbf{r_0}+\mathbf{s}) &= \sum_j I_j^2(\mathbf{r_0}+\mathbf{s}) \\ R_1(\mathbf{r_0}+\mathbf{s}+\mathbf{u}(\mathbf{r_0})) &= \sum_j I_{0;\,j}(\mathbf{r_0}+\mathbf{s}+\mathbf{u}(\mathbf{r_0})) \\ R_2(\mathbf{r_0}+\mathbf{s}+\mathbf{u}(\mathbf{r_0})) &= \sum_j I_{0;\,j}^2(\mathbf{r_0}+\mathbf{s}+\mathbf{u}(\mathbf{r_0})). \end{aligned} \tag{B.6}$$

This allows writing Eq. (B.5) as:

$$\begin{aligned}\mathscr{L}(\mathbf{r_0}) &= \sum_{\mathbf{s}} w(\mathbf{s}) S_2(\mathbf{r_0}+\mathbf{s}) + \xi^2(\mathbf{r_0}) N \langle I\rangle^2 \sum_{\mathbf{s}} w(\mathbf{s}) + \eta^2(\mathbf{r_0}) \sum_{\mathbf{s}} w(\mathbf{s}) R_2(\mathbf{r_0}+\mathbf{s}+\mathbf{u}(\mathbf{r_0}))\\ &\quad - 2\xi(\mathbf{r_0})\langle I\rangle \sum_{\mathbf{s}} w(\mathbf{s}) S_1(\mathbf{r_0}+\mathbf{s}) - 2\eta(\mathbf{r_0}) \sum_{\mathbf{s},\mathbf{j}} w(\mathbf{s}) I_j(\mathbf{r_0}+\mathbf{s}) I_{0;\,j}(\mathbf{r_0}+\mathbf{s}+\mathbf{u}(\mathbf{r_0}))\\ &\quad + 2\xi(\mathbf{r_0})\langle I\rangle \eta(\mathbf{r_0}) \sum_{\mathbf{s}} w(\mathbf{s}) R_1(\mathbf{r_0}+\mathbf{s}+\mathbf{u}(\mathbf{r_0})),\end{aligned} \tag{B.7}$$

where $N = \sum_{\mathbf{j}}$ is the total number of diffuser positions. Furthermore, for a normalised window w, we get $\sum_{\mathbf{s}} w(\mathbf{s}) = 1$.
Further definition of:

$$\begin{aligned} L_1(\mathbf{r_0}) &= \sum_{\mathbf{s}} w(\mathbf{s}) S_2(\mathbf{r_0}+\mathbf{s})\\ L_2 &= N\langle I\rangle^2\\ L_3(\mathbf{r_0};\mathbf{u}(\mathbf{r_0})) &= \sum_{\mathbf{s}} w(\mathbf{s}) R_2(\mathbf{r_0}+\mathbf{s}+\mathbf{u}(\mathbf{r_0}))\\ L_4(\mathbf{r_0}) &= \langle I\rangle \sum_{\mathbf{s}} w(\mathbf{s}) S_1(\mathbf{r_0}+\mathbf{s})\\ L_5(\mathbf{r_0};\mathbf{u}(\mathbf{r_0})) &= \sum_{\mathbf{s},\mathbf{j}} w(\mathbf{s}) I_j(\mathbf{r_0}+\mathbf{s}) I_{0;\,j}(\mathbf{r_0}+\mathbf{s}+\mathbf{u}(\mathbf{r_0}))\\ L_6(\mathbf{r_0};\mathbf{u}(\mathbf{r_0})) &= \langle I\rangle \sum_{\mathbf{s}} w(\mathbf{s}) R_1(\mathbf{r_0}+\mathbf{s}+\mathbf{u}(\mathbf{r_0})),\end{aligned} \tag{B.8}$$

leads to the simplified equation:

$$\mathscr{L}(\mathbf{r_0}) = L_1 + \xi^2 L_2 + \eta^2 L_3(\mathbf{u}(\mathbf{r_0})) - 2\xi L_4 - 2\eta L_5(\mathbf{u}(\mathbf{r_0})) + 2\xi\eta L_6(\mathbf{u}(\mathbf{r_0})). \tag{B.9}$$

The **recipe** to determine the local displacement $\mathbf{u}$ of the speckle pattern at each pixel $\mathbf{r_0}$, which is directly related to the refraction angle of the X-rays induced by the presence of the sample, is the following:

1. Construct the terms S_1, S_2, R_1 and R_2 from the measured sample and reference speckle patterns using Eq. (B.6).
2. Use these to construct the terms L_1, L_2, L_3, L_4, L_5 and L_6 with Eq. (B.8).
3. Determine analytic solutions for ξ and η.
4. These give directly the transmission $T = \xi + \eta$ and dark-field signal $v = \frac{\eta}{\xi+\eta} = \frac{\eta}{T}$.
5. Construct the cost function $\mathscr{L}$ from L_1, L_2, L_3, L_4, L_5 and L_6 as in Eq. (B.9) using the values for ξ and η.
6. Perform sub-pixel minimisation of $\mathscr{L}$ with respect to the displacement $\mathbf{u}$.
7. The horizontal and vertical refraction angles α_x and α_y can then be determined from the value of $\mathbf{u}$ obtained in the previous step via the relations $\alpha_x = u_x p_{\text{eff}}/z$

and $\alpha_y = u_y p_{\text{eff}}/z$, where u_x and u_y are the horizontal and vertical components of the speckle displacement $\mathbf{u}$, respectively, p_{eff} is the effective pixel size and z the propagation distance.

To achieve point 3 of above recipe, we determine the minima of $\mathscr{L}$ in Eq. (B.9) with respect to the variables ξ and η. Setting the partial derivations of Eq. (B.9) with respect to ξ and η equal to zero, delivers a linear system of equations that can be solved analytically:

$$\begin{aligned} 2\xi L_2 - 2L_4 + 2\eta L_6 &= 0 \\ 2\eta L_3 - 2L_5 + 2\xi L_6 &= 0, \end{aligned} \tag{B.10}$$

which can be reorganised to:

$$\begin{aligned} \xi L_2 + \eta L_6 &= L_4 \\ \eta L_3 + \xi L_6 &= L_5. \end{aligned} \tag{B.11}$$

Rewriting this in matrix form gives:

$$\begin{bmatrix} L_2 & L_6 \\ L_6 & L_3 \end{bmatrix} \begin{bmatrix} \xi \\ \eta \end{bmatrix} = \begin{bmatrix} L_4 \\ L_5 \end{bmatrix}. \tag{B.12}$$

Matrix inversion delivers the solutions for ξ and η:

$$\begin{bmatrix} \xi \\ \eta \end{bmatrix} = \frac{1}{L_2 L_3 - L_6^2} \begin{bmatrix} L_3 & -L_6 \\ -L_6 & L_2 \end{bmatrix} \begin{bmatrix} L_4 \\ L_5 \end{bmatrix}, \tag{B.13}$$

or:

$$\begin{aligned} \xi &= \frac{L_3 L_4 - L_5 L_6}{L_2 L_3 - L_6^2} \\ \eta &= \frac{L_2 L_5 - L_4 L_6}{L_2 L_3 - L_6^2}. \end{aligned} \tag{B.14}$$

References

Engelhardt M, Kottler C, Bunk O, David C, Schroer C, Baumann J, Schuster M, Pfeiffer F (2008) The fractional Talbot effect in differential x-ray phase-contrast imaging for extended and polychromatic x-ray sources. J Microsc 232(1):145–157

Hipp A, Willner M, Herzen J, Auweter S, Chabior M, Meiser J, Achterhold K, Mohr J, Pfeiffer F (2014) Energy-resolved visibility analysis of grating interferometers operated at polychromatic X-ray sources. Opt Express 22(25):30394–30409

Weitkamp T, Diaz A, David C, Pfeiffer F, Stampanoni M, Cloetens P, Ziegler E (2005) X-ray phase imaging with a grating interferometer. Opt Express 13(16):6296–6304

Curriculum Vitae

Marie-Christine Andree Zdora

Research Fellow
School of Physics & Astronomy
University of Southampton
Highfield Campus, Southampton SO17 1BJ
United Kingdom.

Education

06/2015–01/2020	**Ph.D. in Physics**, University College London, UK and Diamond Light Source Ltd, UK Thesis title: "X-ray phase-contrast imaging using near-field speckles" Supervisors: Dr. Irene Zanette, Prof. Pierre Thibault, Prof. Christoph Rau
10/2011–05/2015	**Master of Science (Physics)**, Technical University Munich, Germany Specialisation in Biophysics Master's project at Chair E17 for Biomedical Physics (Prof. Franz Pfeiffer) Thesis title: "X-ray Multimodal Imaging Using Near-Field Speckles"
03/2012–06/2013	**Master of Medical Physics**, University of Sydney, Australia Master's project at Chair for Applied Physics, University of Sydney & Royal Prince Alfred Hospital, Radiation Oncology, Sydney (Prof. David McKenzie, Prof. Natalka Suchowerska) Thesis title: "Nanoparticle enhancement of organic scintillators—Improving the energy dependence of scintillation dosimeters"
10/2008–09/2011	**Bachelor of Science (Physics)**, Technical University Munich, Germany Specialisation in Biophysics Bachelor's thesis at Chair E17 for Biomedical Physics (Prof. Franz Pfeiffer) Thesis title: "Evaluation of staining protocols for improved soft tissue contrast in absorption-based microCT"
09/1999–06/2008	**Abitur** (High School Certificate), Maximiliansgymnasium Munich, Germany High School Certificate: Abitur, Grade: 1.0 (High Distinction) Top of class.

M.-C. Zdora, *X-ray Phase-Contrast Imaging Using Near-Field Speckles*, Springer Theses, https://doi.org/10.1007/978-3-030-66329-2

Further Research Experience

12/2010–02/2012	**Student research assistant**, Technical University Munich, Germany, Chair E17 for Biomedical Physics
04/2010–10/2010	**Student research assistant**, Technical University Munich, Germany, Chair E21 at FRM II, Garching
06/2009–07/2009	**Student research assistant**, Ludwig Maximilians University Munich, Germany, Chair for Biophysics.

Teaching Experience

02/2016–03/2016	**Lab demonstrator** for first-year Physics electronics lab, University College London, UK
04/2014–08/2014	**Supervisor** for Physics Bachelor student, Technical University Munich, Germany
03/2013–06/2013	**Lab demonstrator** for first-year Physics lab course, University of Sydney, Australia.

Awards and Scholarships

07/2019	**Stjepan Marcelja Visiting Fellowship** to visit the Department of Applied Mathematics, Australian National University, Canberra, Australia
04/2019	**Runners-Up prize** at Research Images as Art / Art Images as Research competition for the contribution "Glimpse into a mouse kidney", awarded by the UCL Doctoral School, London, UK
08/2018	**Werner Meyer-Ilse Memorial Award** for exceptional contributions to the advancement of X-ray microscopy
06/2018	**Second poster prize** at the Third Annual Workshop on Advances in X-ray Imaging at Diamond Light Source, Didcot, UK
07/2017	**Second poster prize** at the Second Annual Workshop on Advances in X-ray Imaging at Diamond Light Source, Didcot, UK
05/2016	**Award of Congressi Stefano Franscini (CSF)** for Best Contribution at Processes in Combined Optical and Imaging Methods conference in Ascona, Switzerland
06/2015–01/2020	**UCL Ph.D. Studentship for postgraduate studies** at the Department of Physics & Astronomy and Diamond Light Source Ltd (joint studentship)
09/2014	**Grant of the French Association for Crystallography (AFC)** to attend XTOP 2014 conference in Villard de Lans, France
02/2012–11/2012	**Postgraduate Scholarship of the German Academic Exchange Service (DAAD)** for studies at the University of Sydney, Australia
declined	**TUMExchange scholarship of Technical University Munich, Germany** for postgraduate placement at Queensland University of Technology, Brisbane, Australia.

06/2008	**Max-Planck award of Maximiliansgymnasium Munich** for outstanding academic high-school achievements
06/2008	**DPG award (German Society for Physics)** for outstanding achievements in high-school Physics
06/2008	**DMV Abitur award of the German Mathematics association** for outstanding Abitur (final high school exams) achievements
06/2008–05/2015	**E-fellows scholarship** for outstanding Abitur achievements.

Publications

First-Author Publications

1. M.-C. Zdora, P. Thibault, W. Kuo, V. Fernandez, H. Deyhle, J. Vila-Comamala, M. P. Olbinado, A. Rack, P. M. Lackie, O. L. Katsamenis, M. J. Lawson, V. Kurtcuoglu, C. Rau, F. Pfeiffer, and I. Zanette, "X-ray phase tomography with near-field speckles for three-dimensional virtual histology," Optica **7**(9) 1221–1227 (2020).
2. M.-C. Zdora, I. Zanette, T. Walker, N. W. Phillips, R. Smith, H. Deyhle, S. Ahmed, and P. Thibault, "X-ray phase imaging with the unified modulated pattern analysis of near-field speckles at a laboratory source," Appl. Opt. **59**(8), 2270–2275 (2020).
3. M.-C. Zdora, P. Thibault, H. Deyhle, J. Vila-Comamala, W. Kuo, C. Rau, and I. Zanette, "Advanced X-ray phase-contrast and dark-field imaging with the unified modulated pattern analysis (UMPA)," Microsc. Microanal. **24**(S2), 22–23 (2018).
4. M.-C. Zdora, "State of the art of X-ray speckle-based phase-contrast and dark-field imaging," J. Imaging **4**(5), 60 (2018).
5. M.-C. Zdora, P. Thibault, H. Deyhle, J. Vila-Comamala, C. Rau, and I. Zanette, "Tunable X-ray speckle-based phase-contrast and dark-field imaging using the unified modulated pattern analysis approach," J. Instrum. **13**(05), C05005 (2018).
6. M.-C. Zdora, I. Zanette, T. Zhou, F. J. Koch, J. Romell, S. Sala, A. Last, Y. Ohishi, N. Hirao, C. Rau, and P. Thibault, "At-wavelength optics characterisation via X-ray speckle- and grating-based unified modulated pattern analysis," Opt. Express **26**(4), 4989–5004 (2018).
7. M.-C. Zdora, P. Thibault, T. Zhou, F. J. Koch, J. Romell, S. Sala, A. Last, C. Rau, and I. Zanette, "X-ray Phase-Contrast Imaging and Metrology through Unified Modulated Pattern Analysis," Phys. Rev. Lett. **118**(20), 203903 (2017).
8. M.-C. Zdora, P. Thibault, C. Rau, and I. Zanette, "Characterisation of speckle-based X-ray phase-contrast imaging," J. Phys.: Conf. Ser. **849**(1), 012024 (2017).

9. M.-C. Zdora, J. Vila-Comamala, G. Schulz, A. Khimchenko, A. Hipp, A. C. Cook, D. Dilg, C. David, C. Grünzweig, C. Rau, P. Thibault, and I. Zanette, "X-ray phase microtomography with a single grating for high-throughput investigations of biological tissue," Biomed. Opt. Express **8**(2), 1257–1270 (2017).
10. M.-C. Zdora, P. Thibault, J. Herzen, F. Pfeiffer, and I. Zanette, "Simulations of multi-contrast X-ray imaging using near-field speckles," AIP Conf. Proc. **1696**(1), 020016 (2016).
11. M.-C. Zdora, P. Thibault, F. Pfeiffer, and I. Zanette, "Simulations of X-ray speckle-based dark-field and phase-contrast imaging with a polychromatic beam," J. Appl. Phys. **118**(11), 113105 (2015).

Co-authored Publications

1. C. Rau, D. Batey, S. Cipiccia, X. Shi, S. Marathe, M. Storm, A. Bodey, and M.-C. Zdora, "New imaging opportunities at the DIAMOND beamline I13L," Proc. SPIE **11112**, X-Ray Nanoimaging: Instruments and Methods IV, 111120L (2019).
2. C. Rau, M. Storm, S. Marathe, A. J. Bodey, M.-C. Zdora, S. Cipiccia, D. Batey, X. Shi, S. M. L. Schroeder, G. Das, M. Loveridge, R. Ziesche, and B. Connolly, "Fast Multi-scale imaging using the Beamline I13L at the Diamond Light Source," Proc. SPIE **11113**, Developments in X-Ray Tomography XII, 111130P (2019).
3. P. Vagovič, T. Sato, L. Mikeš, G. Mills, R. Graceffa, F. Mattsson, P. Villanueva-Perez, A. Ershov, T. Faragó, J. Uličný, H. Kirkwood, R. Letrun, R. Mokso, M.-C. Zdora, M. P. Olbinado, A. Rack, T. Baumbach, J. Schulz, A. Meents, H. N. Chapman, and A. P. Mancuso, "Megahertz x-ray microscopy at x-ray free-electron laser and synchrotron sources," Optica **6**(9), 1106–1109 (2019).
4. P. Bidola, J. Martins de Souza e Silva, K. Achterhold, E. Munkhbaatar, P. Jost, A.-L. Meinhardt, K. Taphorn, M.-C. Zdora, F. Pfeiffer, and J. Herzen, "A step towards valid detection and quantification of lung cancer volume in experimental mice with contrast agent-based X-ray microtomography," Sci. Rep. **9**(1), 1325 (2019).
5. C. Rau, M. Storm, S. Marathe, A. J. Bodey, S. Cipiccia, D. Batey, X. Shi, M.-C. Zdora, and I. Zanette, "Multi-scale imaging at the diamond beamline I13," AIP Conf. Proc. **2054**(1), 030012 (2019).
6. C. Rau, M. Storm, S. Marathe, A. J. Bodey, S. Cipiccia, D. Batey, X. Shi, M.-C. Zdora, I. Zanette, S. Perez-Tamarit, P. Cimavilla, M. A. Rodriguez-Perez, F. Döring, and C. David, "Multi-Scale Imaging at the Coherence and Imaging Beamline I13 at Diamond," Microsc. Microanal. **24**(S2), 256–257 (2018).
7. I. Teh, D. McClymont, M.-C. Zdora, H. Whittington, K. Gehmlich, C. Rau, C. A. Lygate, and J. E. Schneider, "Validation of Diffusion Tensor Imaging in Diseased Myocardium," Proceedings of the Joint Annual Meeting ISMRM-ESMRMB 2018 (2018).

8. M. P. Olbinado, J. Grenzer, P. Pradel, T. De Resseguier, P. Vagovič, M.-C. Zdora, V. A. Guzenko, C. David, and A. Rack, "Advances in indirect detector systems for ultra high-speed hard X-ray imaging with synchrotron light," J. Instrum. **13**(04), C04004 (2018).
9. I. Teh, D. McClymont, M.-C. Zdora, H. J. Whittington, V. Davidoiu, J. Lee, C. A. Lygate, C. Rau, I. Zanette, and J. E. Schneider, "Validation of Diffusion Tensor MRI Measurements of Cardiac Microstructure with Structure Tensor Synchrotron Radiation Imaging," J. Cariov. Magn. Reson. **19**(1), 31 (2017).
10. C. Rau, A. Bodey, M. Storm, S. Cipiccia, S. Marathe, M.-C. Zdora, I. Zanette, U. Wagner, D. Batey, and X. Shi, "Micro- and nano-tomography at the DIAMOND beamline I13L imaging and coherence," Proc. SPIE **10391**, Developments in X-Ray Tomography XI, 103910T (2017).
11. S. Marathe, M.-C. Zdora, I. Zanette, S. Cipiccia, and C. Rau, "Comparison of data processing techniques for single-grating x-ray Talbot interferometer data," Proc. SPIE **10391**, Developments in X-Ray Tomography XI, 103910S (2017).
12. C. Rau, U. H. Wagner, M. Ogurreck, X. Shi, D. Batey, S. Cipiccia, S. Marathe, A. J. Bodey, M.-C. Zdora, I. Zanette, M. Saliba, V. S. C. Kuppili, S. Sala, S. H. Chalkidis, and P. Thibault, "The imaging and coherence beamline I13L at DIAMOND (Conference Presentation)," Proc. SPIE **10389**, X-Ray Nanoimaging: Instruments and Methods III, 1038905 (2017).
13. G. Schulz, C. Götz, M. Müller-Gerbl, I. Zanette, M.-C. Zdora, A. Khimchenko, H. Deyhle, P. Thalmann, and B. Müller, "Multimodal imaging of the human knee down to the cellular level," J. Phys.: Conf. Ser. **849**(1), 012026 (2017).
14. A. Khimchenko, C. Bikis, G. Schulz, M.-C. Zdora, I. Zanette, J. Vila-Comamala, G. Schweighauser, J. Hench, S. E. Hieber, and H. Deyhle, "Hard X-ray submicrometer tomography of human brain tissue at Diamond Light Source," J. Phys.: Conf. Ser. **849**(1), 012030 (2017).
15. J. Romell, T. Zhou, M.-C. Zdora, S. Sala, F. J. Koch, H. M. Hertz, and A. Burvall, "Comparison of laboratory grating-based and speckle-tracking x-ray phase-contrast imaging," J. Phys.: Conf. Ser. **849**(1), 012035 (2017).
16. T. Zhou, M.-C. Zdora, I. Zanette, J. Romell, H. M. Hertz, and A. Burvall, "Noise analysis of speckle-based x-ray phase-contrast imaging," Opt. Lett. **41**(23), 5490–5493 (2016).
17. G. Schulz, C. Götz, H. Deyhle, M. Müller-Gerbl, I. Zanette, M.-C. Zdora, A. Khimchenko, P. Thalmann, A. Rack, and B. Müller, "Hierarchical imaging of the human knee," Proc. SPIE **9967**, Developments in X-Ray Tomography X, 99670R (2016).
18. A. Khimchenko, G. Schulz, H. Deyhle, P. Thalmann, I. Zanette, M.-C. Zdora, C. Bikis, A. C. Hipp, S. E. Hieber, G. Schweighauser, J. Hench, and B. Müller, "X-ray micro-tomography for investigations of brain tissues on cellular level," Proc. SPIE **9967**, Developments in X-Ray Tomography X, 996703 (2016).
19. S. E. Hieber, C. Bikis, A. Khimchenko, G. Schulz, H. Deyhle, P. Thalmann, N. Chicherova, A. Rack, M.-C. Zdora, I. Zanette, G. Schweighauser, J. Hench, and B. Müller, "Computational cell quantification in the human brain tissues based

on hard X-ray phase-contrast tomograms," Proc. SPIE **9967**, Developments in X-Ray Tomography X, 99670K (2016).

20. C. Rau, U. H. Wagner, J. Vila-Comamala, A. Bodey, A. Parson, M. García-Fernández, A. De Fanis, Z. Pešić, I. Zanette, and M.-C. Zdora, "Micro- and nano-imaging at the diamond beamline I13L-imaging and coherence," AIP Conf. Proc. **1741**(1), 030008 (2016).
21. I. Zanette, M.-C. Zdora, T. Zhou, A. Burvall, D. H. Larsson, P. Thibault, H. M. Hertz, and F. Pfeiffer, "X-ray microtomography using correlation of near-field speckles for material characterization," Proc. Natl. Acad. Sci. U.S.A. **112**(41), 12569–12573 (2015).
22. T. Zhou, I. Zanette, M.-C. Zdora, U. Lundström, D. H. Larsson, H. M. Hertz, F. Pfeiffer, and A. Burvall, "Speckle-based X-ray phase-contrast imaging with a laboratory source and the scanning technique," Opt. Lett. **40**(12), 2822–2825 (2015).
23. I. Zanette, T. Zhou, A. Burvall, U. Lundström, D. H. Larsson, M.-C. Zdora, P. Thibault, F. Pfeiffer, and H. M. Hertz, "Speckle-based X-ray phase-contrast and dark-field imaging with a laboratory source," Phys. Rev. Lett. **112**(25), 253903 (2014).

Talks and Posters at Conferences and Seminars

- 10/2020 **Invited seminar talk** (online) at Swiss Light Source, Paul Scherrer Institute (PSI), Switzerland
- 10/2020 **Invited workshop talk** (online) at the workshop "Soft Matter, Health, and Life Science" at PETRA IV, DESY, Germany
- 10/2020 **Invited seminar talk** (online) at the Australian and New Zealand Optical Society
- 03/2020 **Invited seminar talk** at the Australian National University, Canberra, Australia
- 03/2020 **Invited seminar talk** at Monash University, Melbourne, Australia
- 02/2020 **Invited seminar talk** at the Francis Crick Institute, London, UK
- 10/2019 **Invited conference talk** at XNPIG 2019, Sendai, Japan
- 09/2019 **Contributed conference talk** at Workshop on Coherence at ESRF-EBS, Grenoble, France
- 03/2019 **Invited conference talk** at TMS Annual Meeting 2019, San Antonio, USA
- 03/2019 **Invited conference talk** at symposium on "Hard X-Ray Imaging of Biological Soft Tissues" at the Francis Crick Institute, London, UK
- 01/2019 **Contributed conference talk** at IMXP 2019, Garmisch-Partenkirchen, Germany
- 12/2018 **Invited seminar talk** at Physics Department, Technical University Munich, Germany
- 10/2018 **Invited seminar talk** at CFEL/DESY, Hamburg, Germany
- 08/2018 **Contributed conference talk** at XRM 2018, Saskatoon, Canada
- 06/2018 **Poster presentation** at Third Annual Workshop on Advances in X-ray Imaging at Diamond Light Source, Didcot, UK
- 04/2018 **Invited seminar talk** at ESRF, Grenoble, France
- 01/2018 **Contributed conference talk** at IMXP 2018, Garmisch-Partenkirchen, Germany

11/2017	**Contributed conference talk** at Zeiss XEN meeting, London, UK
09/2017	**Contributed conference talk** at ICXOM24, Trieste, Italy
09/2017	**Contributed conference talk** at XNPIG 2017, Zurich, Switzerland
07/2017	**Poster presentation** at Second Annual Workshop on Advances in X-ray Imaging at Diamond Light Source, Didcot, UK
01/2017	**Contributed conference talk** at IMXP 2017, Garmisch-Partenkirchen, Germany
10/2016	**Invited seminar talk** at Swiss Light Source, Paul Scherrer Institute (PSI), Switzerland
08/2016	**Poster presentation** at XRM2016, Oxford, UK
06/2016	**Contributed conference talk** at Coherence 2016, Saint Malo, France
06/2016	**Invited workshop talk** at Diamond Light Source Science Away Day, Oxford, UK
05/2016	**Contributed conference talk** at Processes in Combined Digital Optical and Imaging Methods Applied to Mechanical Engineering, Ascona, Switzerland
01/2016	**Contributed conference talk** at IMXP 2016, Garmisch-Partenkirchen, Germany
09/2015	**Poster presentation** at ICXOM23, Long Island, New York, USA
09/2015	**Poster presentation** at Diamond User Meeting, Diamond Light Source, Didcot, UK
01/2015	**Invited seminar talk** at the Centre for X-ray Tomography, Ghent University, Belgium
09/2014	**Contributed conference talk** at XTOP 2014, Villard de Lans, France
06/2014	**Invited workshop talk** at Science 3D meeting at DESY, Hamburg, Germany
03/2014	**Contributed conference talk** at DPG Frühjahrstagung of the German Society for Physics in Mainz, Germany.

Additional Skills

Programming Skills

Experienced user of Python.

Computer Skills

Confident user of LateX, MS Word, MS Excel, MS Powerpoint, Inkscape, Blender.

Language Skills

English: Fluent oral and written
German: Native speaker
Italian: Basic skills oral and written.